Adults' Mathematical Thinking and Emotions

The perpetual concerns about the teaching and learning of mathematics, and its use in work and everyday life, relate to a number of issues:

- doubts about the transferability of school learning to outside settings
- declining participation in A-level and higher education mathematics courses
- under-representation of certain groups, such as females, in mathematics study
- the aversion of many people to mathematics generally.

This book addresses these issues through investigating the following:

- the ways in which numerate thinking and performance of adults are context-related
- the inseparability of thinking and emotion, and the consequent ways in which mathematical activity is emotional, and not simply cognitive
- the understanding of mathematics anxiety in psychological, psychoanalytic and feminist theories
- social differences in mathematics performance, anxiety and confidence.

Drawing on works such as those of Jean Lave and Valerie Walkerdine, and ideas from poststructuralism and psychoanalysis, this volume shows ways of reconceptualising current debates in mathematics education, including its psychological and sociological aspects. It points to ideas for practical applications in education and training, such as clarifying the problems with the transfer of learning, and countering mathematics anxiety. It also illustrates a number of ways of fruitfully combining quantitative and qualitative methodologies in educational research.

Jeff Evans is Principal Lecturer in Social Statistics at Middlesex University, and teaches social and business statistics and research methods.

Studies in Mathematics Education Series
Edited by Paul Ernest
University of Exeter, UK

The Philosophy of Mathematics Education
Paul Ernest

Understanding in Mathematics
Anna Sierpinska

Mathematics Education and Philosophy
Edited by Paul Ernest

Constructing Mathematical Knowledge
Edited by Paul Ernest

Investigating Mathematics Teaching
Barbara Jaworski

Radical Constructivism
Ernst von Glaserfeld

The Sociology of Mathematics Education
Paul Dowling

Counting Girls Out
Girls and mathematics
Valerie Walkerdine

Writing Mathematically
The discourse of investigation
Candia Morgan

Rethinking the Mathematics Curriculum
Edited by Celia Hoyles, Candia Morgan and Geoffrey Woodhouse

International Comparisons in Mathematics Education
Edited by Gabriele Kaiser, Eduardo Luna and Ian Huntley

Mathematics Teacher Education
Critical international perspectives
Edited by Barbara Jaworski, Terry Wood and Sandy Dawson

Learning Mathematics
From hierarchies to networks
Edited by Leone Burton

The Pragmatics of Mathematics Education
Vagueness and mathematical discourse
Tim Rowland

Adults' Mathematical Thinking and Emotions
A study of numerate practices
Jeff Evans

Adults' Mathematical Thinking and Emotions

A Study of Numerate Practices

Jeff Evans

London and New York

First published 2000
by RoutledgeFalmer
11 New Fetter Lane, London EC4P 4EE

Simultaneously published in the USA and Canada
by RoutledgeFalmer
29 West 35th Street, New York, NY 10001

RoutledgeFalmer is an imprint of the Taylor & Francis Group

© 2000 Jeff Evans

Typeset in Times by
Curran Publishing Services Ltd, Norwich

Printed and bound in Great Britain by
St Edmundsbury Press, Bury St Edmunds, Suffolk

British Library Cataloguing in Publication Data
A catalogue record for this book is available from the British Library

Library of Congress Cataloging in Publication Data
Evans, Jeff, 1944–
Adults' mathematical thinking and emotions: a study of numerate prac-
tices/Jeff Evans.
   p. cm. — (Studies in mathematics education series; 15)
Includes bibliographical references and index.
1. Mathematics–Study and teaching (Continuing)–Psychological
aspects. 2. Adult learning. I. Title. II. Series.
QA11.E93 2000
510`.71   5—dc21   00-032306

ISBN 0–750–70913–8 (hbk)
ISBN 0–750–70912–X (pbk)

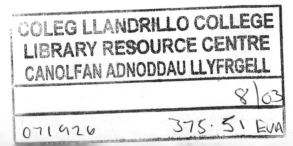

Contents

11 **Conclusions and Contributions**   **225**

Figures

Tables

Series Editor's Preface

Mathematics education is established world-wide as a major area of study, with numerous dedicated journals and conferences serving ever-growing national and international communities of scholars. As it develops, research in mathematics education is becoming more theoretically orientated. Although originally rooted in mathematics and psychology, vigorous new perspectives are pervading it from disciplines and fields as diverse as philosophy, logic, sociology, anthropology, history, women's studies, cognitive science, linguistics, semiotics, hermenutics, post-structuralism and postmodernisrn. These new research perspectives are providing fresh lenses through which teachers and researchers can view the theory and practice of mathematics teaching and learning.

The series 'Studies in Mathematics Education' aims to encourage the development and dissemination of theoretical perspectives in mathematics education as well as their critical scrutiny. It is a series of research contributions to the field based on disciplined perspectives that link theory with practice. The series is founded on the philosophy that theory is the practitioner's most powerful tool in understanding and changing practice. Whether the practice concerns the teaching and learning of mathematics from preschool to adulthood, teacher education, or educational research, the series offers new perspectives to help clarify issues, pose and solve problems and stimulate debate. It aims to have a major impact on the development of mathematics education as a field of study in the third millennium.

In the past decade one of the most influential and controversial developments in mathematics education has been the emergence of theories foregrounding the social dimensions of learning. This has been prompted by work in social psychology, sociology, anthropology and the 'rediscovery' of Vygotsky by the mathematics education research community. The resultant controversy between social and constructivist views of learning can be seen throughout the research literature, such as Burton (1999) and Ernest (1994), two earlier volumes in this series.

Some of the key questions addressed in this debate include the following. Can learning be regarded solely as an individual cognitive activity of the learner? Does the social context shape learning in any profound way? To what extent does the uniqueness of the individual learner, including her/his subjectivity, affect in the form of emotions, attitudes, beliefs and values, out-of-class experience and knowledge come into her/his learning? Is the assumption that what is learned in school can unproblematically be 'transferred' to and utilised in out-of-school activity justified?

These are all important questions. Each of these questions is addressed here, and this book not only provides a map of current work but also gives some of the potentially most fruitful answers available to date. One of the unique features of the book is the way in which it combines multiple perspectives to provide these insights into learning, activity and being. The research looks at both cognitive and affective dimensions of learning, and combines both quantitative and qualitative research methodologies. It employs both sociocultural and language/semiotic based approaches, including structuralist and poststructuralist perspectives. Although overtly about adults' responses to and performances in mathematics, a full understanding of the data requires deep theoretical reflections and frameworks, and takes us, the readers, through a variety of exciting domains illuminating the relationships between the following:

- practical mathematics and its contexts
- mathematics performance and social difference
- mathematical affect, emotion and mathematics anxiety
- knowledge and learning in context and the possibility of transfer
- social theories of learning, including sociocultural approaches and situated cognition
- practices, discourses and semiotic relations
- theory, pedagogy and professional practice.

Jeff Evans draws on a wide literature base, including such disparate thinkers as Freud, Lacan, Lave, de Saussure and Walkerdine. The result is a richly theorised and original position. The most significant innovation of the book is the deep and multiply theorised perspectives on the research topics and issues addressed.

Another important feature is the rationale for the careful deployment of quantitative and qualitative approaches, in a complementary way. The combination of the questionnaires analysed statistically and semi-structured interviews analysed using semiotic and psychoanalytic insights will be of great interest to researchers in mathematics education.

The book also offers a critical perspective on a number of 'hot' issues in education, including the concept of numeracy, mathematical performance, 'transferable skills', citizenship, and social difference and disadvantage. It concludes with valuable reflections for changing practice, including making contexts for learning more student friendly, developing citizenship though shared activities, and extending the scope of numeracy to be empowering in learners' lives. The conclusion, together with the rest of the book, will be particularly welcomed by teachers who subscribe to the philosophy of this series, that theory is the practitioner's most powerful tool. Likewise, the book's consummate interdisciplinarity, drawing on ideas from mathematics education, psychology, anthropology, sociology, women's studies, semiotics, and poststructuralism, embodies the spirit of the series.

Paul Ernest
University of Exeter, March 2000

Author's Preface and Acknowledgements

One of the messages of this work is that what may appear as individual learning, thinking and performance are better seen as social accomplishments, depending on cultural and collective resources. The production of this book is no exception. I appreciate the quality of the earlier work of many people (some of it discussed here). I also want to acknowledge the encouragement, support and help I have had from colleagues, friends, groups and communities.

At Middlesex University, the Mathematics and Statistics Group, the Business School, the Research Committee, and the Computing, Library and Media Services staff have supported this work in many ways. I also appreciate my connections with colleagues in the mathematics education, the statistics and the social science communities, in Britain and internationally. I would like especially to mention the Research into Social Perspectives in Mathematics Education group, the British Society for Research into the Learning of Mathematics, the International Group for the Psychology of Mathematics Education, Adults Learning Maths – a Research Forum, the Project on Mathematical Thinking and Learning (funded by the Portuguese agency, JNICT) and the Radical Statistics Group.

An earlier version of this work was submitted as a doctoral dissertation at the University of London Institute of Education. I thank my supervisors, Harvey Goldstein and Valerie Walkerdine, for sharing with me their specialist knowledge in that earlier project, and for their willingness to discuss its more general and interdisciplinary aspects, in a way that has provided a platform for my continuing development of the ideas. I also value my ongoing contact with the Mathematical Sciences group at the Institute, an environment that is both exciting intellectually and supportive.

From among the many who have contributed to the final form of this project, by commenting critically on parts of the book, or by helping with crucial stages of production, special thanks are due to: Jeannie Billington, Mark Coulson, Katherine Crawford, Celia Hoyles, Steve Lerman, Joao-Filipe Matos, Candia Morgan, Monique Polins, Florian Sala, Linda Santimano, Peter Sneddon, Chris Stanley, Ingrid Thorstad and Carol Anne Wien. In particular, I am grateful to Tom Wengraf for an impressive job of commenting on the lion's share of chapters at a crucial moment; to Paul Ernest for much good advice and long-term faith in the project; and especially to Anna Tsatsaroni for her constant intellectual and emotional support, in this and other projects.

I also want to express my appreciation for the massive contribution from the students of the (then) Polytechnic. Over 900 gave twenty minutes each to complete the questionnaire, and twenty-five gave up to a further hour to be interviewed about their experiences and thinking about mathematics. I hope that their willingness to participate will be repaid, at least in part, by the contribution of this study, and others, to the challenging task of making the learning and the use of mathematics, much less mystifying and much more satisfying.

I thank the following for their kind permission to reproduce specific material:

* the National Institute of Adult Continuing Education (England and Wales) for Figures 2.1 and 2.2;
* Springer-Verlag and Elizabeth Fennema for Figure 3.1;
* Taylor and Francis Group for Figure 6.1; and
* Brigid Sewell for Figures A2.1 to A2.5.

Jeff Evans
London, February 2000

1 Introduction: Mathematics, the Difficult Subject

There seems to be a need for a more technological bias in science teaching that will lead towards practical applications in industry, rather than towards academic studies. Or to take other examples, why is it that such a high proportion of girls abandon science before leaving school?

(Callaghan 1976: 5)

There is considerable evidence that those who give up mathematics do so because they perceive it as 'hard' or 'boring', that is having little significance for them.

(Brown 1995)

I began this study of adults' mathematical thinking and emotions during an exciting period for mathematics education. The findings of an enquiry into mathematics teaching in the UK (Cockcroft Committee 1982) had recently reported, and there was a growing concern with adult 'numeracy'. Many in the research communities were discussing exciting and impressive research on the ways that people's mathematical thinking might be different in different settings; for example, when doing school problems, working in street markets, or shopping in supermarkets (e.g. Carraher et al. 1985, Lave et al. 1984). In addition, a few writers on both sides of the Atlantic were discussing mathematics anxiety, and other kinds of feelings about mathematics (e.g. Tobias 1978, Buxton 1981, Nimier 1978).

In this chapter, I outline the background to the study in terms of policy concerns and research emphases in play when I began. I indicate the most important aspects of the conceptual basis of my work, and briefly describe my use of quantitative and qualitative methodologies.

Perpetual Concerns about Mathematics Learning and Use

Concerns over the mathematics curriculum and mathematics teaching have been a continuing feature of debates in education throughout this century (see e.g. Howson 1983, McIntosh 1981). These issues occasionally flare up as crises, around several flashpoints, including:

- Assertions that students fail to apply their mathematics learning from school to the workplace, to other 'everyday' settings, or to other subjects: *the problem of application or 'transfer'*.
- Decreases in already low levels of participation in A-level and higher education (HE) mathematics courses – and allegedly lower standards of preparation among those accepted to study mathematics in HE (London Mathematical Society 1995): *the participation and preparation problems*.
- The apparent under-representation of particular groups – especially females – in mathematics study: *the inclusiveness problem*.
- Evidence of the perception of mathematics as 'hard', 'boring', anxiety-provoking, even hateful: *the affective problem*.[1]

These four areas of concern are interrelated. For example, both the inclusiveness and affective issues affect the participation problem, and affective issues are also likely to affect students' commitment, and hence their preparation. Sometimes, action on one area undermines another; for example, the attempt to combat many students' lack of confidence and alienation, by connecting mathematics with what appears to be 'accessible' – typically, based in everyday contexts – risks disconnecting it from its roots in science and technology, which some argue is what is genuinely useful (Noss 1997). In particular, I shall argue in this book that affective issues are implicated widely in the learning and doing of mathematics, including the 'transfer' or application of such learning in other contexts.

The first epigraph heading the chapter displays concerns about all of the first three issues raised. The then Prime Minister James Callaghan's Ruskin College speech in October 1976 launched the 'Great Debate' in education, and arguably opened the way for the Conservative Party's educational reforms of the 1980s and 1990s. These led to construction of a national curriculum and testing framework, which looks set to remain in place well beyond 2000.

Despite the time elapsed and the changes occurring, both in the educational world and outside, since the speech, many current commentators still appear to share a traditional view of 'mathematical ability'. It is seen as involving a set of abstract cognitive 'skills', which can be applied to perform a range of tasks, in a variety of practical contexts. This is considered to take place through a relatively straightforward process of transfer. In any formal educational system, the issue of transfer of learning is clearly of major importance, whether in the 'application' of school knowledge in contexts outside the school, or in the 'harnessing' of outside learning to help with school aims. (Though the latter process has so far received less attention, it is bound to be increasingly crucial in systems where 'mature students' or 'recurrent education' are encouraged.) In this traditional view, performance is usually measured by the number of correct responses to a set of test items, of the type often used in schools. However, the possibility that correct performance could be produced merely by rote learning has led mathematics educators to stipulate that correct performance should be produced through 'real understanding' (Skemp 1976).[2]

In response to concerns expressed by the Prime Minister and others, in 1977 the Labour government announced its decision to establish an Inquiry into the teaching of mathematics 'with particular regard to its effectiveness and intelligibility and to the match between the mathematical curriculum and the skills required in further education, employment and adult life generally' (Cockcroft Committee 1982). The Committee's report focused on a number of issues, including how to promote the application of school mathematics in practical everyday life, drawing on research which it had commissioned. This is not to say that the interest in 'practical mathematics' began with Cockcroft: there were already a number of projects with a similar focus.[3] However, Cockcroft used the term 'numeracy' to mean the use of basic mathematical operations with confidence in practical everyday situations – and also to emphasise that the 'transfer' from abstract mathematics to everyday applications might not be as straightforward as the traditional view suggested.

Three major developments in further education and adult education during the 1980s and 1990s reinforced the interest in 'numeracy' – though, as we shall see, the meaning of this term varies across contexts:

1   the mushrooming of adult literacy and 'basic skills' courses
2   the development of 'Access' courses, providing adults with intensive preparation and a 'second chance' alternative entry route for higher education
3   the development of vocational qualifications, such as GNVQs and NVQs.

Thus numeracy became important in several areas of educational policy-making. In particular, the adult literacy movement led to an awareness, starting among tutors, that a lack of 'numeracy' existed as a widespread problem, somewhat independently of illiteracy. Public discussions resulted, concerning the handicaps, for individuals and for society, resulting from this lack of numeracy (Evans 1989a).

Besides signalling their avowedly more practical focus, calling courses 'numeracy' probably served to deal with many students', and teachers', negative feelings towards mathematics – lack of confidence, anxiety and dislike. Many students perceive there to be fairly strong boundaries between the mathematics that they have met at school or college, and the other activities that make up 'real life' (or indeed other academic disciplines), even when the latter make substantial use of quantitative, spatial or other 'numerate' ideas and strategies. One might suspect that both the sense of 'boundaries', and the negative feelings, will be implicated in any possibilities of transfer between academic mathematics and practical activities.

Research in mathematics education and psychology since the 1950s has often produced standard findings on social differences in performance in school mathematics: namely, gender differences in favour of males, and social class differences in favour of the middle classes. In the 1970s concern about gender differences increased markedly: most (but not all) researchers accepted their existence (at least by the early teenage years in the USA and UK). In the USA, many attributed them to gender differences in 'participation' or course-taking. These issues of subject choice in turn were explained by *attitudes* and *affect*, that is, by gender differences in feelings towards mathematics: of enjoyment/dislike, confidence/anxiety, beliefs about usefulness and difficulty, and so on. However, by the mid-1990s, gender researchers began to argue that the situation was changing, so that gender differences in school-level performance, and even at university, are disappearing (for example Keitel et al. 1996, but see also Fennema 1995).

Nevertheless, partly in order to explain gender differences in mathematics performance, much research on affect and attitudes has been carried out with
schoolchildren since the mid-1970s, especially in the USA. Much of this research
on affect has focused on '*mathematics anxiety*'. Anxiety as a concept has its basis
in Freud's work which places crucial emphasis on the possibility that anxiety may
be 'latent' or unconscious. Psychology took on board these concepts, but the stress
on conscious, observable phenomena in mainstream American psychology from the
late 1940s onwards led to an emphasis on 'manifest anxiety', which was considered
observable and quantifiable. Later research, using broadly the same methodology,
studied college students (largely in the USA), especially with regard to gender and
other differences in reported anxiety, and the relationship between mathematics
anxiety and performance.

One of the reasons that 'mathematics anxiety' has received most attention among
affective factors has been the continuing attention paid on both sides of the Atlantic to
Sheila Tobias's *Overcoming Math Anxiety* (1978) and Laurie Buxton's *Do You Panic
about Maths?* (1981). These and other researchers have shown how some of the ways
that school and college mathematics have traditionally been taught lead to negative
emotions: its emphasis on abstraction, individualism, and speed have spawned
feelings of boredom, isolation, and anxiety. Widely-held myths like the 'one right
method' have often led to humiliation and anxiety (lest one's own, perhaps 'illicit',
methods be found out); myths about the 'one right order' to learn mathematics and the
possession – or lack – of a 'mathematical mind' encourage fantasies about 'starting
back at the beginning', and/or a slavish dependence on teacher (or text).

Despite the wealth of research undertaken in recent years, there are still gaps in
our conceptions and our knowledge. There remain confusion about the idea of
transfer of mathematical (and other) learning from school to outside contexts, and
disappointment about students' ability to accomplish this. In any case, how should
we specify the *context* of learning and thinking?

There is still relatively little research on the mathematical thinking of adults,
especially in the UK. Further, much of the research that has been done on adults has
not paid much attention to *affect* or *emotion*, though it does focus more nowadays
on contexts outside the academic.

Therefore, the research reported here studies thinking and affect, and their inter-
relationship, among a group of adults who were involved in college mathematics
and who also were experienced in a wide range of practical activities that might
provide a context for the use of ideas recognisable as 'mathematical'.

Developing Ideas and Methodologies for the Study

The main study reported in this book is about 'mathematical' thinking in various
contexts – college and out-of-college – by adult students. Now what is 'mathematical'
is not straightforward. A recent symposium characterised mathematics variously as:

- a discipline in its own right
- a collection of skills for wide application
- a set of 'thinking tools'

- a set of principles and techniques for modelling
- a powerful language for sharing and systematising knowledge
- a part of our cultural heritage (Hoyles et al. 1999).

There are certainly differences of view on this issue, which cannot be discussed fully here.[4] In this book, in describing mathematical thinking, I aim to emphasise the idea that it is *thinking in context*.

In this study, *adults* are understood as people of a range of ages, who:

- participate in a substantial range of activities and social relations, normally including some outside the home, school or college
- have at least the opportunity for paid or voluntary work
- are conscious of having social or political interests.

Thus, I would include many 'adolescents', and virtually all 'mature students', who in the UK are 21 or over, and usually have previous work or child-care experience.

Adults' lack of numeracy cannot fully be explained by their weakness in school mathematics, as measured by school tests for several reasons:

1   Adults use numerate ideas in a variety of social contexts, which could not be anticipated by school mathematics teaching.
2   Adults may lose certain school mathematics skills through lack of use.
3   Society's definition of a minimum adult level of competence may change.[5]
4   Lack of confidence may be as important a problem for adults as lack of knowledge.
5   The errors made by adults may differ from those made by children – sometimes because adults' greater knowledge and experience may make simple problems more complex (Withnall et al. 1981).

I began my work with the broad problem: How do adults develop, or fail to develop, mathematical thinking, and confidence in it? I developed a simple model which aimed to explain this by several factors or processes:

- socialisation depending on the individual's position in class, gender and other terms
- critical incidents in one's history of learning mathematics
- affective characteristics, which I saw as representing an 'internalisation' of the experiences from the first two sets of determinants.

I aimed to attempt to produce (or gain access to) information that would allow me to describe the 'level' of numeracy in the population of adults in the UK, or a reasonably representative sub-population; and to relate this to respondents' backgrounds, experiences, and feelings about mathematics. At the start, I was interested in the population of adults at large, though the practical constraints of doing empirical research soon led me to choose a population that would allow a compromise between representativeness and convenience (see Chapter 2).

I considered that using a methodology including both quantitative and qualitative aspects would be the best way to meet my objectives. I aimed to employ a suitably designed questionnaire, analysed using quantitative methods and modelling, to allow a general description of the study population, and of the relationships between social, affective and performance variables. I aimed to combine this with 'qualitative' (less structured) interviewing, so as to elicit descriptions of the sorts of everyday activities where subjects might (or might not) use numerate reasoning; to describe in more detail students' critical incidents and experiences; and to illuminate their more fluid perceptions, goals and interests. The interviews would also allow me to describe how adults might re-emerge from their difficulties with numeracy to have a 'second chance'. Together questionnaires and interviews would provide an appropriate methodology.

As the research developed, it became clear that I would need to develop my concepts more fully, and to deploy alternative theoretical models. In particular, the concepts of the *context* of mathematical activity, *performance* in problem-solving, *mathematical affect*, and *mathematics anxiety* needed rethinking. In response both to these conceptual challenges, and to the emerging limitations of the questionnaire results, I revised my aims for both the quantitative and the qualitative parts of the research.

Thus the broad aim of the book is to contribute to an understanding of numerate activity among adults, both in and out of school/college, and how it is developed, or constrained. I will argue not only that the notion of context is basic to the understanding of mathematical thinking, but also that the context must be conceived more broadly than in much earlier work, and as an integral aspect of activity. Further, in order to understand mathematical thinking in context, I shall also show that it is essential to understand affect, emotion, and feelings.

The broad aims of the book relate to specific objectives, as follows. The aim of understanding the development of mathematical thinking and performance led to:

- description of levels of numeracy among a selected population of adult students, comparing them with those of the national population
- investigation of social differences in performance, and in affective / emotional responses to mathematics
- critical consideration of a number of findings about mathematics performance and emotion, which have been widely considered to be valid; for example, 'Males perform better than females' and 'Women have more maths anxiety'.

Concern with a broader conception of the context of thinking and its grounding in practice led to:

- a review of the conceptions of 'context' used in mathematics education and psychology, assessing them conceptually and through empirical research
- proposing a methodology for describing the context on the basis of the practice(s) 'called up' by subjects when they are presented with numerate problems
- contributing to a reformulation of the notion of 'transfer of learning'.

The aim of understanding the role of affect and emotion in mathematical thinking led to:

- documenting the range of emotions described in relation to mathematics and numbers by members of the sample
- illustrating the contribution of insights from psychoanalysis in the study of mathematical affect, including anxiety, and mathematical thinking
- examining the relationship of thinking and emotion, both within samples of adults, and for particular subjects.

In addition, I aimed to produce:

- illustrations of ways of using quantitative and qualitative methodologies, and especially ideas for fruitfully combining them, in educational research
- recommendations for improving mathematics education, and the development of numeracy.

Overview of the Book

In Chapters 2 to 5, I describe the conceptual basis, methodology and findings of the quantitative part of the study. This phase uses a notion of the context of mathematical thinking that is consistent with that used at the time I began the overall study: that the context of a mathematical problem is determined basically by its wording and format. In Chapter 2, I consider the different ideas of numeracy or 'practical mathematics' used in several key British studies of adults. I also attempt to compare the levels of numerate performance of the students in my sample with those of a national survey done at about the same time. In Chapter 3, I draw on the literature on gender and social class differences in mathematics performance, to develop my model, and I summarise the results of the survey relating to such performance differences. Statistical modelling is used, so as to control for the effects of other variables that may be 'confounded' with gender and class. Chapter 4 takes up the discussion of affect, and emotion, in particular, the development of psychological notions of anxiety and of mathematics anxiety. I present results from my survey on the links between anxiety and performance, and on the 'dimensions' of mathematics anxiety. This part of the study concludes with an Interlude in Chapter 5 which aims to evaluate the findings of the quantitative strand, and to point to ways to build on them.

The discussion up to this point suggests that the notion of context I began with might need to be reconsidered. Therefore, Chapter 6 considers a range of ways of conceptualising the context, in studies of practical, out-of-school mathematics, and the implications for the possibility of transfer of mathematical knowledge. Chapter 7 seeks to reclaim the conception of mathematics anxiety from the limitations imposed on it in most psychological and educational research since the 1950s; in particular, the contribution of psychoanalysis is discussed, as is its combination with ideas of poststructuralism in the work of Valerie Walkerdine and others. These two chapters provide the basis for a set of themes to be addressed in the qualitative strand of the study, which is discussed, along with the methodology for the interviews, in Chapter 8. In Chapters 9 and 10, the results of this strand are presented, first as cross-subject analyses (including all of the subsample) and then in individual case studies. In these analyses I propose a notion of the context of mathematical thinking that can be captured by the idea of *positioning in practices*, and reconsider situated cognition and gender differences in terms of this idea. I also argue for the importance of the unconscious in understanding emotion connected with mathematical thinking and learning.

Chapter 11 presents conclusions and summarises the contributions of this study to theory, methodology, and mathematics education pedagogy and practice.

2 Mathematical Thinking in Context among Adults

While in the 1960s and early 1970s the preferred version of school mathematics tended to favour 'abstract' algebraic approaches . . . the dominant orthodoxy since the time of the Cockcroft Report of 1982 has favoured the teaching and learning of mathematics within 'realistic' settings.

(Cooper and Dunne 1998: 117)

The launch in 1957 of *Sputnik*, the world's first satellite in space, by the Soviet Union, shocked the USA and other Western industrial nations into a re-examination of their educational systems. In mathematics this led to an emphasis, via the 'New Maths', on abstract and axiomatic approaches, as indicated by the epigraph from Cooper and Dunne (see also Cooper 1985).

However, in recent years there has been heightened concern that mathematics should be taught and learned *within 'practical'* or *'realistic'* settings. No doubt this has been partly because of one of the problems highlighted in Chapter 1, namely the alleged inability of young workers to 'transfer' their school mathematics to work contexts, leading to renewed complaints against the educational system. These concerns were acknowledged and assessed by the Cockcroft Report (1982), and its emphasis on 'numeracy' has been seen as a response to these issues.

Nevertheless, 'practical mathematics' or numeracy can be understood in different ways, as can the related idea of the 'context' of mathematical activity.[1] I examine some of these differences here, and consider the conceptual bases and broad findings of the key studies done in recent years on numeracy in the adult population of the UK.

I also describe the methodology of my own survey of adult students, and the beginnings of my conceptual map, and I present results comparing the level of performance of the members of my sample with that found in one of the national surveys.

Conceptions of Practical Mathematics and the Context

The aim of emphasising practical applications of school mathematics raises the question of what is different about non-school contexts. Until the early 1980s, the exploration of 'context' in mathematics education seems to have been confined to describing the effect on performance of variations in the wording and format of word-problems and logical exercises (for example, Wason's 'selection problem') at school; see Bell et al. (1983). However, context has recently been discussed in a much more wide-ranging way, as I discuss later (see also Chapter 6).

Here I consider the notions of context, and related views on the applicability of mathematics, within two views competing over at least the last twenty-five years. These may be called the 'numerical skills' and the 'functional' approaches, the latter articulated by the Cockcroft Report (1982).

'Numerical Skills' Approaches

One approach pre-dating Cockcroft considered that the practical and vocational 'applicability' of school and college mathematics came from learning numerical skills in an abstract way. In a discussion of the number skills needed for work and for citizenship, Glenn (1978) emphasised 'an integrated set of mental skills and understanding' that could be defined only by listing its component skills. He presented a set of behavioural objectives, for example:

- Add without error using an algorithm any two multi-digit numbers.
- Multiply any decimal by an integer of one or two digits.

Here the context is usually unmentioned, or at most vaguely specified as, for example, 'a suitable context' (Glenn 1978: 126–31). Thus this view assumes that if the 'basics' are learned properly in their abstraction, they can be applied in any appropriate context at will.

Another view which emphasises the learning of a set of *abstract skills* in arithmetic calculation, which are assumed to be applicable in a range of practical contexts, is that of Ruth Rees and George Barr (1984, 1985); their work is based on research on mathematics learning in diverse areas of vocational education. Teaching adults numerical skills which 'transfer' means exposing 'students to different contexts which have *the same mathematical content*' (Rees and Barr 1984: 195; my emphasis). Examples of contexts are: electrical work, painting and decorating, and O-level Mathematics at school.[2] The challenge for tutors is 'to bridge the gap between using computational skills in pure form and their application in real-life settings' (ibid. 1985: 3). Thus they seem to consider the mathematical task or content to be neatly *separable from the context*, and to be unaffected by it.

For Rees and Barr, the context of a task is played down in importance. It can be captured by *the wording of the problem*, and they say almost nothing about the *conditions* in which the various groups studied actually did perform. For them, a problem such as '8 divided by 0.16' can be said to have a 'pure' context, or no context at all (ibid. 1984: 177–8). Rees and Barr are proponents of the idea that mathematics 'skills' can basically be considered in their pure form, as *context-independent*.

Thus what can be called a 'numerical skills' approach to learning mathematics and its 'practical' applications normally exhibits the following features:

1   An emphasis on learning a set of abstract basic mathematical skills in calcu-
    lation and manipulation of numbers.
2   A related notion that it is possible to have 'the same' mathematical task in
    different contexts.
3   A consequent view of task and context as neatly separable for analytical
    purposes.
4   A resultant tendency to downplay the importance of the context.
5   A view of 'transfer' to other subjects and to everyday living and working, as
    relatively straightforward. That is, as long as the mathematics is learned
    properly, in its abstraction, the applications will follow.

This set of views has been challenged by several other approaches, one of which was
the Cockcroft Report (1982).

The Cockcroft Report and 'Functional' Numeracy

The Cockcroft Committee's brief (from James Callaghan's Labour government in
1977) was to consider 'the match between the mathematical curriculum [in schools],
and the skills required in further education, employment and adult life generally'
(Cockcroft Report 1982: ix). The Committee considers the 'mathematical needs of
everyday life', and proposes the idea of 'numeracy' as the ability to cope confidently
with these (ibid. paras 32–5). However, the idea of numeracy needs definition.

The Committee recalls the 1959 Crowther Report's very broad concept of
numeracy, as the 'mirror-image of literacy', including familiarity with the scientific
method, thinking quantitatively, avoiding statistical fallacies, and so on.[3] In contrast,
Cockcroft reports that 'numeracy' was mentioned by many submissions to them, but
mostly in the narrower sense of being 'able to perform basic arithmetic operations'.
Taking an intermediate position, they want 'numeracy' to comprise not only this
latter ability, but also to the ability to *make use of* arithmetic operations *with confidence* in *practical everyday situations*, so as to cope with the latter (ibid. para. 38).

Thus Cockcroft defines the word 'numerate' to imply the possession of two
attributes:

•   An 'at-homeness' with numbers, and an ability to make use of mathematical
    skills which enables an individual to cope with the practical mathematical
    demands of everyday life.
•   Some appreciation and understanding of information which is presented in
    mathematical terms, for instance in graphs, charts or tables, or by reference to
    percentage increase or decrease.

The Report argues that together these imply an appreciation and understanding of
some of the ways that mathematics can be used as a *means of communication*, for
example, to represent, to explain and to predict (ibid. para. 39 and Chapter 1).

Related to the 'at-homeness' with numbers, they also stress the importance of a
'feeling for number' which permits sensible estimation and approximation and which
enables straightforward mental calculation to be accomplished' (ibid.: para. 33).

There are several noteworthy aspects of this definition. First, attitudes, as well as skills, are considered important: confidence and familiarity ('at-homeness') count, as well as competence. Second, the criterion for which skills are important is practical: namely, relevance to the context of the person's everyday life. Third, their notion of numeracy includes the appreciation of numerical information, as well as the use of techniques, and this appreciation is implicitly critical.

The reception given to the Report shows that Cockcroft succeeded in opening up a space for 'numeracy' in traditional discourses about mathematics education. By choosing a previously little-used word, the Committee signified their view that the mere 'skills of computation' were not sufficient, but rather the confident use of mathematical operations in practical everyday situations was essential to their aims. The use of a word different from 'mathematics' also implies that the 'transfer' of skills from abstract school maths to everyday applications is not as straightforward as a 'numerical skills' approach – sketched in the previous subsection – might suggest.

Similarly, the research study on adults commissioned for Cockcroft urges a 'functional approach', rather than 'formal manipulation' (ACACE 1982: 57). It reports on a survey of almost 3,000 adults investigating 'the national level of adults' mathematical ability' (see next section), and on a more intensive study of about 100 adults done by Brigid Sewell.

Sewell used the term 'functional mathematics' to emphasise the practical usefulness of mathematics, in distinction to the 'abstraction' of its concepts, as used by mathematicians (Sewell 1981: 1). An important feature of her study was her constructing problems in practical contexts; these she aimed to represent by the careful use of wording, tables, graphs, and facsimiles of maps, pay slips, electricity bills, and so on.

This discussion points to an important distinction. An emphasis on numerical skills, the ability to perform basic arithmetic operations, as in the approaches discussed in the previous subsection, may be called *proficiency numeracy*. In contrast Cockcroft's (and Sewell's) emphasis on the ability to make use of basic arithmetic operations with confidence in practical everyday situations might be labelled *functional numeracy*.[4]

The functional view was developed further in adult education during the mid-1970s adult literacy campaigns, when adult numeracy problems became evident in their own right. Again, the expression 'functional numeracy' is used to emphasise its practical nature (as in 'functional literacy'). However, Withnall et al. define numeracy as 'efficiency in mathematics relative to the tasks to be achieved' (Withnall et al. 1981: 30). This relativity might suggest that there is no definition in abstract terms that will be generally valid for *everyone*. Others are concerned at the way the idea of functional numeracy is sometimes seen as applicable to all students at all times:

> To present a student with a worksheet on costing a holiday in Wales when in fact he or she is going to Majorca, or is unemployed and is going nowhere, is a travesty of the functional approach. . . . To be truly functional the work must derive from the student.
>
> (Riley 1984: 2)

This suggests the need for a distinction in functional numeracy, depending on whether the relevant capabilities are seen as able to be defined as *relatively general* across a society or group – or as needing to be specified for *a particular individual*. This distinction helps to differentiate the conceptual basis of the adult numeracy studies discussed in the next section.

The distinction between numerical skills proficiency and functional numeracy provides the basis for differentiating two types of performance items – 'abstract' or school mathematics and practical mathematics – in my survey of adult numeracy.

Recent British Surveys of Adult Numeracy

Since the time of Cockcroft, there have been several large-scale representative surveys, aiming to assess adults' levels of numeracy or practical mathematics in the UK. They also can help to estimate inequalities in this area related to gender, social class and age (Evans 1989a).

The national survey of adults 'mathematical ability' done for Cockcroft (ACACE 1982), conducted by Gallup, was based on a representative quota sample of almost 3,000 adults in February 1981.[5] All ten questions were meant to test everyday or 'practical' mathematics.[6] Overall, six of the questions had to do with spending money in shopping or eating out; see for example Questions 10 (25 per cent off) and 3 (10 per cent tip) in Figures 2.1 (p. 18) and 2.2 (p. 20) respectively. Another question, about reading a rail timetable, was clearly 'practical'. Two others, about the meaning of 'inflation' and about a graph depicting temperature changes, also *seemed* practical. On the other hand, a question asking whether 'three hundred thousand' or 'a quarter of a million' was greater, without giving any units, seemed rather more 'abstract'.

This national survey provides useful information on the likely 'practical mathematical' capabilities of British adults (based on responses given in a street interview with a stranger). The most important results are as follows:

- The questions on simple operations, 'percentage of' and graphs were answered correctly between 68 per cent and 88 per cent of the time, but Question 10 (deducing, from a percentage reduction, what fraction of the original price has to be paid), the timetable question and the inflation question were answered correctly only 64 per cent, 55 per cent and 40 per cent of the time respectively.
- Men performed better than women, and the difference – between 2 per cent and 20 per cent, depending on the question – was largest on the 'abstract' comparison of large numbers, the timetable question and Question 3 (calculating a 10 per cent tip on a restaurant bill)).[7]
- The young, especially the 25–34 age group, generally did better, and the over-65s least well, though the difference was less on 'money' questions.
- Social classes AB (professional and intermediate occupations) did best and classes DE (semi-skilled and unskilled) did least well, with the differences greatest for the questions on reading the timetable, tipping, inflation and reading graphs.

The results are somewhat discouraging. Only two questions had more than 80 per cent of the sample answering correctly. Two of the three worst-answered questions (see earlier) involved percentages, though it is not clear that that was the reason since Question 10 (25 per cent off) was also 'tricky' (percentage *reduction* implying fraction of price *to be paid*), and the inflation question involved a complex concept. However, Sewell's associated interview study (1981) found many proclaiming their inability to understand percentages (Cockcroft Report 1982: para. 26).

The complex concept needed to understand the inflation question requires the respondent to make a distinction between the *level* of prices and their *rate of increase*. Over half the sample answered incorrectly. This illustrates the importance of numeracy, broadly defined, not only for work and everyday transactions, but also for the understanding needed for effective 'citizenship'.[8]

The national survey produced differences in performance across social class, gender, and age groups that were not unexpected. Further, an implication of the results described is that differences by gender or age, say, may vary widely depending on the mathematical content of the questions, and on their context – for example, abstract, money, or timetables.

The results of this national survey are compared with those from the survey used in my study, later in the chapter.

Another study, done by the Adult Literacy and Basis Skills Unit (ALBSU), also used a functional notion of numeracy, but one related to a conception of a particular individual's needs (see previous section). The data came from the fourth follow-up of the National Child Development Study (NCDS), which interviewed some 12,500 23-year-olds in 1981 (Hamilton and Stasinopoulos 1987).[9] Respondents were asked *self-rating* questions: whether they had had problems with numeracy ('number work' or 'basic maths'), and with literacy ('reading', 'writing' and 'spelling') since leaving school.

This method of measurement was used in preference to assessing performance on examples of everyday problems (cf. the ACACE study reported earlier). The drawback of the latter approach, according to the authors – echoing the quotations from Withnall et al. (1981) and Riley (1984) – was that the competences measured 'vary from task to task and may not be relevant or representative of the problems faced by people outside the community (or time) they were originally designed for' (Hamilton and Stasinopoulos 1987: 8). Thus no standard assessed questions on numeracy were included.

The most important results here are as follows:

- Some 5 per cent of the sample reported 'problems' with numeracy ('number work' or 'basic maths'), compared with 10 per cent for literacy.
- Of the 5 per cent reporting 'problems' with numeracy, over a quarter reported 'difficulties' in everyday life arising from these problems, for example, difficulties at work, getting jobs, or in household management.
- There were no gender differences in reporting numeracy problems, but, of those reporting problems with numeracy, five times as many men as women had attended classes (Hamilton and Stasinopoulos 1987).

The first and the last results are unexpected. The first – that only 5 per cent have 'numeracy problems' – is a more optimistic result about adult numeracy in the UK than that for the 16–24 age group in the ACACE study (done at about the same time). The latter showed a 73 per cent average of correct answers (on eleven questions), with the 5th percentile at only three questions correct; a very poor score indeed. This suggests that ALBSU's respondents self-ratings of their numerical skills may have been over-optimistic, or else that the two studies are based on very different conceptions of numeracy. One source of invalidity for the ALBSU study is that self-ratings were made (presumably) on the basis of many different perceptions of what are 'problems', 'difficulties', and especially 'number work' or 'basic maths'.

Later studies addressed the problem by using both types of measure with the same sample. Ekinsmyth and Bynner's (1994) study was based on a representative sample survey in early 1992 of 1650 members of the 1970 British Cohort Study (see note 9) at age 21. It used both a *self-assessed* indicator of 'numberwork' difficulties, and also a half-hour *'objective' assessment* of 'numeracy skills'; the latter included fourteen quantitative questions clustered around eight everyday tasks, such as setting a video recorder, considering a hire purchase arrangement, reading a table of statistics, similar in style to the ACACE's general measure, but more demanding.

Here the numbers with low *assessed levels* of numeracy, as defined by the study's standards, were 55 per cent for numeracy, compared with only 19 per cent for literacy, a striking difference! Yet only 4 per cent reported 'difficulties with numberwork', and, even among those with low numeracy assessment scores, the vast majority did not report numberwork difficulties. The researchers explain this as follows: 'large numbers of both sexes are either unaware of their poor skills, or are aware of them but fail to acknowledge that they have a problem' (Ekinsmyth and Bynner 1994: 24). Also, 'reporting a difficulty depends on a number of factors, not least the importance of the skill to the individual' (ibid.: 40). Thus both kinds of appraisal – standard assessment and self-reporting – are essential to understand the adult's position.

This study also illuminates the apparent anomaly of there being no gender difference in self-reporting of difficulties in the ALBSU study, again observed by Ekinsmyth and Bynner. Here men with low scores on assessed numeracy were twice as likely as women with the same level of scores to report difficulties.

Overall, the problem with the validity of self-report measures, and the related anomalies in the findings, led me to prefer *general, assessed* outcome measures for my survey, in the style of the ACACE study.

Conceptual Map and Research Questions

Here I begin to develop the conceptual basis of the survey phase of the study. I introduce the 'conceptual map', which highlights the key variables and the expected interrelationships among them. It is also used to guide the formulation of research questions and hypotheses, the specification of indicators for key variables, and the data analysis. I begin by describing decisions about the research setting and the cohorts of students to be studied.

Research Setting and Sample of Adults

In beginning my research, I had to decide between studying a sample of adults from the population at large, or a sample of adult students in tertiary education. The former would have the advantage of being more representative, but there would likely have been difficulties in contacting them, and the costs of travel and so on would have been great. The students would be more accessible, and would also be a more appropriate group for studying the differences, and connections, between school or college mathematics, and 'practical maths' or numeracy. Therefore, I opted for an appropriate group of tertiary students. My criteria were that the group should have a good proportion of members aged beyond 18–21 and having experience of paid work or raising a family before college entry – in addition to the group's being convenient and accessible.

I took the opportunity to conduct my research at one particular polytechnic, where I had been teaching for some years. The polytechnic had (and, as a 'new university', continues to have) a relatively high proportion of 'mature' students (21 years of age or over). Many of these were returning to study after some years' experience of work or child-care.

Within the polytechnic, I chose to study entrants to the BA Social Science (BASS), and the two-year Diploma in Higher Education (DipHE), where students specialised in subjects ranging across the arts and sciences (including mathematics and statistics).[10] For these two courses over 60 per cent of entrants were mature students, and up to half of these were admitted without 'traditional' qualifications (i.e. two A-levels).

Doing my research at this polytechnic, and with these samples of students, had a number of advantages, with respect to social variety in the sample, and convenience and access to respondents and data:

1   These groups of students were relatively gender-balanced, and presented a wide range of ages, social class and ethnic backgrounds, and post-school experiences.
2   I could ask colleagues for access at lectures to groups of students to reply to my questionnaires – with greater hope of cooperation.
3   I could ask students for interview, with less risk of flat refusal,[11, 12] as I was already known to all students in the BASS first year, and the first year courses in Maths/Quantitative Methods on both BASS and DipHE were generally acknowledged by students to be taught in a challenging, but humane way.
4   I had access to course records, so that I could check, say, the numbers registered for particular courses.

Of course, there were possible disadvantages. My position as lecturer in Mathematics and Statistics, and my role as Maths coordinator for BASS Year 1, risked 'reactive effects', especially in the interviews. This threat was addressed by producing a 'reflexive account' for each student cohort generally, plus one for each interviewee; see Chapters 8 and 10.

Overall, I felt that the choices of the particular polytechnic, and of the BA Social Science and DipHE courses, were appropriate, because they included a richly varied group of mature students. Using the latter as a 'working population' for the research would allow me to produce a sample that would be as representative of the population of adults in the UK as any face-to-face higher education institution could provide at that time.

Developing a Conceptual Map

The main outcomes of interest in the study were *mathematics performance*, and mathematical affect, especially *mathematics anxiety*. In this chapter, I focus on the performance variables; the model is developed further in Chapters 3 and 4, where social and affective variables, respectively, are brought in.

Because of my interest in the context of mathematical thinking, I attempted at the beginning to divide performance items into 'abstract', and 'practical' types, depending on the overt context (wording and format) of the problems. I call these *'school mathematics' (SM)*, and *'practical maths' (PM)* types of items / contexts, respectively, in discussing the survey phase of the study, in Chapters 2 to 5. Other senses in which I use these terms will be specified as the idea of context is developed in the book.

The 'Performance Scale' included ten abstract or 'school maths' questions, and twelve questions that were 'practical'[13] in the sense of relating to a context, briefly described within the question. The SM subscale included one item from the ACACE survey (see pp. 13–15), on comparing the size of two large numbers, and the PM subscale included nine others; thus all the ACACE questions (except for that on reading a rail timetable) were re-used in the polytechnic study (see Table 2.2, p. 20). The remaining performance items were constructed, with the aim of creating 'parallels' between the SM and PM subscales, in terms of the mathematics topics involved and difficulty levels. The PM items can also be classified by the practical context specified, the main ones being spending money (six questions) and interpreting opinion poll results (two questions); these were considered to avoid giving an advantage to either gender, or to any social class or age group.

The number correct on each of these subscales formed indices of performance on 'school maths' items and 'practical maths' items respectively. (For more detail on the questions used, see Evans 1993, Appendix Q4.)

Research Questions

My review of the literature led to a number of research questions or hypotheses, to be addressed in the polytechnic survey. Sometimes it was possible to formulate fairly precise hypotheses about what results were expected, but in other cases (where the literature or my own ideas were less specific), it was possible only to indicate broad research questions, or foreshadowed problems. In the quantitative part of the study, seven research questions will be addressed. In this section I shall formulate the first two research questions, and indicate the results I expected; in the next section, I shall consider the results. The first two research questions concern: first, overall levels of performance in the polytechnic adult student samples; and, second, the specification of contexts of performance, as 'school mathematics' or 'practical mathematics'.

Research Question 1: Overall Levels of Performance in the Polytechnic Samples

I found disappointing the levels of performance in the national sample reported by ACACE (1982), and later those found by Ekinsmyth and Bynner (1994) and by Basic Skills Agency (1997; see note 12). So I expected that the polytechnic students would perform at substantially better levels than the adults included in the ACACE's national survey. The indicators were comparable since the Performance Scale in the polytechnic questionnaire used all but one of the ACACE questions (see earlier). The basis for my expectation that the polytechnic students would perform better can be seen by comparing the profiles of the two samples (see Table 2.1).

Table 2.1 uses indicators for social variables in the polytechnic survey that have not yet fully been discussed, such as social class (see Chapter 3). We can see that the polytechnic students were younger, and appeared to be more middle class (using parents' occupational class at least).[14] As entrants to higher education, they would have much better general educational qualifications than the national sample. All these characteristics were associated with higher performance on the national survey (see 'Recent British Surveys of Adult Numeracy', earlier). Thus, despite the slightly lower proportion of men among the students, I expected the polytechnic sample results to be superior, both overall, and for each individual question.

Research Question 2: Specification of Contexts of Performance

As already indicated, my assumption for this part of the study was that performance items could be divided into school maths (SM) and practical maths (PM) types, and that this division would be indicative of performance in different *types of context* – abstract and practical, respectively – represented by the wording and format of the items.

In analysing the data, I focused first on the correlation between the overall scores on PM and SM. If these subscales were 'measuring the same thing', the correlation should in theory be close to 1 (though allowance must be made for measurement error). Conversely, if, as I expected, SM and PM were measuring

Table 2.1 Comparison of Profiles of the National Sample (ACACE) and Polytechnic Sample

Sample	National (n = 890)	Polytechnic (n = 935)
Gender (% women)	52%	59%
Age (years) mean	43	24
range	16–65+	17–64
Social class (%MC / %WC)	38% / 62%	
Parental social class		58% / 30%*
Students' own class		38% / 23%*

*   based on cohort 3 sample (n = 291), and not including mixed parental social class, non-response, 'don't know' and 'not applicable'

Source: for National Sample, ACACE (1982)

different types of performance, that correlation should be substantially less than 1. I also aimed to compare the performance on individual PM items, taken from ACACE and based in a 'money context', with that on those SM items constructed (see above) to be 'comparable' to them in difficulty and mathematical operation(s) used. My expectations that success rates would be higher in the PM 'money' items than in the abstract SM items were based on earlier research by Lave et al. (1984) in the USA and by the Assessment of Performance Unit mathematics team:[15]

> We know that money contexts get higher success rates than other contexts using decimals because, in the former, decimals can be avoided by thinking of x.yz as x pounds and yz pence, or xyz pence. [Thus 4.18 can be thought of as £4 and 18 pence or 418 pence.]
>
> (Derek Foxman 1988: personal communication)

Results related to these research questions are given in the next section.

Survey Results for Performance and Context

The fieldwork phase of the survey is described in Appendix 1. In each of the three years of the study, there were samples from the BA Social Science (BASS) and the Diploma in Higher Education (DipHE); the DipHE samples included students taking mathematics and statistics (Quantitative Methods) and, after the first year, another group not doing so. Overall, in the eight subsamples, 935 students completed the questionnaire.[16] After inspection of the separate results (Evans 1993), the subsamples were pooled for most of the analyses presented in this chapter. Nevertheless, allowance was made for cohort (year) and course differences when full modelling of the combined effects of all factors such as gender, social class, age and qualification in maths was done for outcomes such as performance and mathematics anxiety (see Chapters 3 and 4).

Overall Levels of Performance

When we compare the number of questions correct on average for the national (ACACE) and the polytechnic samples, we find the expected difference – one full question out of ten – in favour of the polytechnic students (Table 2.2). This is a substantial difference.

I also considered differences on particular questions, seeing the different questions as indicators of a range of practical numerical skills. The results for individual questions in Table 2.2 suggest that the polytechnic scores were higher on eight of ten questions; the two exceptions were:

1   Question 10 (national)/Question 6 (polytechnic) on the fractional amount of the original price you would expect to pay in a shop advertising '25 per cent off marked prices'

2   Question 3 (national)/Question 18 (polytechnic) on calculating a 10 per cent tip.

Table 2.2 Comparison of Results of the National Sample (ACACE) and Polytechnic
Sample

Question no. National (Poly)	Percentage correct		
	Polytechnic (n = 935)	National (n = 2890)	Difference (Poly–National)
Question 1 (1) (total cost of snack)	95	88	+7
Question 2 (2) (multiply costs)	89	74	+15
Question 3 (18) (10% tip)	65	72	–7
Question 4 (3) (abstract comparison of nos)	81	77	+4
Question 5 (4) (cost division)	88	68	+20
Question 6 (–) (reading timetable)	(not used)	55	—
Question 7 (14) (inflation)	67	40	+27
Question 8 (5) (price reduction)	87	70	+17
Question 9A (19) (temperature graph)	93	87	+6
Question 9B (20) (temperature graph)	87	72	+15
Question 10 (6) (25% off)	64	64	0
Total Correct			
Mean	8.1	7.1	
Standard Deviation	1.68	n.a.	

Source: for National Sample, ACACE (1982)

Figures 2.1 and 2.2 show the formats of the two pairs of questions.

In trying to explain these two anomalous results, it seems useful to look more
closely, not only at the differing characteristics of respondents, and the more
obvious differences in question format, but also at the different contexts (broadly
understood) of the two surveys: their *settings* and their *practices of administration*.
These may help to explain differences in performance more generally.

The national survey numeracy questions formed part of the Gallup Social Survey,
based on street interviews in February 1981. The ten numeracy items were divided
into clusters of three or four and interspersed with clusters of other items asking for
social and political opinions.[17] The polytechnic survey questionnaires were

distributed at the end of mass lectures, during the first or second week of the students' course (see Appendix 1 for further on these settings). Within the polytechnic questionnaire, the performance questions were preceded by the Experience Scale, and followed by the Situational Attitude Scale (mathematics anxiety items).

As for administration practices, most importantly, the polytechnic Performance Scale was *timed* (ten minutes). Further the polytechnic survey was completed by the student, whereas the national questions were administered by the interviewer. This could have made a substantial difference to the results on certain questions, where the interviewers might have provided 'cues' to respondents about how to respond. Such cues, often subconsciously emitted and received, can help the respondent to formulate an answer, or to correct a slightly erroneous answer (see later). It is even possible that the interviewers may have provided more explicit help, as they would have done for other sorts of responses within the same questionnaire.[18]

I now consider each pair of performance questions with anomalous results, in turn.

First, Question 10 (national) / Question 6 (polytechnic): these two questions concerned a 25 per cent reduction in price; for their formats, see Figure 2.1.

The percentage correct was the same for both (64 per cent) samples. However, this question was one of the three least well answered on the national survey, and

---

Question 10   CARD 10

25% OFF

ALL MARKED PRICES

**If you saw this sign in a shop, would you
expect to pay:**
        **A half, or
        three-quarters, or
        a quarter, or
        a third**
    of the original price?

6. **25% OFF
ALL MARKED PRICES**

If you saw this sign in a shop, would you expect to pay:
    (a) a half, or
    (b) three-quarters, or
    (c) a quarter, or
    (d) a third
of the original price?

ANSWER..

Figure 2.1 Comparison of the Format of Question 10 on the National Survey and Question 6
on the Polytechnic Survey

Sources: for National Survey question, ACACE (1982); for Polytechnic Survey question, Appendix 1

had the lowest percentage correct on the polytechnic survey. The reason *both* groups did poorly might be that solving the question requires several steps and it appears slightly 'tricky': if you simply convert '25 per cent off' to 'a quarter', you get the wrong answer, as you must continue your reasoning to obtain 'three quarters' of the original price (correct). About a third of both samples gave the (wrong) answer; in the case of the students, this was usually 'a quarter' (32 per cent of total responses); the incorrect answers were more varied among the national respondents. This suggests that many of the students may have been rushed by the time limit into not fully completing what was one of the more complex problems posed; thus it would explain their unexpected 'lack of superiority' over the national sample, who had no time limit, as well as a relatively sympathetic interviewer at hand.

Second, Question 3 (national)/Question 18 (polytechnic): These two questions concerned a 10 per cent tip; for their formats, see Fig. 2.2.

Here, the national question presented the costs of four dishes summed to a total on a card resembling a restaurant bill, whereas the polytechnic question presented only a total amount. This difference may have had a impact on the way respondents perceived the questions. The polytechnic question may simply have reminded the students of school maths questions, whereas the form of the national question may have allowed respondents to call to mind the activity of eating out at a restaurant. Again it is difficult to know for certain, but some support for this idea is given by the actual responses to the two questions; see Table 2.3.

Perhaps the most striking feature of Table 2.3 is the substantial proportion of the polytechnic sample – about one in seven – who gave the answer '37.2p'. This is just the sort of 'over-precise' answer we would expect from those who have 'called up' school mathematics in responding, rather than recognising the sort of practical tipping situation that the national question wished to evoke, where '37.2p' has no

Question 3		
	CARD 3	

This is a restaurant bill. If you wanted to leave a 10% tip, how much would the tip be?

Soup	.35p
Main course	£2.20p
Sweet	.68p
Coffee	.30p
Total	£3.53p

18. Suppose you go to a restaurant and the bill comes to a total of £3.72. If you wanted to leave a 10% tip, how much would the tip be?

ANSWER...

Figure 2.2 Comparison of the Format of Question 3 on the National Survey and Question 18 on the Polytechnic Survey

Sources: As for Figure 2.1

Table 2.3 Comparison of Results of Question 3 for the National Sample and Question 18 for the Polytechnic Sample

Sample response	Percentage for each response	
	Polytechnic (n = 935)	National (n = 2,890)
'37p', or '38p' (*correct* – Poly) '35p', or '36p' (*correct* – National)	65	72
'37.2p' (*over-precise* – Poly) '35.3p' (*over-precise* – National)	13	19
Other, wrong	10	
No answer, 'don't know'	11	9
Total	99	100

Note: for the format of the questions, see Figure 2.2

Source: for National Survey results, ACACE (1982)

meaning. It is also the sort of answer that may have been revised in the national survey, as a result of cues from the interviewer. It is also noticeable that the proportion not answering was now slightly greater among the polytechnic students than in the national sample. This is likely to be because the tipping question was presented towards the end of the Performance Scale (Question 18 out of 24), and some of the students appeared to be running out of time by that point.[19]

These comparisons show the potential influence on responses of factors like setting and administration practices, as well as question format. I have shown here how there might be outcome differences between the two surveys due to such factors, besides those that are due to sample differences, but also that it would be difficult to estimate their relative strengths of these influences.

This discussion of the settings and administration practices – or social relations – of surveys and tests suggests additional features that might be included in a fuller characterisation of the *context* of responses to questions such as these.

Specification of Contexts of Performance: School Mathematics and Practical Maths

There is little point in comparing the overall levels of performance on school mathematics (SM) and practical maths (PM), as we cannot be sure of the relative difficulty of items from the two. In addition, the time limit on the completion of the Performance Scale could be expected to take a greater toll on the PM score, since, of the last seven questions included in either subscale score, all but one (Question 16) were included in the PM subscale (see note 19).

As indicated in the discussion of research hypotheses earlier, we would expect the correlation between the SM and PM scores to be substantially less than 1, if the two were 'measuring different things'. In the event, the observed correlation between school maths (SM) performance and practical maths (PM) performance was 0.55 for the whole sample (n= 935). Since this value is still moderately large, it is not really small enough to provide unambiguous evidence that the two sets of items are *not* measuring the same thing.[20]

As for comparisons of success rates on individual PM 'money' items, with that on 'comparable' SM items, see Table 2.4.

The expectation that success rates would be higher in the PM 'money' items than in the abstract SM items is refuted for three of the four comparisons shown in Table 2.4, with the exception of the subtraction items. These are shown in Figure 2.3.

The difference favouring the PM subtraction item (87 per cent correct) over the SM (78 per cent) here does suggest that money contexts get higher success rates than abstract ones involving decimals, as suggested by the quotation from Foxman earlier (p. 19). However, the expected advantage is not observed for the other three pairs of questions. This is understandable, for this reason: though constructed to parallel the PM items expressed in pence, the SM questions involve no decimal calculations (see Appendix 1).

Thus, both the relatively low observed correlation between overall PM and SM scores, and the inconsistent differences in success rates on individual items, fail to give clear confirmation to the expectation that PM and SM subscales might be measuring different types, different contexts, of performance. That is, they raise a question as to whether the 'practical maths' items really are situated in a practical context. (Alternatively, the second finding may question whether practical contexts actually 'help' problem-solving performance.)

Indeed, there are several reasons to consider that the two subscales might actually be 'measuring *rather similar* things'. First, when the questions on the two subscales are inspected, the practical maths items (despite the origins of most in the ACACE 'functional mathematics' scale) appear very similar to the sorts of questions done at school, and therefore the PM score may not have measured anything very different from the SM. Second, the two sets of questions were completed by the students at the same time and *under the same conditions*, that is, the same setting and administration practices; indeed the school maths and the practical maths questions were roughly alternated in order in the Performance Scale.

The analysis of this result raises the question as to whether my strategy for

Table 2.4 Comparison of Performance of Polytechnic Sample between Practical ('Money') Maths and School Maths Types of Questions

| Type of context Type of operation | *Percentage correct* | |
	Practical mathematics/ money	*School mathematics/ abstract*
Addition	(Qu. 1) 95	(Qu. 7) 98
Subtraction	(Qu. 5) 87	(Qu. 8) 78*
Multiplication	(Qu. 2) 89	(Qu. 10) 91
Division	(Qu. 4) 88	(Qu. 9) 90**

Notes:
*   For Cohort 3 version of item only
**   For Cohort 1 and 3 version of item only[21]

5. If you bought a raincoat in the 'summer sales' reduced from £44 to £29.50, how much would you save? (PM 'money' type)

8. 56 − 23.5 = ... (SM 'abstract' type)

Figure 2.3 Comparison of the Practical ('Money') Maths and School Maths Types of Subtraction Question in the Polytechnic Survey

specifying differences between the contexts of performance through different wordings of items in the questionnaire has been successful. The implications are taken up in the conclusion to the chapter.

Conclusions

One of the problems highlighted in Chapter 1 was that of how mathematical thinking and 'performance' might depend on the context. Conceptions of the context of doing mathematics in turn have implications for our ideas of what mathematics in practical situations or numeracy would be like, and for the possibilities of transfer of learning across contexts, for example, from school to everyday and work settings.

I have discussed two clusters of views so far on the context and transfer of mathematical thinking; these may be called 'numerical skills proficiency' and 'functional numeracy' approaches. The proficiency approach emphasises the learning of a set of abstract mathematical skills, in calculation and manipulation of numbers. In contrast, the functional approach emphasises not only the ability to perform basic calculations, but also to do so 'with confidence in practical everyday situations'. This is the idea of numeracy, as propounded by the Cockcroft Report (1982).

Both approaches see the context as rather obviously defined, based on 'natural' settings – such as shopping, electricians' work, or GCSE Mathematics at school – and as able to be read off from the wording etcetera of the problem or task. However, the functional approach generally aims to specify the context of using mathematics more fully than the proficiency approach. In research, this may mean going so far as to attempt to describe the 'mathematical needs' of contexts like 'everyday life' and 'employment' (Cockcroft Report 1982: chs 2, 3); in problem construction, it often means heightened attention to wording, formatting, diagrams, and so on.

Concerning transfer, the proficiency approach assumes that transfer from educational situations to a wide range of everyday and work situations will occur in a largely straightforward way as long as teaching and learning are done 'properly', while the functional position is more cautious about the conditions under which transfer can be expected to occur. Nevertheless the distinction is one of degree, not of kind. Both approaches hold to the idea that one can be sure of having 'the same' mathematical content in different contexts, that the mathematics is not changed by the setting where it is used.

These two approaches have continuing influence in contemporary mathematics education policies and practices. Indeed, they have often been in competition in the last twenty years. If functionalist approaches received a boost from Cockcroft in the 1980s, the numerical skills approach has seemed to regain the ascendancy in the 1990s. In particular, the latter approach can be seen to have expanded to relate to cognitive skills in vocational training (Wolf 1991), key skills in higher education, and the National Numeracy Strategy being promoted in UK schools (Numeracy Task Force 1998; see also Chapter 11).

These two approaches are significant for the research design of my study. Because of their somewhat differing views on context, they provide a basis for my distinguishing the performance dimensions of abstract or school mathematics (SM) and practical mathematics (PM): proficiency approaches would emphasise the former and functional approaches the latter.

In reviewing several pioneering surveys of adults in the UK, I found disagreement concerning the desirability of generality in measuring numeracy. After considering their conceptions of numeracy, indicators for it, and results, I adopted for my survey the idea of numeracy (or practical mathematics) as relatively general across British society, and as measurable by a standard set of questions *assessed* by the researcher, as used by the ACACE (1982) study for Cockcroft and by Ekinsmyth and Bynner (1994), rather than the more individual conception of numeracy, measurable by *self-report* indicators, as in the analysis of the National Child Development Study cohort at age 23 (Hamilton and Stasinopoulos 1987).

Thus I produced a set of performance items, some worded in 'school' (SM) and some in 'practical' (PM) contexts, the latter including most of the ACACE questions. This was done so as to have an indicator of numeracy that would apply across my sample, and to be able to make comparisons of my respondents' results with those of the relevant national survey.

My survey was undertaken with three years' entrants to social science, arts and science courses at a London polytechnic (n > 900). Besides the performance items, data were produced on social variables (gender, social class, age), experience with mathematics and numbers, and affective variables (see Chapter 4).

The basis of a conceptual map is proposed in this chapter, and will be developed further in the next two. Two research questions have been addressed here. First, how would the performance of students entering higher education compare with that of the ACACE's national sample? Second, is the method used here – of distinguishing between school mathematics (SM) and practical maths (PM) contexts, on the basis of wording and format of items – adequate?

For research question 1, as expected, the overall performance level among the students was higher than that on the national survey, except for two of the items. These anomalies were explained by differences, not only in item format, but also in the settings and administration practices, or *social relations*, between the two surveys.

On research question 2, the results raised questions about the assumption that the SM and PM items might be measuring different things, that is, performance in different contexts. The basically identical conditions in which the two types of items were answered heightened these doubts.

These findings have major implications for the study:

- The distinction between school maths and practical maths contexts of performance, as operationalised so far, needs to be deployed cautiously in the remaining analyses of the survey results. In particular, the separate analysis of SM and PM performance outcomes may not always be called for. Questions over the reliability of the PM scale (especially from the operation of the time limit) may limit its use as an outcome measure.
- This chapter suggests considering elaborations of, or alternatives to, the narrow characterisation of the context of mathematical thinking and performance, as simply the wording and format of problems or tasks. Thus 'school mathematics' might be taken to mean the mathematics that is taught and learned in the school, comprising curriculum, pedagogy and assessment; this would include the setting and the social relations of what would be seen more broadly as an *activity* or *practice* rather than simply as a type of performance item. In this view, the idea of 'practical mathematics' is more complicated, since it can be seen as an element of a large number of activities, such as shopping, playing cards, and many kinds of work. (I shall return to the discussion of context and 'transfer' in Chapters 5 and 6.)

The studies reviewed in this chapter point to the relationship of social differences (due to gender, social class or age) to performance outcomes. They also attest to the importance of affective variables, notably 'confidence'. These issues relate to the problems of 'inclusiveness' and 'affect' raised in Chapter 1. Thus, in the next two chapters, I investigate whether there might be social differences in performance or affective variables (especially mathematics anxiety levels), and what might be the impact of mathematics anxiety, confidence and other affective variables on mathematics performance.

3   Mathematics Performance and Social Difference

I dropped mathematics at 12, through some freak in the syllabus. . . . I cannot deny that I dropped maths with a sigh of relief, for I had always loathed it, always felt uncomprehending even while getting tolerable marks, didn't like subjects I wasn't good at, and had no notion of this subject's appeal or significance. The reason, I imagine, was that, like most girls, I had been badly taught from the beginning: I am not really as innumerate as I pretend, and suspect there is little wrong with the basic equipment but I shall never know. . . . And that effectively, though I did not appreciate it at the time, closed most careers and half of culture to me forever.

(Margaret Drabble, *The Guardian* 5 August 1975: 16)

Differences in academic achievement across social groups have been a long-standing concern among educators, policy-makers and the public. However, until recently, the literature on academic achievement and social class, and that on achievement and gender, have tended to be somewhat separate. A concern about social class probably predominated in the 1950s and 1960s, at least in the UK, but a sensitivity about gender differences, especially in mathematics and science education, was in the ascendancy in the 1970s and 1980s. In the 1990s, people have tended to focus on several dimensions of difference: gender, social class and ethnicity.[1]

Here I use ideas from several literatures as a basis for developing my 'conceptual map' of plausible influences on differences in school mathematics and practical maths performance among adults, and for articulating research questions to be investigated. I also present findings from my survey on gender and social class differences in performance.

Gender Differences

In recent years, gender has been a central and vibrant concept, both in mathematics education and in social research generally.

Two national studies of adults reviewed in the previous chapter (ACACE 1982, Ekinsmyth and Bynner 1994) used a functional idea of numeracy and showed substantial gender differences, in performance scales similar to my 'practical maths' (PM) scale. In addition, the recent International Numeracy Survey (Basic Skills

Agency 1997) also reported gender differences in the UK subsample on a set of items of which a majority were of the proficiency type, that is, akin to my 'school mathematics' (SM) items. Thus, on the basis of the previous surveys of adults, there was reason to investigate gender differences in my sample.

Gender differences in mathematics performance at school have been observed in many societies during recent times, and these continue to be monitored by researchers (such as Keitel et al. 1996). Here I do not present a detailed review of recent findings in this area: suffice it to say that the situation is dynamic and differs across countries and cultures (Hanna 1996), and that recent results suggest that the male 'advantage' in school mathematics performance, once somewhat pronounced, is lessening or disappearing in many advanced industrial societies, at least.[2]

Here the main focus is on the ideas underlying the basic model I use to explain differences in performance due to gender – and other influences – in the survey. These ideas can be situated by sketching a brief history of the beginnings of gender and mathematics research in the USA and the UK. Gender differences have been a concern in such research in the USA since at least the 1970s, even longer than in the UK, where concerns with social class differences predominated well into the 1970s. In the USA, Elizabeth Fennema (1979) and others interpreted the gender differences in national standardised tests as indicating mainly that young men had taken more mathematics courses at school than young women; this pointed to the importance of controlling for *participation* or course-taking, when comparing performances. At the same time researchers such as Fennema and Sherman (1976) emphasised the role of 'affective factors', both as influences on participation and performance, and also as influenced by social variables – for example, perceptions of parents and teachers – that were themselves linked with gender.

In the UK, both school-level findings – for example, those from the APU (Foxman et al. 1985) – and statistics for higher education (Cohen and Fraser 1991) in the 1980s showed a pattern of gender differences similar to those of the USA. However, these results were contested at the time, for example by some of the contributions to Burton (1986), and by Walkerdine et al. (1989). Walden and Walkerdine themselves turned their attention to the way girls' good performance was discounted in classrooms, for example by the tendency for girls to be entered less often than boys for the higher status, more 'demanding' examinations, that is, GCE-O levels (rather than CSEs), and A-levels (Walden and Walkerdine 1985: 44–5), at least until the advent of the GCSE examination in the late 1980s.

Fennema and her colleagues developed their later version 'generic' model (Fennema 1989), to include 'mediating learning activities' (MLAs) between affective factors and academic performance generally (for example, course-taking, mathematics problem-solving); MLAs include: choosing to solve a novel problem, 'thinking independently', and persisting (Fennema and Peterson 1985). In addition, external variables are considered to influence both affect and MLAs; see Figure 3.1.

There are a number of points to note about this model. First, variables such as innate 'abilities' are omitted, on the grounds that 'the examination of biological variables, which are often believed to be unchangeable, would not offer much help in achieving equity in education' (Fennema 1985: 304). Second, affect is considered as

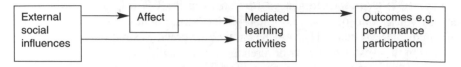

Figure 3.1 Fennema's Generic Model for Relating Social Influences, Affect and
          Mathematical Outcomes

Source: adapted from Fennema (1989)

important, because of its influence on outcomes like performance and course-taking,
via 'mediating learning activities'.

Third, external social variables are understood to include the influence of home,
school, and the community. The home and the school are considered to exert influences on the individual student through others' aspirations, expectations, beliefs and
attitudes – especially those of parents and teachers respectively – and through the
student's perception of these (for example, Chipman and Wilson 1985). Community
factors include a range of influences from peers to cultural 'stereotypes', such as
beliefs in 'mathematics as a male domain'.

To summarise, the studies briefly reviewed here have emphasised the relevance
of affective and social influences on mathematics education outcomes, and the
importance of controlling for mathematics course-taking (in US studies, at least).
Yet it is important to note that most of the studies referred to have been at school
level, and in the USA. At the time that I began my study there was a relative dearth
of information on gender (and other) influences on 'preparation' for, and performance in, mathematics courses for non-specialists in higher education,
particularly in the UK, despite non-specialists normally being the majority on
mathematics courses in higher education.[4] Such gaps pointed to questions that
could be investigated in the present study.

The Fennema model provides a stimulus for the type of conceptual map I develop
for the analysis of the survey results in this and the next chapter. In particular, the
studies mentioned here and related literature reviewed in Chapter 4 , provide the
basis for my investigation both of social differences like gender and of affective
variables, as important influences on mathematics performance among adults.

Social Class Differences

This discussion of social class differences is dissimilar to that on gender for two
reasons. First, social class differences have not had nearly as much attention (as
gender) in recent mathematics education research, but they remain important in
social research generally (Reid 1998). Second, one of my aims here is to clarify the
notion of social class, and the indicators that might be used both in the questionnaire
and in the interviews.

The literature on social class differences in educational performance generally is
immense: the reader is referred to reviews such as those by Reid (1981, 1998) which
indicate the wide-ranging and persistent advantages for those from the middle or

'non-manual' class over those from working class or 'manual' backgrounds, in the UK. Reid gives a basic definition of social class as 'a grouping of people into categories on the basis of occupation . . . [since] occupation is a good indicator of the *economic* situation of a person and a family' (Reid 1981: 6–7; my emphasis). The indicator for social class most used in social and educational research in the UK until recently has been the Registrar General's social class categories (Table 3.1), based on groupings of occupations according to 'general standing within the community' (Reid 1981: Ch.2).[5, 6]

One of the problems in measuring social class has to do with the unit of measurement. When an indicator for the *family or household unit's* social class is needed, the work reviewed by Reid normally uses the occupation of the 'head of household' – with 'default' definition as the senior male – but this risks minimising the role of the mother in fostering educational attitudes and actions.

British research relevant to mathematics learning focusing on social class includes the National Child Development Study, whose cohort members' numeracy at age 23 was examined earlier (see Chapter 2). Using the Registrar General's occupational classes (Table 3.1) as the indicator, they found, for example, substantial differences (between non-manual, skilled manual and unskilled manual groupings) in a standardised mathematics test at age 16, which represented a widening of differences found at ages 7 and 11 (e.g. Fogelman 1983). The APU mathematics studies (see Chapter 2), using an indicator for social class only *for the school as a unit*, showed a decrease in performance across schools as the proportion of pupils on free school meals increased (Foxman et al. 1985).

More recently, Boaler (1997), also using Registrar General's classes, argues (using results from very small numbers of students) that social class can be related, in some schools at least, to decisions made on 'setting' students into 'ability groups' for teaching, which itself is one of several influences on final performance. Cooper and Dunne (1998, 2000) use a scheme due to Goldthorpe and others (for example, Erickson and Goldthorpe 1993), also based on occupations, grouped on the basis of 'dominance' or authority in the workplace into three classes:

1 'service' (cf. Registrar General's I and II)
2 'intermediate' (cf. Registrar General's IIIN, plus supervisors, technicians, farmers and small proprietors)
3 'working' class (cf. Registrar General's IIIM, IV and V, plus agricultural workers).

Table 3.1 Registrar General's Social Class Categories

I	Professional, etc. occupations
II	Intermediate
III(N)	Skilled Non-manual
III(N)	Skilled Manual
IV	Partly Skilled
V	Unskilled

Source: OPCS (1980)

Cooper and Dunne use this indicator to analyse differences in children's responses to 'esoteric' and 'realistic' items, broadly similar to my school mathematics and practical maths items respectively.

Rather than social class, American social research tends to prefer 'socio-economic status' (SES), but has generally used either concept much less than British research. SES may be defined as 'the position that an individual or a family occupies with reference to the prevailing standards of *cultural* possessions, effective income, material possessions and participation in group activity in the community' (White 1982: 462–3; my emphasis).

In one of the few recent reviews focusing on the effects of social class (as well as ethnicity and language group) on differences in mathematics achievement, Secada (1992) offers several possible reasons for the under-emphasis of social class in US studies. First, social class seems to be a less 'salient' characteristic of students than ethnicity, gender or language background. Second,

> it is as if social class differences were inevitable or that, if we find them, the results are somehow explained. It even seems different to claim that lowered achievement is an effect of social class than to claim that it is an effect of ethnicity.
>
> (Secada 1992: 640)

This suggests that US education researchers (in mathematics, at least) feel less optimistic that the negative effects of social class can be combatted and compensated for, than they do for gender or ethnicity, or perhaps that there is less political pressure for them to do so. Further, an earlier review, Reyes and Stanic (1988) suggests that those studies that have included SES are difficult to reconcile because of the use of differing indicators.

However, bringing in indicators of social class (or SES) which include parents' education underlines a concern, not only with economic aspects, but also with cultural facets of social class positions. This suggests a multi-dimensional concept, which would require a number of survey questions to measure it. However, because of the constraints on the questionnaire for this study – namely, it would need to be completed by the student him/herself, in a limited time – I realised that it would not be possible to ask sufficient questions for indicators of multiple dimensions of social class. Thus, for the questionnaire, I based my social class indicator on occupation (for further details, see Appendix 1).

In interviews, however, it would be appropriate to use a broader conception of social class that includes not only *economic* aspects – such as access to wealth, living standards and orientation to money, but also *cultural* ones – such as attitudes to work and to education, beliefs about what can be known and how you know it, views on the meaning of 'growing up' and attaining 'independence', and resources for exercising power or manifesting anxiety (cf. Walkerdine et al. 1989). That is, we could include, in our conception of social class, elements relating to identity-formation or *subjectivity*. This would support my concern with affective and emotional issues.

Conceptual Map and Research Questions

The conceptual map used in this study is derived from Fennema's 'generic' model, referred to earlier (see pp. 28–30). The previous chapter focused on the school mathematics (SM) and practical mathematics (PM) performance variables. Here I relate gender and also social class, as social structural variables, to performance. Finally the next chapter brings in the affective variables, especially 'mathematics anxiety', and presents a full version of my conceptual map.

Much of the US research on gender differences has used the concept of 'participation' or the number of mathematics courses taken, both as a determinant of performance, and as an outcome in its own right. However, in the UK, most students had been taking mathematics courses throughout the first five years of secondary school, even before the advent of the National Curriculum in the late 1980s. Therefore I focus instead on the type of exam passed, the higher status GCE O-level versus CSE at 16+, as well as the level: none, 16+, A-level at 18+. Thus the concept of participation gave way to that of *qualification*, and its role in my analysis is mostly as an influence on mathematics performance in the survey, that needs to be *controlled for*, rather than as an outcome.

I considered further 'social variables' referred to by the generic model, for example, parents' expectations, beliefs and attitudes, or, more precisely, students' perceptions of these (see pp. 29–30). However, the space and time constraints in the questionnaire ruled out including them but, since parental attitudes and so on have been shown repeatedly to relate to social class (for example, Fogelman 1983), I aimed to include some of the former's effectivity via the social class measures in the questionnaire.[1]

My inclusion in the study of 'mature' students suggested that the conceptual map should include not only the student's parental social class (at the beginning of secondary school), but also the student's 'own' occupational social class, where appropriate, at the time of entering college.

I was interested also in *age* as a basic variable, that, like qualification, should be controlled for. Since 'mature' students were seen as having special strengths and needs, I considered age might be a 'proxy' for, that is might substitute for, measures of experience in practical, out-of-school activities requiring numerate thinking.

In this chapter, the third and fourth of the seven research questions addressed by the survey will be investigated:

Research Question 3: gender differences in school mathematics (SM) and practical mathematics (PM) performance; and
Research Question 4: social class differences in SM and PM performance.

Research Question 3: Gender Differences in School Mathematics and Practical Mathematics Performance

At the beginning of my study, many studies of older secondary pupils had reported differences in favour of males, on 'school mathematics' type items. Therefore I expected uncontrolled gender comparisons on SM scores in my study to favour men.

For practical maths items, the results for the ACACE (1982) numeracy questions, which formed part of my PM scale, had showed clear gender differences on most questions. On the other hand, my enlarged PM scale was based on activities familiar to both genders, so there was less reason to expect one group to perform better than the other (see Chapter 2).

Of course apparent gender (or other) differences might result from 'selection effects', or *confounding* of gender differences with others. Here, for example, women students might be either less well-qualified in school mathematics than men, or the women might be older, and hence might have lost more of their familiarity with school mathematics; in this case, it might 'actually' be the age, rather than the gender differences, that makes the difference in performance.

Indeed I did expect to find gender confounded both with qualifications and with age, for the students in this study. First, neither O-level nor A-level Mathematics were required for entry to the courses studied (for most subjects, anyway) and hence the greater success in these exams by males, in the population at large at that time, was likely to be evident here. Second, in my samples, an even higher proportion of the women than of the men were mature students. These are reasons why I needed to control for differences in mathematics qualification, and in age, when considering gender differences.

There might also be *interaction effects* among two (or more) factors; that is, the joint effect of two factors might be more (or less) than the simple 'additive' sum of the two individual effects. For example, if there were greater barriers to a woman's taking and passing O-level or A-level Mathematics at the time of her schooling, the advantage in performance (on the survey) associated with having the qualification might be greater for a woman than for a man; this would result in a 'qualification by sex interaction effect' for performance.

Research Question 4: Social Class Differences in School Mathematics and Practical Mathematics Performance

In this account of the study, research questions and results are presented only for parental social class.[8] The two main categories used were 'middle class' and 'working class' with a residual 'mixed' category (see Appendix 1). Following previous research, I expected students from middle class backgrounds to perform better on school-type questions than working class students. For practical problems, I expected a difference in the same direction, though perhaps not as large.[9] Again, selection effects need to be controlled for, in terms of qualification in mathematics and age.

Now, to the results.

Survey and Modelling Results for Gender Differences in Performance

Here uncontrolled results, comparisons across relevant subgroups, and results based on the full regression models (including affective variables), are presented. The latter results therefore anticipate the discussion of the full conceptual map and the modelling procedures, given in the next chapter.

Taking uncontrolled performance results first, *both* men and women have high mean scores in the whole sample: between 8 and 9 correct out of 10 questions for SM performance, and between 8.5 and 9.5 correct out of 12 for PM performance (see Tables 3.2 and 3.3). Put another way, the tasks on both performance scales were not very difficult for either men or women in the Polytechnic sample. However, the uncontrolled results do show higher scores for men on both measures. The sample estimates of gender differences are substantial – about three-quarters and half of a question for SM and PM performance, respectively – and both are statistically significant ($p<.001$). However, we should control both for qualification in mathematics and for age differences.

Taking qualification in mathematics, 12 per cent more males than females had passed O-level or A-level Mathematics (considered 'high' qualifications) in the whole sample (57 per cent – 45 per cent, derived from Table 3.2); this is moderately substantial, as expected. This difference seems to have resulted largely from the fact that the percentage of men who had passed O-level Mathematics is 10 per cent higher than that among women, and the percentage who had done CSE is 6 per cent lower (Evans 1993). So it seems to support claims made in the literature (before the advent of GCSE exams) that there was a greater tendency for boys to be entered for the more demanding O-level in Mathematics, and for girls to be entered for CSE (Walden and Walkerdine 1985). Thus it is important to control for qualification in mathematics, as well as for age, when considering gender differences, *especially in school mathematics performance* in this sample.

I examined gender differences in performance while controlling for age and qualification in mathematics, in two ways. First, by examining the mean performance scores for the subgroups defined by the eight possible combinations of gender (two), qualification (two levels: O/A-level Mathematics and others) and age (two groups: 18–20 years, and 21+). Second, by modelling the relationships of the set of possible 'predictors' with SM and PM, using analysis of variance and multiple regression procedures, which *simultaneously* control for all the variables entered into the models.

The sets of mean subgroup performance scores are shown in Tables 3.2 and 3.3 for school mathematics performance, and for practical maths performance, respectively.

Overall, we can see the following:

Table 3.2 Differences in School Mathematics Performance by Gender, Age and Qualification in Mathematics: Means of Subgroups for Whole Sample (n = 863*)

Qualification/age	Males	Females	Difference
High/Young (18–20)	9.27 (n = 98)	8.85 (n = 118)	0.42
High/Mature (21 +)	9.24 (n = 104)	8.67 (n = 111)	0.57
Low/Young (18–20)	8.54 (n = 35)	8.33 (n = 69)	0.21
Low/Mature (21 +)	8.19 (n = 117)	7.23 (n = 211)	0.96
Whole sample	8.83 (n = 354)	8.07 (n = 509)	0.76

* Students for whom gender, age or qualification scores were not available were excluded

Table 3.3 Differences in Practical Maths Performance by Gender, Age and Qualification in Mathematics: Means of Subgroups for Whole Sample (n = 863*)

Qualification/age	Males	Females	Difference
High/Young (18–20)	9.85 (n = 98)	9.20 (n = 118)	0.65
High/Mature (21 +)	9.63 (n = 104)	9.15 (n = 111)	0.48
Low/Young (18–20)	8.46 (n = 35)	8.65 (n = 69)	-0.19
Low/Mature (21 +)	8.71 (n = 117)	8.31 (n = 211)	0.40
Whole sample	9.27 (n = 354)	8.72 (n = 509)	0.55

* Students for whom gender, age or qualification scores were not available were excluded

1   For both measures, the highest score was for young (18–20), high-qualified (O-level or A-level Mathematics) males.
2   For both measures, the lowest score was for 'mature' (21+), low-qualified females.
3   Men outperform women, for both performance measures, for all subgroups, except for practical maths performance among the young low-qualified.
4   High qualifications in school mathematics appear to be an advantage, for SM and also for PM performance, for all age / gender groups.
5   Being young (18–20) is also a substantial advantage (average difference > 0.25 question) only among the low-qualified – especially among women for SM – but not among men for PM performance.

From Tables 3.2 and 3.3 it does appear that there are 'interaction effects' among gender, age, and qualification in mathematics for both SM and PM performance. This can be seen from the fact that differences between average male and female performances are not constant for the subgroups differing on age and qualification in mathematics.

However, the most striking finding from this analysis of partly controlled results (that is, on the basis of gender, qualification and age only) is the relatively low performance of the *mature female* group who are also *'low-qualified' in mathematics*. This is especially striking for school mathematics performance, where the difference between this subgroup and all others was large: between one and two questions. However, this result may depend on the effects of other (for example, social class or affective) variables, so far uncontrolled, so it needs confirmation from the full model.

The second method for controlling other variables used the modelling procedures discussed more fully in the next chapter. Here I present results based on the full models (including not only social, but also affective variables).

I begin with the results of modelling school mathematics performance, for the whole sample. The model used here includes qualification in mathematics, age, 'maths test/course anxiety' (one dimension of the MARS mathematics anxiety scale), confidence (self-rating) in mathematics, controlling for these and other variables and allowing for relevant interaction terms. Using this more precise analysis, the difference for younger students (aged 18–20), about one sixth of a question, is no longer statistically significant, and that for mature students (21+), just over half a question, is borderline (see Table 3.4).[10]

Table 3.4 Gender Differences in School Maths Performance: Uncontrolled Group Means (n = 933) and Estimates Controlled for Qualification in Mathematics, Age, etc. in Multiple Regression Model, for Whole Sample (n = 837*)

| | *Average number of questions correct (out of 10) (standard deviation/standard error)[11]* | | |
	Men	Women	Difference
Uncontrolled	8.78	8.07	0.71[12]
	(1.53)	(1.69)	(0.11)
Controlled			
Young (18–20)			0.16
			(0.17)
Mature (21+)			0.57
			(0.24)

* Only cases for which no variable score was missing were included in the regression analyses

Thus we can see that the uncontrolled gender difference in school mathematics performance, which is substantial (almost three-quarters of a question out of ten, and statistically significant at the 0.001 level) becomes much less substantial after controlling for qualification in mathematics, age, and so on.[13] This happens basically because gender is 'confounded with' (or correlated with) qualification in mathematics and other factors which are also related to performance in the school mathematics subscale. This shows that it is important to control not only for mathematics course-taking or qualification, but also for age, affective variables, and relevant interactions, and the multiple regression modelling allows us to do this.

A more informative way to allow for possible sampling error, rather than testing for statistical significance, is to produce *confidence intervals* for the estimates of the effect. For example, a 95 per cent confidence interval for the value of the gender difference for younger (18–20) students is between 0.50, that is, half a question in favour of males, and –0.18 of a question, just under one-fifth of a question in favour of females.[14] The estimate of the gender difference for mature (21+) students was 0.57, with standard error 0.24. In this case, the confidence interval was 0.09 to 1.05, or between one-tenth of a question and just over one question in favour of males. The fact that the lower bound of the confidence interval is just above zero shows that the gender effect for mature students is only of borderline statistical significance.

We can evaluate this model by checking the value of R-squared, the proportion of variation in the outcome, SM performance, that is accounted for by its relationship with the set of 'predictors': gender, age, qualification, affective variables and so on. Here it is 29 per cent: this is not large, but reasonably acceptable.[15]

I turn to the modelling of practical maths performance. The uncontrolled estimate of the gender effect was 0.57 in favour of males, just over one-half of a question (see Table 3.5).

The estimate of the gender effect as a result of the PM modelling is 0.27 of a question in favour of males, a reduction of the order of one-half due to the

Table 3.5 Gender Differences in Practical Maths Performance: Uncontrolled Group Means (n = 933) and Estimate Controlled for Qualification in Mathematics, Age, etc. in Multiple Regression Model, for Whole Sample (n = 853*)

	Average number of questions correct (out of 12) (standard deviation/standard error)[11]		
	Men	*Women*	*Difference*
Uncontrolled	9.24	8.67	0.57[12]
	(1.91)	(1.93)	(0.13)
Controlled			0.27
			(0.13)

* Only cases for which no variable score was missing were included in the regression analyses

controlling of other variables (and about the same order of reduction as for the gender effect in the SM performance modelling). From the value of the standard error of the estimate given in Table 3.5, the 95 per cent confidence interval for the gender effect can be calculated as being between 0.01 and 0.53 of a question. Thus the estimate of the gender effect is statistically significant, but only at the borderline 5 per cent level. And, anyway, one-quarter (0.27) of a question is not a very substantial difference.

In addition, the value of R-squared for the PM performance model for the whole sample is only 17 per cent (some 12 per cent less than the R-squared for the SM model). This value is low, and indicates that a great deal of the variation (83 per cent) in PM scores remains 'unexplained' by the model, that is not 'accounted for' by the relationship of PM performance with the set of predictor variables. For this reason, no further modelling results will be presented in this chapter for practical mathematics performance.

Survey and Modelling Results for Social Class Differences in Performance

Since social class questions were included only in the cohort 3 questionnaire, the data used here comes from that cohort only, and results are presented for parental social class only (see note 8). Uncontrolled differences related to this social class measure were observed for school mathematics performance and for practical maths performance in the expected direction: students with middle class parents scored higher than those with working class parents. However, the differences were not substantial – both less than one-quarter of a question (and not statistically significant) (see Table 3.6).

An estimate for the parental social class effect based on a performance model is available only for SM performance (because of the low value of R-squared for the PM performance model). Its size increases very slightly after controlling for mathematics qualification, gender, age, and relevant affective variables. However, it is still only about a quarter (0.28) of a question. This gives support to my original hypothesis – that students from middle class backgrounds would perform better than working class students – but only weak support, since the

Table 3.6 Parental Social Class Differences in School Mathematics Performance and Practical Maths Performance: Uncontrolled Group Means and Estimate for School Maths Performance Controlled for Qualification in Mathematics, Gender, Age, etc. in Multiple Regression Model, for Cohort 3 (n = 217*)

	Average number of questions correct (out of 12 and 12, respectively) (standard deviation/standard error)[11]					
	School mathematics performance			*Practical maths performance*		
	Middle class parents	*Working class parents*	*Difference*	*Middle class parents*	*Working class parents*	*Difference*
Uncontrolled	8.58 (1.70)	8.36 (1.73)	0.22 (0.26)	9.23 (1.73)	9.13 (1.71)	0.10 (0.27)
Controlled			0.28 (0.21)			N/A

* Cases for which any variable score was missing were excluded from the regression analyses[16]

estimate of parental background effects is not statistically significant, and hence might be due to a quirk in my sample.

Conclusions

This chapter focuses on differences in performance – both in 'school mathematics' and (to a lesser extent) in 'practical maths' – related to gender and to social class. Age and qualification in school mathematics are used basically as controlling variables. It was not possible to study ethnicity in this survey.

Rather than reviewing the research findings on gender differences, which vary across cultures and over time, the focus here is on producing a model which might explain any differences in mathematical outcomes (such as problem-solving performance) related to gender and other social differences, and which will be extended to include affective factors in the next chapter. My model is developed from Fennema's (1989) 'generic' model of educational outcomes.

Social class differences have been less investigated (than gender differences) in mathematics education over the last twenty years, but they remain important in social research generally. Thus, I aimed to take account of social class here. This raises the problem of how to measure it. UK studies have tended to use parental (often father's) occupation, often using the Registrar General's categories, whereas US studies have tended to prefer socio-economic status, aspects of which can include some (or all) of parental occupation, parental education, and sometimes parental income. I have used the Registrar General's categories.

Turning to the results, the gender differences expected (in favour of males) were found for school mathematics (SM) performance, and also for practical maths (PM). However, more of the males had successfully completed O-level or A-level courses in Mathematics, and so mathematics qualification (previous courses taken), and also age, needed to be controlled for. Initial analysis also suggested the existence of interaction effects. Both of these points underlined the importance of multiple regression modelling. This produced estimates of gender effects from the models for the whole sample, for both SM and PM performance, which were reduced by at least one-half, as compared with the uncontrolled differences. Further the modelling results were more specific. For SM performance, the gender difference in favour of men continued to hold only for mature (21+) and not younger (18–20) students. For PM performance the gender difference in favour of men was no longer substantial (one quarter of a question), and was only of borderline statistical significance. The effect of a more carefully controlled and comprehensive analysis is that the gender differences reported here are by no means as clear cut as those reported elsewhere, for example in most of the surveys of adults done since the early 1980s (see Chapter 2).

Social class effects were investigated, using cohort 3 alone (the only one to be asked the relevant questions). Here differences related to parental occupation were in the expected direction: that is, those with middle class parents performed better than those with working class parents. But these differences were not substantial, even before controlling for age and qualification in mathematics. The final estimate of social class effect from the model for SM performance was very little different from the initial uncontrolled value (about one quarter of a question).

Three features of the results reported stand out:

1   The way that initially impressive gender differences were reduced, and made more specific (to subgroups) as a result of using controls: comparing relevant subgroups, and especially using statistical modelling.
2   Social class differences in performance that were generally much smaller than gender differences.
3   The suggestion of a pocket of lower performance among these entrants to higher education who were lacking in O- or A-level Mathematics, mature (21+), and female.

This last finding is relevant in two areas: for research projects like this one, it may provide a basis for selecting students for more intensive interviewing; and for provision in tertiary institutions, it points in specific terms to one (or more) groups that may have particular needs (in this case, related to mathematical background, or numeracy).

The small size of the social class effects was somewhat surprising. There are several possible reasons. The measure for social class, using parental occupation based on the Registrar General's classification, may not have provided a valid measure for relevant differences in household perspectives on education. My collapsing of occupations into two basic groupings – middle class (non-manual) and working class (manual) – may have occluded too much crucial variation. Alternatively, the wording of the relevant question, asking respondents for their father's and their mother's 'paid work when you began secondary school' might have required recall that was too demanding to allow reliable responses (going back at least seven years for younger students and considerably longer for older ones). Some or all of these plausible reasons may help to explain the expected result not being observed. However, the interviews will afford opportunities to assess and distinguish social class positions differently.

In the next chapter, I explore the conception and measurement of affective variables, especially mathematics anxiety, and their role in my model, both as outcomes to be explained, and as influences on performance.

4   Affect and Mathematics Anxiety

Question: What do you dread as you open your eyes in the morning?
Answer: That I'm still at school and it's double maths.
            (Shona MacDonald, 26, promotions manager,
                in *City Limits*, 23–30 May 1991)

The importance of affective factors in education generally, and in the learning of mathematics in particular, is reasserted periodically. In the 1970s and 1980s, the need to enhance females' participation and performance in mathematics generated interest in affect and attitudes towards mathematics, and especially mathematics anxiety (for example, Fennema and Sherman 1976, Tobias 1978). So, too, did the aim of increasing access to higher education, and/or 'second chances' with mathematics, for students and adults generally. More recently, the need to account for blocks in mathematical problem-solving episodes has seemed to require a more cognitive, more qualitative approach (e.g McLeod 1992).

In this study, from among the range of affective factors discussed in recent years in mathematics education, I emphasise mathematics anxiety. However, competing concepts of anxiety have generated controversy among psychologists and others. There have also been lively discussions of notions of mathematics anxiety, the principal measures to be used, and relevant research results – in several areas, especially

*   the relationship between anxiety and performance
*   social differences (gender, social class) and anxiety
*   whether anxiety is a general trait (or pattern of response) of the person, or whether it needs to be considered as specific to the context: as *test* anxiety, or as *mathematics* anxiety, for example.

I discuss what I call a Model A type of relationship between cognitive performance and affect, and complete the development of the 'conceptual map' for the survey part of the study. I present results on the context-specificity of mathematics anxiety; gender and social class differences in it; and its relationship with performance. One question I shall explore is whether and how such a relationship might be specific to the context of mathematical activity.

Affect and Anxiety

First I consider conceptualisations of the 'affective' area. For Laurie Hart Reyes, it comprises

> students' feelings about mathematics, aspects of the classroom, or about them-
> selves as learners of mathematics . . . [including] . . . general feelings such as
> liking/disliking of math . . . [and] . . . perceptions of the difficulty, usefulness,
> and appropriateness of math as a school subject.
>
> (Reyes 1984: 558)

Recent useful reviews by McLeod (1992, 1994) have suggested dividing affect into three dimensions:

* beliefs
* attitudes
* emotion.

These might be seen to parallel, roughly, Reyes's 'perceptions', 'general feelings', and 'feelings'.

Schoenfeld (1985) describes *belief systems* as conceptions about the nature of mathematics, in particular the constitution of mathematical arguments. A belief normally involves a commitment to some sort of factual statement, though it may also involve a commitment to *valuing* something (for example, Cobb 1986). This suggests that it may be difficult to classify beliefs as either simply cognitive or simply affective.

Laurie Hart characterises *attitudes* in mathematics education as 'any of a number of perceptions about mathematics, about oneself, about one's mother, father or teacher' (Hart 1989: 40), where these perceptions are relatively low in emotional intensity (and are like beliefs). An illustration would be Fennema and Sherman's (1976) 'Mathematics Attitude Scales', which have been widely used in mathematics education research (as discussed later). Hart also refers to a broader definition, widely used among psychologists (for example, Triandis 1971), of an attitude towards an object, as having three components:

1   an emotional (affective) reaction
2   a predisposition to behave favourably or unfavourably towards it
3   beliefs about it.

A somewhat different three-way breakdown of affect from that of McLeod is proposed by Herbert Simon (1982), comprising:

* emotion, which can suddenly interrupt the current focus of attention, as in surprise, fear and anger
* mood, less acute but nevertheless able to influence cognition (such as remembering), as in happiness or sadness
* valuation, which may depend on memories of earlier affect.

Simon's characterisation of affect clearly points to cognitive concerns, and, typically for psychologists, focuses more on 'hot', visceral reactions, compared with educational researchers, who tend to 'mean a wide variety of beliefs, attitudes, and emotions ranging from "hot" to "cold"' (Hart 1989).

McLeod's and Hart's concept of *emotion* is much like Simon's: it is normally of a higher intensity than an attitude, and it may be categorised as positive or negative. Thus anxiety, for example, tends to be seen by psychologists as a 'hot' emotion, while some mathematics educators (e.g. Fennema and Sherman (1976) in their original Mathematics Attitude Scales) treat mathematics anxiety more like a 'cooler' attitude (see 'Mathematics Anxiety, Meaning and Relationships', later in this chapter).

To sum up, there is clearly some overlap between McLeod's (1992) tripartite characterisation of 'affect', and psychologists' definition of 'attitudes'. However, for this study, I shall adopt McLeod's characterisation of the affective in terms of beliefs, attitudes and emotion. These three areas are not always easy to distinguish, but they cover reactions across a spectrum, from 'cold', stable beliefs, to 'cool' attitudes, to 'hot'/intense, more transitory emotions. Indeed, McLeod (1989a) distinguishes several dimensions of variation in affect generally:

1   intensity
2   direction (positive vs. negative)
3   stability.[1]

Dimensions 1 to 3 can be useful in distinguishing the three areas within affect (as earlier), and also different types of affect. Thus, (acute) anxiety and frustration are intense and negative, satisfaction and joy are strong and positive, with liking less intense, but more stable. The third dimension also distinguishes these emotions (relatively transitory) from attitudes, including (chronic) anxiety and confidence; indeed, some aspects of confidence might be considered as beliefs.

There has been a substantial measure of agreement about the affective variables that might be expected to influence thinking and performance in mathematics in older students and adults (Evans 1993: Ch. 3, McLeod 1992). These include:

*   mathematics anxiety
*   confidence, including self-concept, self-efficacy, and locus of control (Weiner 1986)
*   perceived usefulness of mathematics
*   perceived difficulty of mathematics
*   finding mathematics interesting and/or enjoyable.

Nevertheless, I had to select particular affective factors for priority in my survey, and both mathematics anxiety and 'confidence' feature strongly in the literature reviews just cited. Further, my experience with teaching adults suggested them both as powerful and opposite influences in learning.

However, as I continued my literature review, I was struck by how well developed was the literature on anxiety, and on mathematics anxiety, as used both by researchers and by those concerned with intervention programmes. Also I became aware of two pencil-and-paper tests, the Fennema-Sherman Mathematics Anxiety Scale, one of a set of nine 'mathematics attitude' scales, and the Richardson-Suinn Mathematics Anxiety Rating Scale (MARS), which were widely used in research and in intervention in the USA, and which could be adapted. Further, rather more studies recently have used mathematics anxiety, than have used 'confidence' [2]

Therefore, for all these reasons, I gave priority in my conceptual map (used as the basis for the survey) to mathematics anxiety scales over the other affective variables. However, confidence and several other affective variables were also included.

Psychological Conceptions and Measures of Anxiety

The study of anxiety within the discipline of psychology certainly came into its own in psychology after the Second World War: an estimated 5,000 articles or books on it were published between 1950 and 1970 (Spielberger 1972a). Here I consider the predominant conceptions of anxiety, the measures used, and research into the relationship between anxiety and performance.

The *concept of anxiety* is widely agreed (for example, Spielberger 1972a) to have its basis in the work of Freud. Freud's views on anxiety will only be introduced in this chapter (see also Chapter 7), since some crucial aspects of his theory are not taken up by most postwar work in psychology. Freud proposed a characterisation of anxiety as:

- motor 'innervations' or 'discharges'
- a perception of these
- feelings of 'unpleasure' (Freud [1916–17] 1974: 443–8).

Because anxiety is 'unpleasant' and often painful, there is a tendency for it, or for the ideas associated with it, to be *repressed*, to be pushed into the *unconscious*. However, repression does not destroy these contents; rather, they retain their charge, but undergo a transformation. If they 'return' to consciousness, they tend to be found in a disguised or distorted form; in 'normal' people, as jokes, or 'slips of the tongue', or in dreams. (Otherwise, they may return as neurosis or illness.)

One important aspect of Freud's thinking was his distinction between 'manifest' and 'latent' content, in dreams (Freud [1900] 1965), fantasies, or talk: the latent content is the set of meanings – including meanings banished to the unconscious by defences – revealed after analysis of the dream. Thus a 'manifest' expression of no anxiety, or even great confidence, in an interview may actually indicate anxiety. Freud therefore postulated the use of *defence mechanisms* such as repression, or 'reversal into the opposite'; see Chapter 7.

Since in the psychoanalytic view anxiety may be unconscious, it therefore cannot be assumed to be *reportable* by the subject. Nor can it be assumed to be *observable*, in a dependable way. Only some of its symptoms may be observable, and because of defences, these may appear in distorted form: as 'no feeling' at all, or indeed as the opposite of anxiety.

Despite these trenchant features of Freud's work, because of the ascendancy of behaviourism and aspirations to be 'scientific', psychological work after the Second World War almost invariably assumed that anxiety was observable – marked by the naming of the first self-report scale as the 'Manifest Anxiety Scale' (Taylor 1953) – and quantifiable. In the 1950s and 1960s, conceptions of 'general anxiety', 'test anxiety', and a host of more 'specific' anxieties (that is, towards specific objects or contexts) were developed, as well as standardised scales to measure them. During this early period, there was controversy as to whether anxiety might best be represented as a personal characteristic ('anxiety-proneness') or as a 'stimulus', that is, part of the context (see Evans 1993: App. V3).

In one response to the controversies, Spielberger offered a psychological definition of anxiety as 'a palpable, but transitory, emotional state or condition characterised by feelings of tension and apprehension and heightened autonomic nervous system activity' (Spielberger 1972b: 24). This definition appears to take into account the three aspects of anxiety stressed by Freud (see earlier), but it still sees anxiety as 'manifest'. Spielberger's contribution was to develop the distinction between measures of one's *transitory* (or acute) emotional state of anxiety, called the 'A-state', and relatively *stable* (or chronic) individual differences in being anxiety-prone, called the 'A-Trait' (ibid.).

The development of 'cognitive' approaches within psychology was extended to anxiety (for example Beck and Emery 1985), and led to a distinction between two components of test anxiety:

- worry or 'lack of confidence' was seen as *cognitive* concern about test performance, the consequences of failure, the ability of others relative to oneself, etc.
- emotionality was conceptualised as *physiological /* autonomic arousal, reflecting the immediate uncertainty of the test-taking situation (Liebert and Morris 1967: 975.)

Overall, by the beginning of the 1970s, to a great extent 'anxiety' had slipped into meaning 'manifest' or expressed anxiety, and not reporting anxiety was usually taken to mean 'no anxiety'. In psychology, Freud's ideas for understanding anxiety had largely been forgotten.

Indicators for anxiety can be divided into four broad types (Spielberger 1972b: 25; Ortony et al. 1988: 8):

- *physiological:* for example, heart rate, respiration rate, galvanic skin response, 'cortical potentials' in the brain (EEG)
- behavioural: flight, 'freezing', restlessness, abnormally fast or abnormally slow speech
- *clinical:* for example, 'projective' measures like the Thematic Aperception Test, and the Rorschach Test, both of which attempt to elicit fantasy material, and to interpret it
- *self-report:* questionnaires, such as Spielberger's State–Trait Anxiety Inventory (Spielberger et al. 1970), and Test Anxiety Inventory.

Each type of indicator has particular advantages and disadvantages (Evans 1993:

69–70). A self-report 'inventory' can be administered and scored quickly, with no special difficulties for group use. The stability of response given by an individual on different occasions is probably greater than for physiological measures, and the reliability of scoring is greater than for projective measures. However, their use is subject to reactive effects: respondents may adopt 'response sets', such as acquiescence, or aim to give socially desirable answers.

Nevertheless, the only type of measure that could possibly be used for administration as part of a questionnaire to a large group, as in my study, was a self-report inventory. Within a one-to-one interview, behavioural measures might be used or (resources permitting) projective measures, but physiological methods would probably be too intrusive.

In studies of *the relationship between anxiety and performance* (or educational attainment), a major concern was whether anxiety is *simply interfering* or 'debilitating' in its influence on performance, or whether there might be an *'inverted U'* relationship between the level of anxiety ('drive' or 'motivation') and the level of performance, for some forms of learning at least. The simple interference relationship would see performance as decreasing, when anxiety increases; the inverted U suggests that, for each individual and each task, there is an optimal level of anxiety, neither too little to facilitate performance, nor so much as to interfere with it; see Figure 4.1.[3]

By the end of the 1960s, the dominant view was that, for academic performance, high anxiety generally was interfering; therefore, the correlation between anxiety and performance should be generally negative, and higher for 'test anxiety' than for 'general anxiety'. This view is confirmed in the studies at all levels of the educational system reviewed in Caudry and Spielberger (1971), in which there is little reference to the 'inverted U'. There may be several reasons for this. The breakdown of the 'drive theory' branch of the behaviourist research programme in the late 1950s

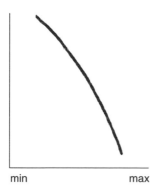

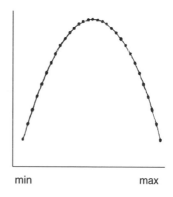

(a) Anxiety as interfering: performance monotone decreasing with increasing anxiety performance

(b) Anxiety may have optimal level for performance: an 'inverted U' relationship performance

Figure 4.1 Two Possible Relationships between Performance and Anxiety

undermined the inverted U's major theoretical underpinning. In addition, it is notable that most research reports published during this time present data in one of two forms:

- differences of mean performances for *two* anxiety groups, high and low; or
- product–moment correlations, of performance with anxiety, essentially *linear* measures of relationship (Caudry and Spielberger 1971).

Such analyses would be unable to confirm or reject a *non-linear* relationship like the 'inverted U'.

This review has highlighted several crucial distinctions as to the way that anxiety can be conceptualised, measured, and related to performance:

- *Manifest* or *latent:* Is anxiety considered always to be available for observation or self-report, or might it be sometimes subject to defence mechanisms that would prevent the subject, and perhaps also an observer, from being aware of it?
- *Chronic* or *transitory:* Is anxiety considered a trait or a state? This distinction is close to that made between 'cool' attitudes and 'hot' emotions (see pp. 43–5).
- Type of *indicator:* Are self-report or other measures used?
- Type of *relationship with performance:* Is this seen as interfering or 'inverted U'?

I shall return to the first issue in Chapter 7. The remaining issues will now be taken up for mathematics anxiety.

Mathematics Anxiety, Measures and Relationships

In the 1950s, the research literature began to refer to 'mathophobia' and 'number anxiety' (for example, Dreger and Aiken 1957). However, it was not until the 1970s that two important new developments occurred, in North America.

First, in the context of a developing women's movement, researchers were seeking explanations for the apparently poorer performance of women in mathematics courses and testing, in terms of factors other than innate ability (see Chapter 3). 'Mathematics participation' and 'mathematics avoidance' (a resistance to taking mathematics courses, beyond compulsory ones) were studied – and as we have seen, this led to an emphasis on affective factors (for example, Fennema 1979) for providing at least part of an explanation for alleged female deficits in performance and participation.

The notion of 'mathematics anxiety' was crucial in this connection. The idea had strong public and media appeal (see for example *Time*, 14 March 1977), and Sheila Tobias's work, including *Overcoming Math Anxiety* (1978), was exceedingly influential in popularising further the notion.

Second, during this same period, a group of 'interveners' (Tobias and Weissbrod 1980) sought to develop techniques for diagnosing and treating the symptoms of maths avoidance, amongst male and female students, especially adults in, or seeking access to, higher education Thus, the concept of mathematics anxiety was widely used, especially in North America, in mathematics workshops and remedial programmes. Also during this period, questionnaires used to measure maths anxiety – for diagnosis and remediation purposes – were produced.

Two measures

The two mathematics anxiety scales most frequently used in recent years have been:

• the Fennema-Sherman Mathematics Anxiety Scale (MAS) (Fennema and Sherman 1976), produced within the 'gender-inclusive' research programme mentioned earlier
• the Mathematics Anxiety Rating Scale (MARS) designed for use in some of the intervention programmes (Richardson and Suinn 1972, Rounds and Hendel 1980).

It is worth comparing the MAS and the MARS, as to their underlying conceptions and as indicators of mathematics anxiety.

The Mathematics Anxiety Scale (MAS) is one of nine domain specific scales which make up the Fennema-Sherman Mathematics Attitude Scales.[4] Fennema and Sherman defined mathematics anxiety as 'feelings of anxiety, dread, nervousness and associated bodily symptoms related to doing math' (Fennema and Sherman 1976: 4). The scale was validated on US secondary school pupils; 'item analysis' procedures used to select the final items aimed to highlight gender differences, and also differences between those taking further mathematics courses in high school, and those not doing so. This means that we should expect that comparisons using the MAS will have a strong tendency to produce 'gender differences' in mathematics anxiety.

If we examine the wording of their items (Fennema and Sherman 1976) with our summary from the previous section in mind, we can note several features. Of the questions, five refer to mathematics (as a school subject) in general, and the rest to mathematics courses, classes, problems, and tests. Thus, the mathematics anxiety appears to be construed as chronic, rather than as a 'transitory' response to an immediately preceding situation. Also, in terms of its relationship with performance, mathematics anxiety is apparently construed as 'debilitating': the twelve items are scored negative for 'anxious' responses and positive for 'non-anxious' responses. A high score is thus indicative of low anxiety.

For the Mathematics Anxiety Rating Scale (MARS), Richardson and Suinn characterise mathematics anxiety as 'involving feelings of terror and anxiety that interfere with the manipulation of numbers and the solving of math problems in a wide variety of situations' (Richardson and Suinn 1972: 551); they were interested in intervention programmes for countering mathematics anxiety (see earlier). To measure mathematics anxiety, Richardson and Suinn produced ninety-eight items which were brief descriptions of behavioural situations, for example, 'adding two three-digit numbers while someone looks over your shoulder', in response to which people were expected to indicate one of five different levels of anxiety from 'not at all' anxious to 'very much'. The MARS was validated with samples of students (mostly female and studying education) at large US Midwestern universities (Richardson and Suinn 1972, Suinn et al. 1972).[5]

We can note several features of the MARS. First, in line with Richardson and Suinn's (1972) observation that 'studies emphasising the identification of different types of anxiety have found that different kinds of anxiety lead to different effects on intellectual performance', the MARS test aims to be situationally specific; the use of factor analysis (see later) aimed to produce even more specific 'factors'. Second, the wording of the questions again suggests that mathematics anxiety is conceived of as chronic or 'trait'. Again, mathematics anxiety is considered as debilitating, rather than facilitating (see Richardson and Suinn's definition earlier).

With regard to their usefulness for questionnaire (self-report) research with adults, we can compare the MAS and the MARS on several aspects:

- range of emotion: the MAS items mention a variety of emotions, 'scare', 'bother', 'worried', 'at ease', 'uncomfortable' and so on, whereas each MARS item asks how much, for each situation mentioned, the respondent is 'frightened by it nowadays'
- context: the MAS items refer only to school mathematics in general, or to mathematics courses, classes, problems and tests, whereas the MARS items refer to a wide range of situations, to do not only with college mathematics (as the MARS was constructed with adults in mind), but also mathematics in a variety of other contexts (see next subsection);
- reliability: the MARS (with ninety-eight items) would tend, other things being equal, to be more be more reliable than the MAS (with twelve items), but more time-consuming (for respondents and researchers)
- symmetry: the set of MAS responses for each item was 'symmetrical', ranging over the standard categories, as

| strongly agree | agree | neither agree nor disagree | disagree | strongly disagree |

whereas the MARS was 'asymmetrical' ranging across

| not at all | a little | a fair amount | much | very much |

and a symmetrical set of responses is, on balance, more likely to yield responses that can be reasonably treated as numerical (interval scale), and hence, for example, can be averaged, at the analysis stage.

On the second point, the MARS is clearly the more appropriate indicator to use for a study of the variation in numerate performance – and in anxiety – *across contexts*. Indeed, it seemed desirable to study the MARS's range of contexts more systematically, for example by using factor analysis to look for underlying dimensions of the various items, as other researchers had done (see later). Concerning the practical implications of research like this, being able to separate out aspects of the context of learning or doing mathematics which cause relatively more anxiety to learners would be valuable for designing 'intervention' programmes. The first and third points suggest using the MARS, though we might want to decrease the number of items. On the fourth point, we attempted to set up a symmetrical scale of responses for the MARS.

For these reasons I chose the MARS as an indicator for mathematics anxiety. I address the issue of the more systematic characterisation of different contexts of mathematics anxiety in the next subsection.

Dimensions of Mathematics Anxiety

In Chapter 2, I discussed distinguishing between 'school mathematics' (SM) and 'practical maths' (PM) performance, on the basis of a notion of context, defined 'naturally' and indicated by the wording and format of the task. Here I investigate whether we might distinguish types or dimensions of mathematics anxiety, also on the basis of their contexts, using two methods:

- a classification of the original ninety-eight MARS items (Suinn 1972), based on my reading of the meaning of their wordings
- reports of factor analyses of responses to MARS items from other studies.

(In 'Survey and Modelling Results', this chapter, I give the results of my factor analysis of responses from the Polytechnic sample to the twenty-six mathematics anxiety items in my questionnaire.)

For my reading, I attempted to categorise the mathematical/numerate activity referred to by each of the ninety-eight MARS items into 'clearly' practical maths, 'clearly' school/college maths, or ambiguous. Here I used the basic notions of 'school mathematics' and 'practical mathematics' developed in Chapter 2.

I found it was possible to classify the mathematical/numerate activity referred to by at least three-quarters of the ninety-eight MARS items more or less clearly into one of two groups: those relating to 'practical maths' situations (twenty-five items), or those relating to 'school/college maths' situations (forty-eight items), including involvement with mathematics classes, textbooks, exercises, tests, exams and so on. The remainder I considered 'ambiguous' (twenty-five items).

Examples of 'clearly practical maths' items would be:

Question 10 Totalling up a dinner bill that you think overcharged you.
Question 47 Reading a cash register receipt.
Question 48 Figuring the sales tax on a purchase that costs more than $1.00.
Question 64 Deciding which courses to take in order to come out with the proper number of credit hours for full-time enrolment.
Question 87 Being responsible for collecting dues for an organisation and keeping track of the amount.

Examples of 'clearly school maths' items would be:

Question 26 Signing up for a math course.
Question 54 Taking an examination (final) in a math course.

Question 72   Being given a homework assignment of many difficult problems which is due the next class meeting.

Question 74   Thinking about an upcoming math test one day before.

Examples of 'ambiguous' items would be:

Question 2   Having someone watch you as you total up a column of figures.

Question 14   Adding up 976 + 777 on paper.

I considered that Question 2 could not be classified straightforwardly, since, though no context is explicitly mentioned, *being watched* while doing a sum tends to happen more at school – and in doing homework (see case studies in Chapter 10) – than elsewhere. I found Question 14 ambiguous in that, though again no context is mentioned, the carrying out of the operation *on paper* tends to be characteristic of school mathematics activity – but not exclusively so.

To summarise the results of the factor analyses I considered (Evans 1993, App. V4), there is a fair amount of agreement in the literature that there are at least two main factors of mathematics anxiety: one related to the use of numbers in *everyday* situations, including those relating to money, and another relating to *academic* mathematics courses, lessons, textbooks, tests and exams. In particular, Rounds and Hendel (1980) found two dimensions or factors, described as follows:

Factor 1:   named *mathematics test (or course) anxiety* by the researchers, correlated substantially with ('loaded on') forty-two of the original items, accounting for 31% of their total variance.[6] About one-third of these reflect apprehension about anticipating, taking, and receiving the results of mathematics tests, while two-thirds referred to activities directly associated with mathematics classes and courses. Examples: Questions 26, 54, 74; see earlier.

Factor 2:   named *numerical anxiety,* correlated substantially with forty-four of the original items, accounting for 8% of the variance. Generally referring to everyday concrete situations requiring some form of number manipulation (e.g. addition or subtraction), slightly over half of these items refer to practical skills necessary for making *money* decisions, a quarter refer to various non-money practical situations, and almost a quarter refer to the use of elementary arithmetic skills with no explicit context of application (Rounds and Hendel: 142). Examples: Questions 10, 14, 48, 64.

Rounds and Hendel's work is important because they address the issue of the context where the numerical work is done, and they interpret their results to indicate that anxiety about mathematics may be 'situationally specific and not transituational'. In particular, their work seems to make a distinction – between maths test/course anxiety and numerical anxiety – which can be seen as parallel to the one made between 'school mathematics' and 'practical mathematics' thinking and performance in Chapter 2.

In a more recent 'meta-analysis' of studies on mathematics anxiety, Hembree concluded that mathematics anxiety comprises more than test anxiety, namely 'a general fear of contact with mathematics, including classes, homework and tests' (Hembree 1990: 45). This suggests that Factor 1 above should be thought of as 'mathematics test and course anxiety'.

Mathematics Anxiety, Performance and Social Difference

Most of the research from the 1970s onwards is based on a conception of mathematics anxiety as 'debilitating', that is, as having a *negative* relationship with performance. Two factors in particular seem related to this change, including:

- the ascendancy among psychologists of views of anxiety as generally interfering with performance (see previous section, 'Psychological Conceptions and Measures of Anxiety')
- the rise of the feminist research programme (see earlier) where any gender differences in performance found to exist, in mid- to late adolescence, are explained by differences in mathematics course-taking ('participation'), which are in turn explained by differences in attitudes, including differences in mathematics anxiety: if the latter is seen as debilitating, and if females 'have more of it', and score higher on maths anxiety scales, then this could explain some, at least, of the deficits in female performance.

In a review of 151 studies done largely in the USA, Hembree (1990) finds that almost all studies find an *uncontrolled* negative correlation between mathematics anxiety and performance. However, many of the studies reported suffer from one or more methodological limitations (Evans 1993, Ch. 3):

1   They use measures only of trait or chronic mathematics anxiety, seen as basically stable, and not of 'state' anxiety, which limits their scope for taking account of the varying contexts of mathematics anxiety.
2   Relationships are sometimes examined using only simple differences in means or simple correlation coefficients, without controlling for other variables indicated as relevant by their own and other studies.
3   Correlations are often reported which, though statistically significant, are small in R-squared terms.
4   The use of linear correlation and regression methods by themselves, without examining the 'shape' of the relationship (for example, through scatterplots), do not allow the researcher to address the possibility of non-linear (for example, 'inverted U') relationships between mathematics anxiety and performance.

Points 2 to 4 suggest that multiple regression models, including the possibility of a non-linear component, are called for.

Further points must be remembered:

5   Justifying causal explanations (and hence prescriptions for intervention) is in principle difficult when the research design is non-experimental, even when the data analysis is based on multiple regression models (cf. Reyes 1984).

6   The strength of relationship found may be dependent on *sample characteristics*, such as gender or previous mathematics course-taking, and also on the *other variables included* in the model, for example other affective variables, such as confidence or test anxiety (cf. Llabre and Suarez 1985).

I close this section by summarising a review of results of studies among college students (in the USA) focusing on relationships between mathematics anxiety scores and social differences, such as gender, social class and age (see Evans 1993: Ch. 3).

For gender, the results are mixed. For example, Hembree (1990) finds an overall 'effect size' of about one-third of a standard deviation for gender (that is, women scoring higher), but this is based on a set of individual study results that range over much larger positive, and some negative, values.

On social class, Evans (1993) found 'no dependable differences in maths anxiety . . . in a very limited literature' and 'almost no findings on the relationship between age and maths anxiety'. Betz (1978), in a sample aged between 17 and 34, found that anxiety scores, using Fennema and Sherman's MAS (see earlier) increased with age for two of her three groups.

These findings are taken into account in the development of the 'conceptual map' and the research questions for the study, to which I now turn.

Model A: Conceptual Map and Research Questions

Several types of models have been proposed as a basis for studying relationships between affective variables and cognitive performance in mathematics tests and problem-solving. The one used for the quantitative part of this study might be called an 'individual differences' approach. Such models aim to explain individual differences in performance scores, and in participation (taking mathematics courses) and so on, using measures of affect and other individual differences. Affect in this approach tends to be represented, not by 'hot' emotion, but by 'cool' attitudes: there is a tendency here to see as relatively stable both the cognitive ('performance levels', 'skills', if not innate abilities) and the affective ('personality', 'traits').

This type of model, which I call 'Model A' for relating the cognitive and affective (Evans and Tsatsaroni 1996), is exemplified by Fennema's (1989) 'generic' model (see Chapter 3), and by the conceptual map I built up for the quantitative part of this study (see Figure 4.2).[7] Here social influences socialise the individual, so that values and affect are 'internalised'. In turn, affect is seen as having an influence, (normally one-way), on differences in cognitive outcomes across individuals. The ultimate explanation for both cognitive outcomes and affect comes from social influences, especially differences like social class and gender, which influence affect through beliefs, perceptions, and so on.

Affect is measured by scores on attitude scales or ('trait') anxiety scales such as the Mathematics Anxiety Rating Scale (MARS) (Richardson and Suinn 1972); outcomes by number of mathematics exams passed, scores on standardised tests and so on.

This type of model can be made to include a comprehensive range of affective factors, and of social variables too. The outcome variables are generally quantifiable, and also of interest to teachers, parents and policy makers, such as differences in mathematics performance, or mathematics course-taking.

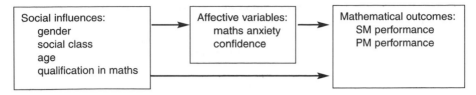

Figure 4.2 Conceptual Map for the Study Using a Model A Approach

In my study, as already indicated, the outcome variables in the main models are the two types of *mathematics performance*. In addition, for some 'intermediate' models (discussed in 'Survey and Modelling Results: Social Differences and Mathematics Anxiety', this chapter), the outcome variables are affective, in the form of *mathematics anxiety*. Just as I divided performance into school mathematics (SM) and practical maths (PM) types (see Chapter 2), I aimed to take account of the context of mathematics anxiety by using items both for 'maths test/course anxiety' (TCA) and for 'numerical anxiety' (NA), as outlined in the previous section of this chapter. Consequently, I was also interested in the relationship between performance and mathematics anxiety in specific contexts.

The conceptual map also includes under 'social variables':

*   'structural' factors such as gender and social class
*   age
*   qualification in school mathematics, as an indicator of course-taking.

In the main versions of my model (Figure 4.2), affective variables can be considered to 'intervene' between these basic social variables, and the performance outcomes. As discussed earlier in this chapter, the affective variables considered worth including in the conceptual map were:

*   mathematics anxiety
*   confidence
*   perceived difficulty of mathematics
*   perceived usefulness
*   perceived interest.

Contexts for usefulness and difficulty were distinguished as 'work', 'everyday', and 'academic' (the latter for difficulty only). However, while I included indicators for all these affective variables in the questionnaire, only results for mathematics anxiety, and to a lesser extent, confidence, are presented in this book.[8]

The last three research questions for the quantitative part of the study concern the following:

Research Question 5: the specification of different contexts of mathematics anxiety.

Research Question 6: gender and social class differences in mathematics anxiety.

Research Question 7: the relationship between performance and mathematics anxiety, both specified as to school or practical context.

These research questions will be developed, and the relevant results presented, in the next three sections.

Survey and Modelling Results: Contexts of Mathematics Anxiety

Research Question 5 concerns the reasonableness of attempting to specify different contexts of mathematics anxiety, and also the adequacy of any results obtained.[9] The section on 'Mathematics Anxiety, Measures and Relationships' (this chapter) introduced the idea of representing mathematics anxiety in different contexts as different 'dimensions', and presented results from earlier research specifying two dimensions, maths test and course anxiety (TCA) and numerical anxiety (NA). In this study, these two dimensions are intended to relate to school and practical contexts respectively.

Here I present an analysis of responses to the set of mathematics anxiety items in the Polytechnic survey. Initially I consider what the correlation between the two dimensions of mathematics anxiety – and their correlations with measures of general anxiety and confidence in mathematics – might tell us about their dimensionality, and hence the possibility of distinguishing two separate contexts of 'mathematics anxiety'. Then I use factor analysis models to explore further the dimensionality of responses from my sample of British students, so as to allow comparisons with earlier US results (see 'Mathematics Anxiety, Measures and Relationships', this chapter) .

Initial Considerations

Research question 5 generates further questions, as follows:

*   Are maths test / course anxiety, TCA, and numerical anxiety, NA, measuring the same type (or context) of anxiety?
*   Are the TCA and NA scales of mathematics anxiety measuring different types of anxiety from general anxiety?
*   How does 'confidence', measured by self-rating of capability in mathematics, relate to TCA and NA?

Here all four variables mentioned are averages of a number of items designed to measure the concept.[10]

I expected each of the three types of anxiety variables to be measuring something different to a greater or lesser extent, so that the correlations of TCA and of NA with general anxiety would both be relatively low, and even the intercorrelation of TCA and NA would not be very high. In contrast I expected the correlation between confidence and TCA at least to be fairly high and negative.

In the results TCA and NA were fairly highly correlated ($r = .68$). This suggested that they were measuring two dimensions of mathematics anxiety that were fairly closely related, but there was still a substantial amount of variation in either dimension that was not associated with the other (54 per cent, since $r^2 = .46$). Thus a factor analysis of the twenty-six items used was likely to produce two or more uncorrelated factors (see later).

However, TCA and NA seemed likely to be measuring a different sort of anxiety from the scale I used for general anxiety, as expected. This is shown by the low correlations of general anxiety with TCA (0.34, $r^2 = .12$), and with NA (0.52, $r^2 = .27$).[11]

Finally, the correlation of confidence with the two dimensions of maths anxiety was of course negative, and was higher for TCA than for NA: $r = -.58$ compared with $r = -.46$. This difference was as expected, since confidence is measured by self-rated capability in *school maths* topics, and TCA refers to maths test and course anxiety. However, the size of the correlation of confidence with TCA was lower than expected; this suggests that they are measuring different types of affect, to some degree.

Factors of Mathematics Anxiety

The aim of factor analysis is to take a set of variables, normally all thought to be measuring 'broadly the same thing', to analyse the intercorrelations among them, and to 'reduce the dimensionality' of the set, to a relatively small number of underlying, relatively general, factors (e.g. Harman 1976). Thus, factor analysis has been used in investigations on whether there might be a 'general type of intelligence', or an 'extraversion personality characteristic'. All factors are *constructions*, based on a quantitative analysis with several decision stages (see later), of a particular data set. Here, the purpose of using factor analysis is as one of several ways of assessing my conceptualising of two or more contexts for mathematics anxiety, parallel to the distinction of school mathematics versus practical maths activity.

Factor analysis, especially in its 'exploratory' form, is often used in an atheoretical, inductivist way: the factors are seen as discovered, rather than as constructed, are named (sometimes somewhat arbitrarily) by the researcher, and are considered to be unproblematically 'real'.[12] However, this need not be: researchers' expectations can be clarified, prior to analysing the data, as can the distinctions to be used, for example in 'naming' the factors at the 'extraction' stage (see later). In addition, the items can be simply *read*, in order to suggest one or more ways of classifying them.

As indicated in 'Mathematics Anxiety, Measures and Relationships' (this chapter), the literature review suggested two dimensions of mathematics anxiety, as measured by the MARS: maths test/course anxiety (TCA) and numerical anxiety (NA). Further, maths test/course anxiety suggested a division into:

- maths test anxiety: anxiety to do with mathematics tests and examinations
- maths course anxiety: anxiety to do with thinking about (and deciding on) mathematics *courses,* being in maths *classes,* and studying *mathematics.*

This latter characterisation of 'maths course anxiety' suggests subdividing it into potentially three further dimensions (cf. Morris et al. 1978).

Thus, prior to the factor analysis, I provisionally classified the thirteen items categorised by Rounds and Hendel (1980) as 'maths test/course anxiety', with the following results (question numbers from the Situational Attitude Scale of the questionnaire, see Appendix 1):

- maths course anxiety: four items: Questions 3 (Enrolling for a course which includes a compulsory mathematics component), 4, 15, 28
- maths class anxiety: six items: Questions 8, 9, 13, and 18, 20, 23 (see later)
- maths studying anxiety: no items in the questionnaire
- maths test anxiety: three items: Questions 25, 31, 35.

Further I was struck by the *active*, immediate character of the situation described in two of the maths class anxiety items, namely:

Question 18 Being asked a question by the teacher in a maths class.
Question 23 Raising your hand in a maths class to ask a question.

This compared with the passive situations described in the other four; for example:

Question 20 Sitting in a mathematics class and waiting for the teacher to arrive.

Next, I classified the thirteen 'numerical anxiety' (NA) items on Rounds and Hendel's basis (1980: 142) of whether the item referred to:

- situations involving 'monetary decisions', in turn divided as to whether they involved:
  - money, immediate exchange: four items: SAS Questions 1 (Determining the amount of change you should get), 14, 22, 24; or
  - money, planning: five items: Questions 5, 17, 19 (Being responsible for keeping track of the amount of subscriptions collected for an organisation), 27, 33.
- other 'practical situations': no items
- use of arithmetic skills 'without a context': four items: Questions 6, 10, 11 (Adding up 967 + 777 on paper), 30.

Thus I produced seven tentative categories of mathematics anxiety item – four nested within the original TCA dimension, and three within NA. These can be regarded as a development of my *hypotheses* about the dimensions of mathematics anxiety that might possibly be constructed in a factor analysis, and an indication of theoretically reasonably ways of naming emergent factors.

There has been a great deal of controversy, especially within psychology and education, about the methods to be used in factor analysis (for example, Harman 1976, McDonald 1970, 1985). There are two basic stages to most factor analyses:

1   *extraction* of a reduced-dimensioned 'factor space' from the larger-dimensional 'variable space'
2   *rotation* of the factor space, so that its underlying dimensions are more easily interpretable in psychological or educational terms.

On the basis that it was desirable in practice to employ several different methods to check whether certain assumptions were important, I decided to use three methods of extraction, and to retain only those factors that were produced independently of method (Harris 1967, but see also McDonald 1970).[13]

Two key concepts are used in interpreting and evaluating the results of a factor analysis. The '*importance*' of each factor extracted by a certain method is given by the proportion of the total variance of all the original items (or variables) that is 'accounted for' by that factor. The *loading* of a factor on each item is basically the correlation of the factor with the item.

Rotation of the factor space can be done to produce factors that are orthogonal (uncorrelated with each other) or 'oblique' (correlated). A choice can be made on theoretical grounds, that is, expectations about whether the factors should be independent or correlated. Alternatively, for each factor 'solution' (that is, factor space extracted), one type of orthogonal rotation and one oblique rotation can be tried, and the results compared, as was done here.[14]

Initially, a four-factor analysis was suggested.[15] Because it produced only three interpretable common factors, the analysis was redone with three factors stipulated. Finally, a two-factor analysis – suggested by the hypothesis of 'two dimensions of mathematics anxiety' – was done. Each analysis began by extracting the relevant number of factors using the three chosen methods (see earlier). For each extraction, the resultant matrix of factor loadings was rotated in two ways, orthogonal and oblique, making six results in all. Only those items that had acceptably high loadings for the majority of the six results are considered in the following summary of the analyses [16] (For more details, see Evans 1993; Section 6.1.)

For the majority (four of six) of the three-factor analyses, the factors found might be 'named' as in Table 4.1. For all the two-factor analyses (and also the remaining three-factor analyses), the solution approached two factors, which might be labelled as in Table 4.2.

What is the meaning of these results for the dimensional structure of mathematics anxiety? In the three-factor solutions, the maths test/course anxiety dimension of Rounds and Hendel was split into two factors: one to do with school or college mathematics situations where *evaluation* (and not only testing) is imminent or likely, and one associated with day-to-day activities of participating in mathematics courses and classes. For numerical anxiety, ten of the thirteen items associated with the planning or spending of money, or with 'unspecified contexts', were associated with one factor. The two-factor solution reproduced Rounds and Hendel's two dimensions of maths test/course anxiety and numerical anxiety – almost.

However, for both analyses, the other three items considered by Rounds and Hendel as 'numerical anxiety', were not straightforwardly classified with '(practical) numerical anxiety'. The item N10 ('having someone watch you . . .') was *always* classified with the 'maths evaluation anxiety' items, while N24 ('figuring out 15 per cent VAT') and N30 ('being given . . . problems involving addition to solve on paper') were for some analyses.

Thus these results support several conclusions. First, two or more dimensions are necessary for describing the types of mathematics anxiety reported by adult

Table 4.1 Three-Factor Analyses: Mathematics Anxiety Items Associated with Each Factor

---

Factor 1: '*maths evaluation anxiety*' (accounting for 40% of the total variance of the original items) – normally 8 to 10 items:
- 'maths test anxiety' – 3 of 3 items: TC25,[17] TC31, TC35;
- 'maths class anxiety/active' – 2 of 2 items: TC18, TC23;
- 'numerical anxiety, context unspecified' – 1 or 2 of 4 items: N10: 'Having someone watch you as you total up a column of figures'; (some analyses) N30: 'Being given a set of numerical problems involving addition to solve on paper';
- 'numerical anxiety, money context, immediate – (some analyses) 1 of 4 items: N24: 'Figuring out VAT at 15% on a purchase which costs more than one pound'.

Factor 2: '*maths course/class anxiety*' (7% of the variance) – normally 8 items:
- 'maths class anxiety/passive' – 4 of 4 items: TC8, TC9, TC13, TC20;
- 'maths course anxiety' – 4 of 4 items: TC3, TC4, TC15, TC28.

Factor 3: '*practical numerical anxiety, mostly in money contexts*' (3% of the variance) – normally 12 to 13 of original NA items, except for: N10 (see above).

Note: For the wording of the items specified, see the Situational Attitude Scale in the Questionnaire, Appendix I

students. Second, the two most important dimensions can be reasonably labelled, so as to suggest

1   school or college contexts
   a   where courses and classes are the basis of social interaction
   b   where evaluation is prominent
2   practical everyday contexts, in many of which money may feature.

There is some indication of the value of subdividing the first dimension into (a) and (b), as indicated. As for the adequacy of this representation, the proportion of total variation (in the original twenty-six items) accounted for by the three-factor solution and the two-factor solution were in both cases about a half (50 per cent and 47 per cent respectively); this is an acceptable proportion. Of course, these conclusions depend on the kinds of mathematics anxiety items originally used in the questionnaire, and on my particular sample of mostly mature students at a British polytechnic.

This factor analysis suggests that consideration might be given to:

Table 4.2 Two-Factor Analyses: Mathematics Anxiety Items Associated with Each Factor

---

Factor 1: '*maths course/class/evaluation anxiety*' (40% of the variance) – including all 13 original TCA items, plus N10, N24 and N30.

Factor 2: '*practical numerical anxiety, mostly in money contexts*' (7% of variance) – 10 to 12 of 13 NA items, except for N10, and sometimes excepting N24 and N30.[18]

Note: For the wording of the items specified, see the Situational Attitude Scale in the Questionnaire, Appendix 1.

- doing some of the regression analyses with three dimensions of mathematics anxiety, instead of two; and / or
- including the three numerical anxiety items singled out above (N10, N24, N30) with 'maths test/course anxiety' for two-dimensional analyses, and with 'maths evaluation anxiety' for three-dimensional analyses.

The implications of the factor analyses for the ideas of 'context' and the context-specificity of anxiety in this study will be discussed in the concluding section.

Survey and Modelling Results: Social Differences and Mathematics Anxiety

In this section I investigate Research Question 6: gender and social class differences in mathematics anxiety. In particular, I want to scrutinise one of the widely-accepted 'myths' discussed in Chapter 1 – that 'women have more mathematics anxiety than men'. In fact the literature review in 'Mathematics Anxiety, Measures and Relationships' (this chapter) produced conflicting results on gender differences in mathematics anxiety. In this study, as in the others reviewed, the anxiety (and other affective) scales are of course self-report measures, which elicit responses which tend to correspond to social norms. For this reason I was prepared for women to report higher levels of both maths test/course anxiety (TCA) and numerical anxiety (NA). Similarly, I expected men might report a higher level of confidence in maths.

As for social class differences in mathematics anxiety and other affective variables, the literature gave little guidance. I expected on general grounds that those with middle class parental backgrounds would have lower anxiety about mathematics in the school context at least, that is on the TCA dimension. I also expected the middle class respondents to report a higher level of confidence in mathematics.

In this analysis, despite reservations about Rounds and Hendel's division of items into maths test/course anxiety and numerical anxiety, I continue to use their division as the basis for two dimensions of mathematics anxiety for analyses presented in this chapter – rather than the two slightly different dimensions – or the three dimensions – suggested by my own analyses. This is because:

- Rounds and Hendel's two-dimensional analysis had been subjected to scrutiny and attempted replication in the literature.
- My suggested three-dimensional analysis of mathematics anxiety threatened to become cumbersome and to undermine the key conceptual distinction made in this part of my study between school and college contexts, on the one hand, and 'practical', everyday contexts on the other.[19]

First I consider gender differences in scores on the two dimensions of mathematics anxiety, maths test/course anxiety, TCA, and numerical anxiety, NA, and also in scores on confident self-rating in mathematics (see note 10). I then compare these uncontrolled differences with the estimates of differences based on modelling the relationship of the mathematics anxiety variables, as outcomes, with a set of social variables, used as predictors in a multiple regression model.[20] The same procedure is repeated for differences in parental social class.

The simple (uncontrolled) gender differences in TCA and in NA were in the expected direction – that is, women reported higher levels of both dimensions of mathematics anxiety than men – and both differences were statistically significant (p < .001) (Table 4.3). However, the scores were not indicative of particularly high anxiety: for TCA, the men averaged around 'neither relaxed nor anxious' (scale point 4), while the women averaged midway between that scale point and 'a little anxious' (scale point 5). For NA, the women averaged a little more than, and the men a little less than, scale point 3 corresponding to 'fairly relaxed'.

In order to control for qualification in mathematics, age, and other relevant affective variables,[21] a multiple regression model was constructed, for each of maths test/course anxiety and numerical anxiety.[22]

As Table 4.3 shows, after controlling for mathematics qualification, age, and relevant affective variables the gender difference in TCA is reduced from a little over a half scale point (on a 7-point scale) to somewhat under a half, and the gender difference for NA is reduced from a quarter scale point to about half that much. The difference for TCA remains highly statistically significant (p < .001), whereas that for NA is on the borderline (p < .05). Here the controls do not reduce the size of the gender effect a great deal, especially in the case of TCA.

Confident self-rating in school mathematics also showed a statistically significant difference (p < .001). Men averaged close to scale-point '3' ('fairly capable'), with women about a third of a scale point lower (towards '2' or 'not very capable') (Evans 1993).

The analyses of social class effects were based on the cohort 3, the only one for which social class measures were included in the questionnaire (see Table 4.4). The uncontrolled parental social class differences for both dimensions of mathematics anxiety, TCA and NA, were small; they increase very slightly after controlling for mathematics qualification, gender, age, and relevant affective variables via the modelling. In no case are they very substantial (about a fifth to a quarter of a scale point). However, in all cases, middle class students report higher levels of anxiety – contrary to my expectations, which were based on general ideas, rather than on specific theoretical considerations or earlier results.

Table 4.3 Gender Differences in Maths Test/Course Anxiety and Numerical Anxiety: Uncontrolled Group Means and Estimates Controlled for Qualification in Mathematics, Age, etc. in Multiple Regression Model, for Whole Sample (n = 849)

	Average scale point score (where 4 = neutral point) (standard deviation/standard error)[23]					
	Test/Course Anxiety (TCA)			Numerical Anxiety (NA)		
	Men	Women	Difference	Men	Women	Difference
Uncontrolled	3.94 (1.12)	4.47 (1.22)	0.53 (0.08)	2.91 (0.91)	3.15 (1.00)	0.24 (0.07)
Controlled			0.40 (0.07)			0.14 (0.06)

Table 4.4 Parental Social Class Differences in Maths Test/Course Anxiety and Numerical Anxiety: Uncontrolled Group Means and Estimates Controlled for Qualification in Mathematics, Gender, Age, etc. in Multiple Regression Model, for Cohort 3 (n = 232 and 220, respectively)

	Average scale point score (where 4 = neutral point) (Standard deviation/Standard error)[23]					
	Middle class	Working class	Difference	Middle class	Working class	Difference
Uncontrolled	4.22 (1.29)	4.01 (1.10)	- 0.21 (0.17)	3.09 (1.07)	2.91 (0.92)	- 0.18 (0.15)
Controlled			- 0.23 (0.15)			- 0.27 (0.14)

Further, and also contrary to expectations, there was no parental social class difference in confident self-rating in mathematics.

It is important to consider the values of R-squared for these models. For the models used to consider gender effects, based on the whole sample (all three cohorts), the values of R-squared were 29 per cent for the TCA model, and 19 per cent for NA. For TCA, this is on the borderline of being acceptable (using my 30 per cent criterion; see Chapter 3, note 15). For NA, however, the value is low. This suggests that the 'explanatory power' of the sets of predictors, for the numerical anxiety dimension especially, is lacking. Therefore there remains a great deal of variation in the mathematics anxiety variables – especially numerical anxiety – that was not accounted for by the social variables of gender, age, social class, and qualification in mathematics. Some – but only some – of this variation was accounted for when certain affective variables were brought into the models.[24]

Survey and Modelling Results: Mathematics Anxiety and Performance

Research question 7 concerns the relationship between performance and mathematics anxiety. In general I expected negative correlations between the two dimensions of performance and the two dimensions of mathematics anxiety. More precisely, on the basis of my ideas of context, I expected a higher correlation for school mathematics performance with maths test/course anxiety (than with numerical anxiety), and a higher correlation for practical maths performance with numerical anxiety (than with maths test/course anxiety). I also expected to find a positive correlation of confidence in school maths with SM performance, and a smaller positive correlation with PM performance.

This research question also brings us back to consider regression models for school mathematics and practical maths performance (as in Chapter 3). Here I aim to judge the 'importance' of mathematics anxiety for 'explaining' performance differences – above that provided by the basic social variables – by checking whether mathematics anxiety adds to the 'variance accounted for' in these performance models. Further, despite the lack of support in the recent literature, I aim to explore whether the relationship between mathematics performance and mathematics anxiety could be described as simply 'debilitating' (that is, represented as a monotone decreasing function) or whether there is a moderate level of maths anxiety that is 'optimal' for performance, so that the relationship could be represented by an 'inverted U' (See Figure 4.1, p. 47). The latter would be modelled by a function including a squared term for the anxiety variable.

I found that the interrelationships between the performance types and the anxiety dimensions were much as expected; see Table 4.5.

Here, the linear correlation coefficients of each of the performance variables with the two mathematics anxiety dimensions are negative, as predicted. Further, the correlation of school mathematics performance with TCA is substantially higher than its correlation with NA, as expected. However, the correlation of practical maths performance with NA is only very slightly more than its correlation with TCA.

Confident self-rating in mathematics has the expected positive correlations with both SM and PM performance, and the former correlation is larger than the latter, as expected, given the focus in the confidence items on school mathematics topics. All six of the correlations reported in Table 4.5 are statistically significant ($p < .001$), which means they can be considered larger (in numerical terms) than zero, allowing for sampling variation. However, only two, the correlations of SM performance with maths test/course anxiety, and with confidence, are larger (numerically) than 0.3 (which would account for at least 9 per cent of the variance, in a model for performance).

Turning to the models, the explanatory power of maths test/course anxiety (TCA) and its square, in the model for school mathematics (SM) performance, and that for numerical anxiety (NA) and its square, in the model for practical maths (PM) performance, can be tested in several ways. First, the linear term for TCA was tested for inclusion in the SM performance model, using the specified criterion of statistical significance in the first step of the model-building procedure (see note 22). Next the square of TCA was tested to check that its inclusion in the model led to a statistically significant increase of explanatory power. For SM performance, these two predictor variables on their own produced a value of R-squared of 11.2 per cent. The same procedure for PM performance model again led to NA, and its square, both exceeding the criteria of statistical significance for inclusion, producing an R-squared of 5.6 per cent. This value is less impressive than that for TCA. The importance of the squared term in each case was confirmed when the final models were specified (Evans 1993, Tables 6.5 and 6.6): in both cases, the squared term remained statistically significant (at around the 5 per cent level). In this way, the model provides support for an 'inverted U'

Table 4.5 Correlations Between School Maths and Practical Maths Performance Types, Mathematics Anxiety Dimensions, and Confidence, for Whole Sample (n = 935)

	Pearson correlation coefficients	
Variables correlated	*SM performance*	*PM performance*
Maths test/course anxiety (TCA)	- .32	- .20
Numerical anxiety (NA)	- .25	- .21
Confidence self-rating	.45	.27

model – both for the relationship of SM performance with maths test/course anxiety, and, less convincingly, for that of PM performance with numerical anxiety – and the regression coefficients for the linear and quadratic terms can be estimated.[25]

In the SM performance model, confidence in mathematics was also included. The coefficient for confidence in the model was substantial: three-quarters of a question (the improvement in performance estimated to be associated with each one point increase in confidence score, *other things being held equal*). That is, the confidence variable made a contribution to the model for SM performance, in addition to that made by the mathematics anxiety variable, TCA. Thus the modelling exercise for SM and PM performance shows the importance of including, as predictors, the relevant mathematics anxiety dimension, its quadratic term, and confidence (the latter for SM performance only).

As indicated in Chapter 3, the R-squared values for these performance models was 29 per cent for school mathematics performance, and 17 per cent for practical maths performance. This is reasonably acceptable for SM performance, but low for PM. For both types of performance, it suggests that there are other influences, as yet unspecified.

Conclusions

Affective factors were brought into the conceptual map for several reasons. They were crucial in the models of Fennema and other feminist researchers in mathematics education (see Chapter 3). The discipline of psychology had long emphasised the importance of the cognitive–affective relationship, and I knew as a teacher of the importance of confidence and other facets of affect in mathematical learning and thinking.

However, affect tends to be seen in different ways: as attitudes, as 'cool', in mathematics education, and as emotion, as 'hot', in psychology. McLeod (1992) has made a helpful proposal that affect should be seen to comprise beliefs, attitudes, and emotions; these three areas are ranked in order of increasing intensity and decreasing stability.

Related to these difficulties, there are at least two types of model for relating affect and mathematical thinking and performance. In what I call Model A (or 'individual-difference' models), the links between cognitive and affective are produced by what are seen as causal relationships. Both 'external' social variables and affect influence cognitive outcomes, for example performance scores, in the individual. This type of model is used for the conceptual map on which the quantitative part of this study is based.

In contrast, what I call Model B tends to place more emphasis on affect as emotion, in studying the *process* of particular individuals' attempts to solve a problem, or series of problems (for example, Mandler 1989a). This model, and further alternatives, are discussed in Chapter 7, as part of the basis for the qualitative part of the study.

I give priority to mathematics anxiety in my conceptualisation of the affective domain for the quantitative phase of the study. Therefore the development of the concept of anxiety – from Freud's pioneering contribution (more fully discussed in Chapter 7) through developments in 1950s psychology, to contemporary cognitive emphases – and its influence on thinking and performance, are highlighted. However, other affective variables, notably confidence, are also included in the conceptual map.

As an indicator for mathematics anxiety in a self-completion questionnaire, Richardson and Suinn's Mathematics Anxiety Rating Scale (MARS) was preferred over Fennema and Sherman's Mathematics Anxiety Scale for reasons of reliability and validity (see 'Mathematics Anxiety, Measures and Relationships', this chapter). Further, given my emphasis on the context of mathematical/numerate thinking, a majority of the MARS items include a description (albeit minimal) of the context in which mathematics anxiety is (reported to be) experienced. In addition, the research pointing to a division into 'maths test/course anxiety' and 'numerical anxiety' (Rounds and Hendel 1980) suggested I would be able to measure mathematics anxiety so as to distinguish anxiety in school and college contexts from anxiety in practical, everyday contexts. This was meant to parallel the distinction between 'school mathematics' and 'practical mathematics' discussed in Chapter 2.

Thus, the full conceptual map for the survey (developed over the last three chapters) aims to relate school mathematics and practical maths performance scores to the appropriate dimension of mathematics anxiety – maths test/course anxiety or numerical anxiety respectively – and to social variables: gender, social class, age, and qualification in school mathematics. School mathematics ('abstract') and practical maths performance measures were based on sets of problems (either constructed, or based on the survey of adults (ACACE 1982) done for the Cockcroft Committee). A subset of the MARS items was used to measure mathematics anxiety. In addition, other affective variables were included in the questionnaire: confidence, perception of mathematics as difficult, perception of mathematics as useful, and perception of mathematics as interesting.

The final three research questions of the seven relating to the survey are investigated in this chapter. Research Question 5 asks: Was the idea of specifying two (or more) different contexts of mathematics anxiety reasonable, and was the method adequate? This was investigated in several ways, including exploring the dimensionality of the mathematics anxiety items used in this study. The factor analyses of these items (twenty-six adapted from the MARS) largely confirmed the basic differentiation by Rounds and Hendel (1980) of the items from the MARS scale into 'maths test/course anxiety' and 'numerical anxiety' dimensions, with several important elaborations. First, though some of the solutions produced two dimensions very similar to those of Rounds and Hendel, others produced three dimensions by splitting maths test/course anxiety into one dimension related to evaluation in school or college mathematics situations, and one related to attending mathematics courses and classes, with the numerical anxiety dimension almost unchanged. However, in all of my solutions, one (or, in some solutions, two) items classed by Rounds and Hendel as 'numerical anxiety, context unspecified', but describing a calculation which was 'given' or 'watched' by another, were grouped with the maths evaluation items, as was (sometimes) one item classed as 'numerical anxiety, money', but describing a percentage calculation more often done in school than outside.

Thus the factor analyses have several interesting implications for the discussion of *context* in this study. First, for this particular sample of social science undergraduates, many of them mature students – or indeed for any sample of people at all – they raise the question as to whether *any* item could be considered to have its context 'unspecified'. This recalls that the same question was posed about the possibility of performance items 'having no context' in connection with the numerical skills approach (see Chapter 2). In particular, for the two numerical anxiety items mentioned earlier, the results of the factor analyses suggest that the subjects may have placed them in a definite context, in the sense of responding to them in the same way as for school mathematics situations (perhaps with evaluative overtones). Second, for at least one item, an action or task that first appeared to form part of some 'practical maths' activity might be more accurately described as part of 'pseudo-realistic school maths'; for example, the item 'Figuring out VAT at 15 per cent . . .'.[26] Finally, the importance of a feeling of being evaluated in academic mathematics contexts was indicated by the dominance within one of the factors of items evoking such a feeling.

Research question 6 concerns gender and social class differences, in reported maths test/course anxiety (TCA) and numerical anxiety (NA). For TCA, the models showed the expected differences for gender, even after controlling for the other relevant variables. However, for NA, the gender effect was smaller (less than a quarter of a scale-point), and only on the borderline of statistical significance. Here the controls do not make much difference to the size of the gender effect, in contrast with the analysis for gender differences in performance (see Chapter 3).

For both dimensions of anxiety, the results for social class (measured by parental occupation) were less clear. The modelling (based on the cohort 3 sample only) suggested, as had the uncontrolled results, that students with middle class parents reported higher levels of both dimensions of mathematics anxiety – but the differences were neither large (barely a quarter of a scale point) nor statistically significant.

Even when other affective variables were brought into the models, there remained a great deal of variation in the mathematics anxiety variables – especially numerical anxiety – that was not accounted for by the modelling. This suggests that much of the variation in these dimensions of anxiety might be based in factors that are more difficult to measure, or particular to individuals.

Finally, Research Question 7 concerns whether the relationship between mathematics anxiety and performance could be described better as a monotone decreasing function – indicating a generally debilitating effect of anxiety – or as an 'inverted U' quadratic relationship, suggesting that moderate levels of anxiety might be optimal. A striking result of the use of controls provided by the regression modelling was the support given to the idea of a *quadratic* relationship between school mathematics performance and maths test/course anxiety, and, less convincingly, to that between practical mathematics performance and numerical anxiety. That is, each 'type' or context of performance was related to the relevant type of mathematics anxiety, and the shape of the relationship found in each case was an 'inverted U'. This finding, though not easily explained by the theories so far considered in this study, at least raises a question about those recent theories of mathematics anxiety, which have tended to see anxiety as having a purely debilitating effect on performance.[27] Further work on replicating, and especially explicating, this finding is needed.

In addition to the effects due to the two maths anxiety dimensions, there was a substantial effect (greater than half a question) estimated for the effects of confidence in the model for school mathematics performance. The fact that this effect was produced while the effect of maths test/course anxiety (TCA) was controlled for suggests that mathematics anxiety and confidence (at least as measured here) might have effects on SM performance that are to some extent independent. That is, contrary to suggestions in earlier research (for example, Fennema and Sherman 1976), it may not be appropriate to conceive of confidence and anxiety – despite a high negative intercorrelation – as 'opposites', or as in some way reducible to one another.

This discussion shows several ways that the statistical modelling can prove useful; these include:

- producing more *precise* estimates of effects, for example for gender, separately from other social variables, or for confidence (as discussed earlier), since the model controls for multiple predictor variables or 'influences' simultaneously
- producing estimates that are more *specific* (to certain subgroups of the sample), since the model takes account of interactions among predictors.

Other gains from using statistical modelling include the possibility of:

- producing an estimate of an individual student's 'expected" score (say on SM performance) on the basis of his/her gender, age, qualification in mathematics, etc. that could be compared with the actually observed score to categorise the student as 'performing to expectation', 'overachieving', or 'underachieving' (see Chapter 10 for the use of this idea).

The next chapter provides an overview of the findings, and the further questions raised, in the quantitative part of the study.

5   Reflections on the Study So Far

Since the processes of determination in the social world are subjective in important ways, involving actors' meanings and intentions, the survey researcher has to face the task of measuring these subjective aspects.

(Marsh 1982: 147)

My report on the first part of this study is now complete, based on the survey of adult students' experiences with mathematics, their performance, and their feelings, especially anxiety, about mathematics. The analyses, discussed fully in the last three chapters, have provided insights about the relationships considered to hold among the variables in the conceptual map.

Here I summarise the import of these analyses, relative to each of the seven numbered research questions as to whether they broadly confirm, partially support, or disconfirm my expectations. Some of my expectations were *broadly confirmed:*

1   The superior performance of the Polytechnic students compared with the general population, on a set of 'functional' or practical mathematical problems, with the (instructive) exception of two problems.
5   The possibility of distinguishing 'dimensions' of 'maths test/course anxiety' and 'numerical anxiety' broadly parallel to school and 'practical' contexts, within the set of mathematics anxiety items used.
6   Gender differences in reported maths test/course anxiety (women greater) and in reported confidence (men greater), when other relevant variables were controlled.
7   Observed relationships[1] of maths test/course anxiety, and of confidence, with school mathematics performance, and a less powerful one of numerical anxiety with practical mathematics performance.

Other hypotheses were only *partially supported:*[2]

3   Gender differences in school mathematics performance – but only for mature students (21+) – and in practical maths performance.
6   A gender difference in reported numerical maths anxiety (women greater).

Others were *disconfirmed*:

2   The adequacy of the distinction between the 'contexts' of the items for measuring school mathematics and practical maths performance.
4/6 Any social class differences in performance or in reported mathematics anxiety.

The analysis for 3 also points to a subgroup of this sample – older, female, and 'low-qualified' in school mathematics – scoring particularly low on school mathematics performance, as measured here.

Thus these findings contribute to challenging several ideas which have sometimes been uncritically accepted, in simple form:

- 'Men generally perform better than women at mathematics' – since the differences observed here were at best borderline, and were not found for younger students in school mathematics performance.
- 'Women have more mathematics anxiety than men', since the differences observed were substantial (and clearly statistically significant) only for one dimension of mathematics anxiety.[3]
- 'The context for a mathematics item is adequately given by its wording' (see later).

These findings also suggest issues for further research, including the interview phase of this study. In particular, there remain some basic problems about the conceptual basis of the research so far:

- the conception of the context, and the description of alternative contexts
- ways that thinking and performance might be understood as being 'in context'
- the conceptualisation of anxiety, and how it might be 'in context'
- the type of relationship represented by the models used
- the general quality of the results, based on averages over (e.g. gender) subgroups.

The analysis so far has assumed that the context is specified appropriately by the *wording* and *format* of the particular task for performance, or self-report item for anxiety. I have noted above that the division of mathematics anxiety items into maths test/course anxiety, and 'numerical' (or practical) anxiety dimensions was more convincing than the division into school mathematics and practical mathematics performance types.

In the case of performance, it was difficult actually to make the distinction between 'abstract' (school) and practical performance when examining particular items. Thus, most of my practical maths questions, such as Questions 19 and 20 (reading a graph of temperature changes) and Question 17 (averaging ages of a group of students), might be considered as 'word problems';[4] these, along with the abstract questions, could be found more likely in many school settings, than as part of practical activities for a typical student.

This raises the question whether the school mathematics and the practical maths items were seen by students as based in contexts that were different. It is clearly necessary to investigate *the subjects' perception of the context*.

The problem of distinguishing academic and everyday contexts may not have been so great for the anxiety items, partly because subjects here report on their 'general' and imagined feelings in the situations described which are (mostly) clearly distinct from that of the survey.[5] However, during pilot testing (see Appendix 1), some subjects had claimed that they could not respond properly to some of the anxiety items, because the contexts were not sufficiently clearly specified. Further, of course, an individual's response to his/her reading of a self-report anxiety item describing a situation in words may differ from their response in the actual situation. So we must keep open the question of whether the attitude items specified the desired contexts with sufficient detail and immediacy. Further, with respect to the division between maths test/course anxiety and numerical anxiety, we are depending a lot on factor analysis to 're-create' the separate contexts – though I have examined these results carefully, by comparing them with previous research, and by 'reading' the context of items before accepting the factor analytic classification.

There are closely related questions, also to do with the conceptualisation of the main outcome variables. Performance has so far been measured as counts of items correct, while qualitative information on, say, the problem-solving *strategy* or *methods* used is not available. Affective measures have been seen basically as personal 'traits', which are observable or reportable, and which can be scored quantitatively. In particular, mathematics anxiety is measured indirectly by self-reports of responses to situations described briefly in a questionnaire item. Though convenient to use, this method used on its own may have limited validity and reliability: this suggests using additional methods of measuring anxiety, for example by observing overt behaviour in problem-solving situations, and/or by asking students to describe critical incidents in their mathematical development, in interviews.

In addition, the analysis so far has not yet taken on board a psychoanalytic perspective. In this domain, anxiety plays a special role, such that it may not be observable in any simple way (see Chapter 4). This is because of *defence mechanisms* which may operate to occlude, or to modify, its expression.

Questions about the type of relationship assumed by the statistical models include

- whether, and how, they are considered to represent relationships of causal influence; and
- the direction of the effects.

Regression models as used here aim to represent and assess causal relationships – by controlling for alternative, 'rival' influences, which might be operating to produce the outcomes of interest. However, aiming to assess, say, whether 'gender (or another social factor) affects performance' raises not only the question of whether gender actually has an influence, but also whether any *intervening variables*, such as the actions and beliefs of parents or teachers, might provide a basis for explaining male and female students' differing attitudes and perceptions, and thereby the presumed influence of gender. I shall attempt to study both students' differing perceptions, and (indirectly) the attitudes and so on, of 'socialisers' like family members or teachers, in the interview phase of the research.

Further, assessing the idea that 'mathematics anxiety influences performance' raises the question as to whether the influence is reciprocal. For example, in my study, the presentation of the mathematics anxiety items after the performance items in the questionnaire might have meant, on reflection, that the student's experience with the performance items would affect responses to the anxiety scales. This would imply the possibility of *influence in both directions*, and would call into question the use of basic regression models, as in Chapters 3 and 4.[6] Thus claims about effectivity, and its direction, need to be cautiously formulated in quantitative modelling such as that reported here.

Finally, the differences and relationships reported are general, in the sense of being based on averages across respondents (in subgroups); therefore certain *processes, meanings,* or *critical incidents* in the development of mathematical thinking or affect in a particular subject will have been lost in the generality of the analysis so far. This problem may be partially attenuated by the use of statistical models, which can allow to some extent for particularity, by using finely-grained subgroups, and by analysing, say, performance scores while adjusting for the individual's anxiety and confidence scores, but it is not overcome.

This limitation is reflected in the fact that the regression models reported on here were not particularly powerful, in their own terms, in accounting for differences in performance scores and in mathematics anxiety – since the values of R-squared tended to be low (never more than 30 per cent for models based on the whole sample, though somewhat higher for those based on cohort 3). That is, variation in the performance and mathematics anxiety variables was not well 'explained' by these models. This suggests that the unexplained variation was more specific than social structural factors could account for in these models, or perhaps even that much of it was individually determined.

Therefore, on reflecting on the results so far, I considered that several areas needed further attention:

1   Ways of characterising more fully the context in which a person thinks about, and reacts emotionally to, a mathematical problem, including the social relations and material resources available in the setting, and taking account of the perceptions of the subject.
2   Description of how thinking, affect and the relations between them depend on the context.
3   A description of the involvement of emotion in mathematical thinking, including consideration of the subject's history of learning mathematics, and also the role of the unconscious in the study of mathematics anxiety, and of affect more generally.
4   Description of both the differences, and the similarities, in the ways that particular subjects with similar 'structural positions' (e.g. gender or social class), act in contexts defined by school mathematics, or other numerate practices.

I aimed to develop my ideas in the ways now clarified, through using a set of semi-structured interviews where subjects would confront a series of problems, and would report on, and evince, their responses to talking about mathematics and numbers, and to attempting problem solutions. This would allow description of:

- the setting, including the social interaction of the interview
- qualitative aspects of the subject's thinking (e.g. type of strategy used)
- description of emotional responses, especially anxiety, using not only expressed emotion, but also observing overt behaviour and analysing the subject's talk.

Issues 1 to 4 are reconsidered – beginning with the idea of the context – in Chapters 6 and 7. The design and results of the further empirical work are discussed in Chapters 8 to 10.

6   Rethinking the Context of Mathematical Thinking

Teacher: Keith, if I had eight apples in my right hand and ten apples in my left hand,
what would I have?
Student: Huge hands, Sir!

(Joke in Christmas cracker)

In this chapter, I consider further what might be meant by the context of mathematical thinking. At the same time, I consider various positions on the possibilities of the 'transfer' of learning. The transfer of learning refers in general to the use, in one context, of ideas and knowledge learned in another. This might take one of several forms:

1   the application of knowledge from pedagogic contexts to work or everyday activities
2   the 'harnessing' of out-of-school activities for the learning of school subjects
3   the use of a school subject like mathematics outside of its own domain, in physics or economics.

Other forms of 'recontextualisation', such as the reformulation of academic discourses as school subjects, are related to these forms of transfer; see, for example, Bernstein (1996).[1]

Here I am especially interested in issues around 1 and 2. These are clearly crucial issues, for schooling in general, and especially for mathematics, which is claimed to have wide applicability across the curriculum, and outside the school or college.

In general, questions around the issues of context and transfer are strongly contested, and a variety of views proliferate in educational circles, as well as in psychology and sociology. The discussion has been especially vibrant in mathematics education, where several conflicting approaches have been on offer. Here I present an overview of the views on context and transfer involved in five approaches:

* utilitarian, including proficiency and functional, views
* constructivism
* sociocultural views, including situated cognition

- structuralism
- poststructuralism.

My position is that neither utilitarian or constructivist views, with their simplistic faith in the basic continuity of knowledge across contexts, nor currently popular 'insulationist' views such as the strong form of situated cognition, which claims that transfer is basically not possible, are adequate. Instead, I analyse why transfer is problematical in principle, and undependable in practice. I shall set down my alternative approach to context and 'transfer', drawing on these other views, but aiming to go beyond them.

A Range of Views on the Meanings of Context and the Possibilities of Transfer

In Chapter 2, I discussed two sets of understandings of the context of mathematical thinking. What I called proficiency approaches include behaviourists favouring the use of learning objectives (for example, Glenn 1978), and those emphasising 'numerical skills' (such as Numeracy Task Force 1998). I contrasted these in several ways with functional views (such as the Cockcroft Report 1982).

However, despite some differences in emphasis, these two sets of approaches share several important ideas. A problem or 'task' (or 'skill'), and the mathematical thinking involved in addressing (or producing) it, are considered by both perspectives as able to be described adequately in abstract terms, with little or no reference to the context, simply as 'proportional reasoning' for example. That is, mathematical knowledge is seen as *de-contextualised*. Within these perspectives, mathematics is considered to provide a set of tools that can be used *in essentially the same form* across a variety of different contexts in working and everyday life. This toolbox metaphor justifies considering these two perspectives together, in the context of the discussion in this chapter, as 'utilitarian' approaches (e.g. Noss 1997).

Utilitarian ideas on the transfer of learning, for example from school to everyday situations, are clear: practical tasks embody mathematics, so the mathematics must simply be recognised, and transfer will be relatively unproblematical (if not straightforward), at least in principle, for those who have been properly taught.

The first part of my study shows the limitations of assuming the context of a problem to be indicated by its wording and format, and of attempting to describe the context simply by naming it as 'school mathematics', 'consumer mathematics' and so on. This means that the context is described 'naturally', rather than being *analysed* for its socially constructed qualities (for example, Atweh et al. 1998).

A number of other problems with these views can be signposted. It has been found difficult to describe a task, in ways that are abstracted from the context, so the notion of the 'same mathematical task' in different contexts is highly problematical (Newman et al. 1989). In addition, the methods used by a particular person for addressing the same task have been shown to vary a great deal across different contexts – for example, in terms of the methods of calculation or types of representation (for example, oral or written) used. Also the levels of performance of what appears to be 'the same task' vary dramatically across different contexts (Nunes et al. 1993, Lave 1988).

Not surprisingly, studies focused on the problem of transfer, from school to outside activities, suggest that much teaching has disappointing results in this respect, and students often 'fail' to accomplish it (for example Boaler 1998, Molyneux and Sutherland 1996). One reason may be that people do not 'spontaneously see' the transfer (the task, the goal) which their teachers (or their managers) have in mind. And even if they see it, they may not be motivated to carry it out. In such cases, a researcher may conclude that a 'mathematical' signifier is not recognised as such, whereas it may be recognised, but its mathematical meaning be undermined by competing values related to other discourses, (Dowling 1991), or by affective conflicts (see Chapter 7).[2]

Even a researcher who considers a learner to be accomplishing some sort of 'transfer' of school algorithms to everyday problems may not accept the traditional view that this transfer is straightforward. Thus Saxe rejects the view of transfer as an 'immediate generalisation or alignment of prior knowledge to a new functional context', and prefers to conceive of it as 'an extended process of repeated constructions . . . of appropriation and specialisation, as children repeatedly address problems that emerge again and again in cultural practices' (Saxe 1991b: 235).

Thus, in recent years, several strong alternatives to the traditional view have emerged. These include *constructivism* and *situated cognition*. Constructivism now takes several forms – including radical constructivism (von Glasersfeld 1998), 'interactionist constructivism' (Cobb 1994), and social constructivism (Ernest 1991) – all with strong affinities with the 'classical constructivism' of Jean Piaget, which has had enormous influences on the teaching of school mathematics worldwide.

Constructivism of course has sharp differences – in the crucial areas of pedagogy and epistemology – with utilitarian approaches. Whereas the latter are normally associated with transmission pedagogy, constructivism sees learning as involving the 'assimilation' and 'accommodation' of the learner's schemas as a result of actions on concrete objects (e.g. Ernest 1994a, Jaworski 1994, Noss and Hoyles 1996b, Walkerdine 1988).

In general, constructivist approaches are 'cognitivist' in that they focus on the individual mind/brain as the site for learning. They share this focus with a range of other approaches, including utilitarians. Further, knowledge and learning result from experience within what most constructivists see as a basically stable, objective world; 'culture and community can enter into cognitivist theory only insofar as they are decomposable into discrete elements that can be conceptualised as included in this world of experience.' (Kirshner and Whitson 1997: vii).

The differences and similarities between utilitarian and constructivist approaches can be further analysed, using Muller and Taylor's (1995) distinction between two tendencies in curriculum development and change: insulation and hybridity. Muller and Taylor focus on the idea of boundaries between, for example, school and everyday knowledges, or discourses, or cultures. For them, insulation stresses

> the impermeable quality of cultural boundaries, of textual classification, of disciplinary autonomy. . . . Hybridity, by contrast, stresses the essential identity and continuity of forms . . . of knowledge, the permeability of classificatory boundaries, and the promiscuity of cultural meanings . . . learning to

'cross-over' cultural boundaries is, or should be, the aim of all pedagogy. Questions of judgement and of classificatory integrity take second place to the goal of individual access to learning.

(Muller and Taylor 1995: 257)

Thus, an hybridiser tends to believe both:

* that boundaries between contexts *are* low; and
* that their crossing *should be* straightforward and/or facilitated by teaching.

The two beliefs tend to go together, in the thinking of utilitarians: as already noted, they tend to see tasks in many contexts as being essentially mathematical, and to believe that the transfer of learning to new contexts should be unproblematical. This should be the case both for the harnessing of everyday examples for teaching school mathematics, and for the application of school mathematical thinking in work or other non-school settings.

Similarly for constructivists. Writing in post-apartheid South Africa, Muller and Taylor are concerned to show the unintended consequences of insufficient acknowledgement of the strength of pedagogic boundaries by constructivists. In their context, 'constructivism' involves an infusion of constructivist pedagogy with 'ethnomathematical' social commitments (Volmink 1995); my choosing to focus on this particular form of constructivism means that this discussion may not be generalisable to other forms. The social commitments of ethnomathematics entail political and pedagogical challenges to a school mathematics dominated by academic mathematics, on the grounds that the latter are both 'sharply located' kind of knowledges: Eurocentric, imperialistic, dominated by male values.[3]

For Muller and Taylor, the constructivists (defined earlier) are 'strong hybridisers whose pedagogy assumes a flattening of the everyday/school boundary' (Muller and Taylor 1995: 267–8). Indeed, constructivists in general assume an 'invariance of understanding across settings' (Noss and Hoyles 1996b: 31). Somewhat surprisingly perhaps, this suggests that constructivists share both the assumptions of low boundaries between contexts, and the hybridising commitments, with the utilitarian approaches.

However, there are differences – besides the pedagogic ones – between the two perspectives. As has been argued, the constructivists tend to be liberal and egalitarian between different forms of knowledge, whereas 'utilitarian' approaches are somewhat imperialist in their desire to privilege mathematics as a special kind of (abstract) knowledge, and to apply it widely. Put another way, the latter see the boundary between school mathematics and everyday activities as more permeable to movement from mathematics towards everyday activities than vice versa, because of the greater abstraction of the mathematics.

Nevertheless, in terms of my objective in this chapter of describing different notions of context and 'transfer', the utilitarian and constructivist positions have much in common. Contexts can be largely ignored or simply named, and transfer is, in principle, basically straightforward. For a more satisfactory description of contexts, and acknowledgement of the problems of transfer, I shall have to open up the question of the social nature of contexts.

The Turn to the Social: Sociocultural Approaches and Situated Cognition

A number of approaches have aimed to bring social perspectives into a consideration of the context of mathematical activity. In this section, I consider the work of researchers who draw on the work of Vygotsky, Leontiev, and other Soviet psychologists, nowadays grouped under the banner of 'sociocultural' approaches, and also the work of those who have strongly emphasised the context of thinking, in the 'situated cognition' approach.

The connections and overlap between the two approaches are extensive. Kirshner and Whitson consider sociocultural approaches to be one of two foundations for situated cognition, along with anthropology (Kirshner and Whitson 1997: viii). On the other hand, it is reasonable to include situated cognition as a broadly sociocultural approach. In any case, there are many lines of cross-reference and cross-fertilisation. For example Jean Lave, probably the best known proponent of situated cognition, was a research student of Michael Cole, who, along with Sylvia Scribner and James Wertsch, played a major role in bringing the work of the Soviet psychologists to the attention of English-speakers.

Researchers working in these programmes have produced a range of studies (many of them cross-cultural): on schooling in mathematics (such as Gay and Cole 1967, Brenner 1985); on forms of literacy in everyday life (including Scribner and Cole 1978); and on numerate thinking out-of-school, for example Scribner (1984) on dairy workers, Lave (1988) on shoppers, and Saxe (1991a) on candy-sellers.

Sociocultural researchers address the issue of context through the idea of *socially organised activity*. Scribner explains 'activities' as: 'enduring, intellectually planned sequences of behaviour, undertaken in the service of dominant motives and directed toward specific objects' (Scribner 1985: 199). In sociocultural approaches, the goal-directed quality of activity is emphasised. Scribner draws on Leontiev's analysis to distinguish three levels of activity:

- activities (for example work activities, play activities);
- the goal-directed actions that comprise them, or
- the specific operations by which the actions are carried out (ibid.: 200).[4]

These three levels can be illustrated in Scribner's research on work in a milk processing plant (for example Scribner 1984, 1985). Here, the various occupations employed in the dairy could be considered as socially organised activities. Examples of goal-directed actions would be specific work tasks, such as assembling a customer's order or pricing a list of deliveries; and these would be based on operations such as taking six quarts of skimmed milk from a case, or multiplying a unit price by a quantity.

We need to consider more carefully the middle level of the hierarchy: actions or 'tasks'. In many psychological and educational approaches, cognitive tasks are situations created – normally by researchers, teachers or testers – to elicit behaviour from learners so as to promote learning, or assess it. It is normally assumed that the task can be specified succinctly in writing or orally, and – in traditional approaches at least – that any other aspects of the context can be assumed to be relatively unimportant, for example for the purposes of transfer of learning.

This assumption – that tasks can be identified – clearly has implications for any notion of context. Michael Cole and his colleagues addressed the problem of defining a task, which serves as 'the environment (or context) within which an informant's behaviour can be framed'. A number of rather demanding conditions are formulated for a task to be 'well-defined' (Laboratory of Comparative Human Cognition 1978: 53–4).

It is relatively easy to satisfy these conditions in experimental or assessment situations, because of the control exercised by the researcher or tester.[5] The consequent 'design features' of tests and experiments tend to facilitate the identification of errors (Cole and Traupmann 1979). To provide a contrasting setting, the research group set up an after-school cooking club, where the children's behaviour could be observed and recorded. The goal in the club was different from that of the testing situation: people came together to cooperate on tasks like baking a cake.

The task environment in the cooking club could be specified to some extent. However, Cole and Traupmann were cautious: 'we have failed to produce a general set of rules for identifying the environment–person relations of the sort that we have labelled cognitive tasks' (Cole and Traupmann 1979, see also Newman et al. 1989). For there were a number of problems:

- A task must be understood in the context of the activity or *higher-level goals* that motivate it – that is, as a 'whole task' – although

    in some settings, like the laboratory, the classroom, or wherever there is a hierarchical division of labour, the higher-level goals may not be under the actors' individual control. In other cases, the actors must formulate the instrumental relation between the goal of the task and the higher-level goals they are primarily trying to achieve.
    (Newman et al. 1984: 192; see also Newman et al. 1989)

- The *social relations* of hierarchy, power, etc. leave particular subjects free to attend to, to reformulate, or even to ignore, the task/goal that is the focus of the researcher's attention (Griffin et al. 1982).
- The subject may be attending to more than one task in a particular setting – behaviour is *multiply determined* (Cole and Traupmann 1979).

Thus, in the study of learning transfer,

    If we want to see how exposure to tasks that arise in an institutionalised setting such as the school affects behaviour in other settings (the home, the super-market, the office), we must go to those other settings to determine: (1) if the social organisation . . . there allows for the occurrence of the tasks that we have hypothesized are occurring at our source point; and (2) how people behave in the everyday contexts of occurrence of those tasks.
    (Cole and Traupmann 1979: 42–3)

That is, it cannot be assumed that a task resembling any particular school-type task will actually be performed in any out-of-school context: it must be investigated empirically, probably using some form of ethnographic observation and description.[6]

Situated Cognition: Jean Lave

Jean Lave has produced a broad-ranging series of studies of the use of mathematics by adults in settings outside the school (Lave et al. 1984, Lave 1988, Lave 1997). This focus, and her early use of the concept of 'activity' have provided many links with the sociocultural theorists discussed earlier.[7] However, she is perhaps best known for her championing of 'situated cognition', the idea that knowing, thinking and learning depend in crucial ways on the situation in which they are done. In addition, her work has constituted a powerful critique of the practices and failures of American schooling.

Lave's work has been deservedly influential among mathematics education researchers. In particular, her *Cognition in Practice* (1988) has been taken by many to provide the basis for a third position on context and transfer, which might be called the *strong form* of situated cognition. This position argues that there is a disjunction between doing mathematics problems in school, and numerate problems in everyday life, because these different contexts are characterised by different *structuring resources*, as outlined later. Further, people's thinking is *specific* to these disjoint practices, and settings. Thus aiming for transfer of learning from school or academic contexts to outside ones is pretty hopeless.

In terms of the framework introduced earlier, the strong form of situated cognition could be classed as having an insulating commitment, a belief that boundaries between practices are high – and, against more traditional views, an aim not to privilege mathematics as a special kind of knowledge. Lave considers the concept of learning transfer (especially from academic to practical situations) to be central to the celebration of the claimed superiority of 'scientific' and mathematical thought over the everyday.

In Lave (1988) activity forms what we might call a 'dialectical triple' with 'persons-acting', and the context or situation. Cognition in everyday practice is 'distributed' (see Salomon 1993) or 'stretched over – not divided among – mind, body, activity, and culturally organised settings (including other actors)'. Thus

> the *specificity* of arithmetic practice within a situation, and *discontinuities* between situations, constitute a provisional basis for pursuing explanations of cognition as a nexus of relations between the mind at work and the world in which it works.
>
> (Lave 1988: 1, emphasis added)

This points to the implications of the move from the idea of the isolated mind, as in cognitivism (see p. 76), to that of a 'situated mind' (cf. Cobb and Bowers 1999).

In developing the idea of context, Lave proposes a number of dialectical relations that are meant to transcend the limitations of polarities, such as individual versus society and cognition versus culture. In particular she relates objective and subjective aspects of the context by distinguishing the *arena*, a durable and public context, and the *setting* which is 'malleable' (for example, by varying displays of products for sale) and is as experienced by the person-acting. Thus a supermarket is an arena which offers a setting for a particular shopper's weekly rounds. Lave (1988) also addresses the 'macro' level of the context, in relating the 'experienced world' to the 'constitutive order' of culture, economy and social structure.

She analyses other aspects of the context through the notion of *structuring resources* (1988, Chapter 6). These have as their basis:

- ongoing activities
- social relationships
- subjective experience of problems as dilemmas, thereby producing motivation
- standard crystallised forms of quantity such as money and mathematics.

Thus, when Lave speaks of the 'proportional articulation of structuring resources', she means the interrelation of elements of different practices in shaping action in a particular situation. This allows her to conceive of different relative predominances or 'mixes' of shopping and mathematics, say, in supermarket best buy decisions, and in her 'best buy simulation experiments' (Lave 1988: 99ff.).

From this point onwards, the term 'practice' will be used interchangeably with 'activity' (in Scribner's sense), unless indicated otherwise. It should be noted that a number of authors, including Lave, use the term 'activity' in a rather broad way, without distinguishing activity from action (unlike Scribner).

Lave's discussion of 'transfer' is linked with a critique of the complex set of ('structural functionalist') ideas from anthropology, about learning, the social world, and people's relationships with it. These depict society as reproducing itself by cultural transmission (or socialisation), with knowledge/learning being unproblematically internalised by learners, then transferred or applied in other settings. She considers this an impoverished account of learning and of problem-solving (Lave 1996b).

Lave echoes many of the criticisms of transfer experiments discussed earlier by Cole and his colleagues. The 'problems' presented in such contexts are considered as objective and factual, because they are constructed by experimenters, rather than by subjects, and the experimenters preformulate the correct or appropriate solutions. This has two consequences: a subject who does not take on the problems, or who does not produce the appropriate solution, is deemed to have 'failed', and the researcher is unable to study the fruitful methods the subject may have to deal with certain problems. Therefore, such research fails to describe much observable problem-solving activity.

Along with others, Lave criticises the 'normative' perception of there being 'one correct method' in problem-solving.[8] She notes that people tend towards early formulation of a 'solution shape', with 'gap-closing' in the resolution of 'snags' or dilemmas (Lave 1988: 139–42, 158ff.) The ease with which a person formulates 'solution shapes', as well as the availability to him/her of developed strategies, for everyday numerate problem-solving, suggest ways of giving substance to the idea of familiarity discussed in Chapter 2.

Lave (1997) also relates an individual's thinking to sociocultural aspects, using illustrations from de la Rocha's research (1985; Lave 1988) on numerate problem-solving amongst a small group of Weight Watchers. Sociocultural 'contradictions' (for example between ways of eating for pleasure and eating for appearance) are experienced as 'dilemmas' (e.g. to binge and feel better now, or to diet, so as to look better in future), and both are embedded in social practice. At the lowest level, a 'problem' (such as calculating protein intake options while preparing lunch) is motivated by reference to the dilemmas that make it meaningful (Lave 1997) or else it is merely a 'closed-system puzzle'.

The empirical material for Lave (1988) and Lave (1997, first published in 1990) comes from the Adult Math Project, begun in 1978 (see also de la Rocha 1985, Murtaugh 1985). Lave chose arithmetic as the focus for this research, as it 'has a highly structured and incorrigible lexicon, easily recognisable in the course of ongoing activity' (1988: 5). However, it will be important to examine critically both the 'incorrigibility' (purity) and the 'recognisability' of the arithmetical lexicon in this study.

The research team recruited thirty-five Californian participants, twenty-five to their supermarket study of shopping activity, and ten to a study of dieting. There were thirty-two female and three male participants. (For further on the method, see Lave 1988: Ch. 3.) They produced six types of empirical material for the shopping study which will be the main focus here, including:

- observations of 'best buy' decisions made while shopping, conducted in the supermarket
- a shopping simulation experiment, with twelve 'similar' best buy question, administered in the subject's home
- tests on arithmetic, 'number facts', and 'measurement facts', presented orally
- a standardised test of multiple-choice questions (MCQs).

The results for the shopping study can be arranged in four categories:

1   differences in levels of 'correct' performance between everyday and school (or school-type) contexts
2   differences in methods used between everyday and school-type contexts
3   correlations in performance between everyday and school-type contexts
4   correlations of performance in everyday, and in school-type, contexts with 'school background' measures.

These four categories of results can also be found in the results of the Weight Watchers study (Lave 1997), and in those of Scribner (see earlier) and of Nunes et al. (1993). In general, differences in categories 1 and 2 are taken to suggest discontinuities between thinking in school and non-school contexts. For category 3, higher correlations are expected among the various measures of school and school-type performance, and lower correlations between these measures and those of everyday performance; for category 4, higher correlations are expected for measures of length, recency, or success in schooling with various measures of school and school-type performance, than with measures of everyday performance.

Two of the most important results of the shopping study were:

- the substantial differences in correct performance in the two situations
- an observable contrast between the difficulty subjects had with the arithmetic test, and the familiarity with which they handled supermarket calculations.[9]

For further on the results see Lave (1988) and Evans (1993: Appendix W5).

From these results, Lave's main conclusion is there is a *discontinuity* between numerate performance in the supermarket – measured by the observations and simulations – and performance in 'school-like' mathematics activities – measured by the four 'tests' in the list earlier. She suggests that the discontinuity was due to the context of the activity.

The results from the shopping study appear to confirm Lave's (1988) views on how activity and thinking are constituted in context-specific ways. But there remain several problems.

First, the *categorising* of the six sets of tasks was not all that clear. For example, were the 'measurement facts' questions firmly in the school mathematics category, or intermediate between school-like and practical everyday performance? The idea of different proportions of structuring resources could be used to differentiate the tasks a priori. But instead Lave simply categorises the tasks 'naturally' into two disparate groups.

Second, comparing performance levels in this way assumes that the different performance measures were *of the same difficulty*! There are many pitfalls and circularities in attempting to assess difficulty levels, and Lave reports her own attempts to judge relative difficulties only for the best buy simulation problems. The reader will recall that I avoided comparing levels of correct performance on my school mathematics and practical mathematics scales in Chapter 2, for just this reason.

Third, even if one takes the performance results on their own terms, as Lave does, they can hardly be interpreted to show straightforwardly the discontinuity in performance claimed; see Table 6.1.

Looking at Table 6.1, rather than the neat discontinuity between 'school-like' and shopping performance claimed by Lave, one can as easily see:

- a clustering of very high results for observed shopping decisions and shopping simulations

Table 6.1 Levels of Performance in Different Contexts in Lave's Research

Performance measure	% correct
Observed shopping decisions	98[1]
Shopping simulation	93
Number facts	85
Standardised test	82
Measurement facts	66
Arithmetic session	59[2]

Notes:
1 Percentage of problems *observed*
2 Percentage of problems *completed*

Source: Lave 1988: 56 (Table 6).

- moderately high results for number facts and standardised test performance
- low results for measurement facts and the arithmetic session.

Also, the supermarket observation results should undoubtedly be bracketed as not comparable with the others – since the subject was free to attempt a calculation or not, rather than having the problems *assigned* by the researchers!

Therefore the most celebrated finding of the situated cognition programme of research can be seen to be much less dependable than it first seems. Accepting it on this basis requires accepting a problematical grouping of different tasks, ignoring the issue of possibly differential difficulties across tasks, and agreeing to a contentious reading of the results. However, Lave discusses 'convergent findings' from other researchers, such as Scribner and Fahrmeier (1982), Carraher et al. (1985), Nunes et al. (1993); see also pp. 88–93.

We need to consider several more general issues concerning Lave's (and others') research on the context and situated cognition. These concern:

- the idea of different mixes of structuring resources
- involvement of the subject in 'locating' the context of the problem
- 'reactivity' in the research arrangements.

The postulation of different proportions of structuring resources is a potentially powerful idea. It allows for a whole range of 'mixes' of activity. It raises the problem of how the activity, or mix of activities, that the person is 'acting within' during a given episode is 'determined', as well as how the researcher would know. However, there are several problems about the way it is used in Lave (1988). First, it might have been used with effect when discussing the 'clustering' of the performance results, but was neglected. Second, its use sometimes appears imprecise. Lave's arguing, for example, that her best buy simulation problems attempted at home were structured similarly to the decision-making episodes observed during supermarket shopping (Lave 1988: 114–15) risks underplaying their differences, and investing both with a specious 'naturalness'. This ignores their distinctive aspects as research contexts, in terms particularly of social relations between researcher and subject, especially whether the latter's 'position' is to accept the former assigning problems to be solved. Finally, the idea of a mix of structuring resources sits uneasily with claims of a *disjunction* between contexts, or a discontinuity between performances in different contexts, as made by Lave and other proponents of situated cognition.

Lave (1988: 69) acknowledges the shopper's freedom to choose to recognise a dilemma as a 'problem' – or not – as the children often could in Cole and his colleagues' cooking club. Indeed, Lave's work appears to differ from Scribner's (1985), and from Saxe's and Noss et al.'s (see later), at least partly because of the differing positions of employment and shopping, in the 'constitutive order': the dairy loader normally *has to do something* in response to an order, whereas a shopper seems to be free, to choose an item on their list without doing a calculation, or even to ignore the item. This suggests that issues of *power*, or *authority*, such as the need to obey orders, must be involved in a full description of the context.

In any case, Lave does not appear to extend the idea of the shopper's freedom to include the subject in her analysis for 'locating' (or determining) the context of the problems solved, since she herself categorises each of the six sets as 'school-like', or 'everyday'. This suggests the idea of an 'objective structure' – reflecting perhaps a determinate predominance of structuring resources – from which one can 'read off' the context which applies to *all* subjects' thinking. Against this I argue the need for a 'negotiated' judgement of context, depending partly on *the subject's perceptions* of the demands of the research situation.

Lave rightly warns against the reactivity involved in interviewing or experimental simulations, since being approached in certain ways may make people feel they have to respond in a 'school-like' way. However, there is also danger of reactivity in 'naturalistic' observation (see also Agre 1997). Thus one might wonder about how candid subjects in the Adult Math Project were prepared to be, especially after up to forty hours' contact with the research team. Shoppers accompanied by a researcher may have felt impelled to report as if they were making decisions in a purely rational 'best buy' manner; for example,

> [Buying] the five pounds [bag twice] would be four dollars and 32 cents, versus four dollars and 30 cents [for the ten-pound bag]. I guess I'm going to have to buy the ten-pound bag just to save a few pennies.
>
> (Shopper, quoted in Murtaugh 1985: 190)

Murtaugh interprets this shopper as comparing 'prices for ten pounds of sugar, the quantity she already has decided to purchase'. However, it also seems possible to interpret the formulation 'I guess I'm going to have to . . .' as suggesting that this shopper is attempting to display a 'rational' performance for the benefit of the researcher. This is illustrative of the problems of reactivity, which suggest that many of the 'explanations' by shoppers quoted by Lave and her colleagues require critical scrutiny.

Jean Lave's more recent work (Chaiklin and Lave 1993, Lave 1996a) focuses on describing learning within 'communities of practice', including a consideration of apprenticeship as a model of situated learning (Lave and Wenger 1991). This work is no longer so concerned to stress discontinuities between practices: it acknowledges that no practice could ever be completely closed, and that a community of practice must be understood in relation to other tangential, or overlapping, communities. The approach now consists of identifying communities of practice which are interdependent, and studying the bridges between them, particularly the social relations and identities across them (Lave 1996b). For example, school homework has been examined because it moves 'back and forth' between home and school, and actually also to the bowling alley, the snack bar, and so on (see also Chaiklin and Lave 1993).

Jean Lave's contribution to elucidating the situated characteristics of cognition has been substantial. She has shown how to look for, and to begin to describe, the ways that different contexts may be discontinuous, and may have effects on activity and thinking in them. These ideas need to be included in a 'context-sensitive' conception of thinking and learning. However, it is important to avoid the *cul-de-sac* of the strong form of situated cognition – which threatens to portray a proliferation of differently situated types of mathematical thinking (Noss and Hoyles 1996b), with high boundaries, and no 'overlap' between them.

On the contrary, the differences in practices pointed to by Lave and situated cognition need to be described *and analysed*. This means letting go the assumption that practices and communities of practice can be seen as 'natural', and instead analysing the bases of difference among them. This will involve using some of the dimensions isolated in the notion of 'structuring resources'.

Further, if the researcher is to 'negotiate' his/her judgement of the context, by taking account of the subject's perceptions, the research needs to give space to the subject to voice these. I shall show that this requires consideration, in a systematic way, of the effects of language underlying practices in different contexts. An indication of how such a concern with language might help clarify the idea of practices and boundaries between practices is given by considering certain structuralist and 'poststructuralist' approaches in 'The Turn to Language', this chapter.

First, however, I consider the work of sociocultural researchers who have developed more fully the notion of structuring resources; in particular, that of the representation of mathematical ideas.

The Turn to Representations: Street Mathematics and Computer Microworlds

In this section, I discuss the work of three research teams whose work might be classed broadly as sociocultural, though they are not as closely identified as Jean Lave is with situated cognition. These are the teams led by Geoffrey Saxe; by Terezinha Nunes, Analucia Schliemann and David Carraher; and by Richard Noss and Celia Hoyles.

Geoffrey Saxe (1991a, 1991b, 1994) aims to develop a new level of analysis – activity in a sociohistorical context – where 'culture and cognition are constitutive of each other' (Saxe 1991a: 184). Thus his aims are similar to Lave's. He is responding to what he perceives as the shortcomings of the theories of Piaget and Vygotsky, in terms of their ability to explain how 'the mathematical understandings that have emerged over the history of a cultural group become the child's own, interwoven with the child's purposive problem-solving activities' (Saxe 1991b: 230).

Saxe has studied the development of numerical cognition in a number of cultural contexts: among the Oksapmin in Papua New Guinea, both among children during the advent of western-style schooling and among adults during the establishment of currency relations (rather than barter) for trading goods, and among children working as candy sellers in North-East Brazil, during a time of inflation.[10] Because of the social and historical conditions, the subjects of all these studies needed to learn or 'develop' from changing conditions. As in other studies reviewed here, this research combined phases of intensive observation, and simulation or one-to-one testing.

Saxe's theoretical framework has several components, which can be related to the candy sellers. At the centre are the seller's *emergent goals* (emerging during participation in the practice); these seem to correspond to tasks or actions in the activity/action/operation hierarchy (see above). These are influenced and constrained by four 'parameters' of the context:

- The *activity structure* of the practice, the general motives for participation in selling (cf. 'high-level goals' referred to by Cole et al.) and the types of tasks that must be accomplished.
- Related *social interactions:* for example, wholesalers or fellow sellers (or family members) may provide help in pricing, and customers in making change.
- Cultural *artefacts* (such as currency denominations and inflation), conventions. (for instance, boxing of candy and price-ratio selling prices), and sign forms (such as the number symbols, 'currency arithmetic').
- The individual's *prior understandings.*

Saxe's 'parameters' recall, and indeed parallel, Lave's 'structuring resources', but they are discussed much more systematically, especially the activity structure of the practice. Saxe takes this analysis further than others, in his description of the candy-selling practice as involving the cycle:

purchase ➤ preparing to sell (pricing) ➤ selling ➤ preparing to buy ➤ (etc.)

Of course, it may be easier to discuss issues of 'form' in the case of work practices, because of the relatively strong exercise of authority there (see earlier) – and this form may be even clearer here, with the relatively simple buy/sell structure to the practice.

On the issue of transfer, Saxe is more positive, and less polemical, than Lave (Pea 1990). He is interested in 'the interplay of form and function across cultural practices'. For example, he describes how a seller (aged 13, fifth grade completed) carries out the pricing function in the candy-selling practice, determining the wholesale unit price by trying out several possible values, each time using a standard school multiplication algorithm. He then calculates his profit per candy bar, and hence per box of fifty, using school algorithms for subtraction and multiplication (Saxe 1991a: 61) rather than alternative methods like *decomposition* and *repeated addition*, often found in 'street mathematics' (Nunes et al. 1993).

Saxe also reports examples of *harnessing*, where sellers (aged around 10) adapted calculation methods from selling practices, to attempt school-type problems, and performed better on such problems than non-sellers (Saxe 1994: 154).

Saxe conceives of transfer as 'an extended process of repeated constructions', of 'appropriation and specialization' – that is, adapting ideas and cultural tools – rather than as an 'immediate generalization or alignment of prior knowledge to a new functional context' (Saxe 1991b: 235); his work suggests that repeated attempts may be needed so that transfer can take place. Thus Saxe's position resonates with Nunes et al.'s (1993) claim that subjects use 'pragmatic reasoning schemas', rather than one abstract system (see p. 92).[11]

Like other authors, Saxe includes within the context social (or pedagogic) interactions as supports, but, more than others, he emphasises 'sign forms' or language, and cultural artefacts and conventions. His inclusion of prior understandings as one of his parameters indicates a position where cognition is not completely situated outside of the individual, unlike strongly situated cognition. Concerning transfer, he also points to the importance of some depth of knowing, or familiarity with, the specialised knowledge forms involved in solving practice-linked problems, in order for them to be appropriated and transformed.

The school of researchers led by Terezinha Nunes (formerly Carraher), Analucia Schliemann and David Carraher have studied the everyday practices of various groups of workers around Recife in North-East Brazil, including carpenters (T. N. Carraher 1986), bookies (Schliemann and Acioly 1989), farmers (Abreu and D. Carraher 1989), and fishermen (Nunes et al. 1993, who also provide an overview of these studies).

These studies generally begin with ethnographic description of the work practices (particularly the numerate aspects), followed by studies using a fusion of experimental designs and Piagetian clinical interviews, and drawing on the ethnographies. Here the subject is asked to solve several sets of problems constructed by the researchers. Typically, the first set of problems are familiar from the work context, but are 'beyond' the familiar tasks; later sets of problems require varying 'levels of transfer'; see below. Transfer is measured in terms of correct performance in solving the different types of problems. This hybrid approach avoids the shortcomings of purely ethnographic studies, which cannot describe the use of mathematics operations that do not arise spontaneously in the settings studied, and those of the standard 'transfer experiments' criticised by Lave.

One of Nunes et al.'s early concerns was with social class differences in school mathematics failure (Carraher 1988). When they began their research programme with an inquiry as to how children were actually solving problems outside of school, 'in the street', they found that:

- children out of school had their own distinctive methods for solving 'mathematical' problems; and
- when they were allowed to solve problems in their own ways, many of the social class differences disappeared (Carraher et al. 1985).

Nunes et al. document clear and interesting differences in calculation methods between street mathematics and school mathematics. They describe some of the 'heuristics' that allow people to accomplish arithmetical calculations while conserving meaning; for example,

- decomposition to simplify addition or subtraction
- repeated grouping to simplify multiplication or division
- 'rated addition' for proportional reasoning.

Thus $252 - 57$ becomes $252 - (52 + 5$, as 57 is 'decomposed'$) = 200 - 5 = 195$. And $10 \times 35 = 105$ (i.e. 3×35; a 'well-known result') $+ 105 + 105 + 35 = 350$ (Nunes et al. 1993).

When they compared the children's performances in street contexts with those in school-like testing contexts, the performances on what appeared to be 'the same task' were superior (in terms of correctness) in the street contexts (Carraher et al. 1985). However, talking about 'the same task in different contexts', and seeking to compare cognition and performance across contexts as different as street markets and testing in school settings, would be seen as highly questionable by researchers accepting the 'situatedness' of cognition This is because different contexts could be expected to differ on a number of aspects, such as the setting, the social relations at play and so on. That is, 'like is *not* being compared with like'. This criticism clearly applies to the production of what I call (p. 84) 'the most celebrated finding of the situated cognition programme' by a number of research teams, and not only by Lave.

The problem about whether the ideas 'really are' mathematics is addressed by Nunes et al., using Gerard Vergnaud's theory of concepts (Vergnaud 1988). Here concepts (developing in a learner's mind) are seen always to have three aspects: invariants, representations and situations. 'Invariants' refer to the properties or relations associated with the concept, such as symmetry, commutativity, conservation of equality. 'Representations' are based on the set of symbols (linguistic or non-linguistic) used to communicate or discuss invariants: 'representations always involve keeping some features of the concept in focus, while losing sight of others' (Nunes et al. 1993: 145). 'Situations' make the concept meaningful; this aspect appears to be broad enough to include social situations (such as selling, sporting events, maths classes) and real or imagined problem-situations (e.g. 'times problems', proportionality problems) (D. Carraher 1991: 178). This triple aspect of mathematical concepts allows Nunes et al. to argue that the invariants must be constant across thinking in different contexts, whereas the representation and especially the situation depend on the context.

This idea was used with effect in Carraher et al. (1987). They asked which aspects of the different contexts could account for the differential performance across contexts found in Carraher et al. (1985). One reason they proposed was that the social relations between researcher and researched were different; indeed, the child working as street vendor may well not have realised that the 'customer' was also a researcher. In this study, they controlled for such differences in social relations between contexts by presenting problems in *three situations within one context*: the context was testing-in-school, and the three situations were simulated store problems, word problems, and computation exercises. Problems set in each of these three situations were created using each of the four operations (+, −, x, ÷), matching for difficulty − thereby avoiding the problem of uncontrolled variations in difficulty found in Lave's shopping research (see earlier) − and performances across the three situations were compared.

They first found that the number correct was higher in the simulated store and word-problem situations than for the computation exercises. But this could not be explained merely by the availability of concrete objects for the simulations, since the level of correct performance was equally high for word-problems, where none were used. They then found that correct performance was also correlated with the choice of *procedure* − or type of representation − with 'oral' calculations being done correctly more often than 'written' ones. Further, they concluded that, when the procedure used by the children was controlled for, the differences in correct performance across situations disappeared.

For these researchers, the difference between written and oral procedures is based on their being learned in school and in informal contexts, respectively. This difference also parallels the distinction between a 'manipulation of symbols' approach, and a 'manipulation of quantities' approach (Reed and Lave 1979). The manipulation of symbols approach is based on the memorisation and recall of arithmetic operation facts, and of *algorithms* which use written representation. The

manipulation of quantities approach uses *heuristics*, such as decomposition (for +
and −), and repeated grouping (for x and ÷). Algorithms need to be memorised and
applied relatively rigidly; heuristics are more flexible in general.

Thus they conclude that

> the situations in which arithmetic problems are embedded may have a strong
> impact on how they are solved. This impact is not produced by some peculiarity
> of the testing situation, such as anxiety, but seems to result from the meaning
> that problems have for children when they engage in problem solving.
>
> (Carraher et al. 1987: 95)

and

> The effect of the situation upon the child's performance is mediated by the
> choice of strategy [i.e. procedure].
>
> (Carraher 1988: 5)

Situations that present quantities embedded in certain transactions allow children
to *preserve meaning* in problem-solving procedures, because of the physical quantities that are being quantified (such as money, cars), and because the meaning of the
quantifier itself within the number system (for example ones, tens, hundreds) is
preserved. On the other hand, written, school-based procedures which involve the
manipulation of symbols may lead children to focus not on physical quantities and
preserving meaning, but on written numbers and rules; since these rules may be
designed to convey meaning in abstract ways such as through place-value, they may
be experienced by the learner as being associated with a loss of meaning. This may
explain some learners' apparent willingness to accept results that would be recognised as absurd by anyone who was 'controlling for meaning'.

Thus, Nunes et al. seem to have reformulated the problem of transfer. Though the
invariants may be the same between school maths addition and totalling prices in the
market, since the situations and the representations are different, the concepts are
not strictly the same (Carraher 1991). Thus neither the gap in performance between
market and testing contexts (Carraher et al. 1985), nor that between store simulation
or word problem situations, on the one hand, and computation situations, on the
other (Carraher et al. 1987), provide a refutation of the possibility of transfer of
learning. Rather, the differences arose because the subjects perceived that different
procedures were 'called for' in the different contexts or situations (Carraher et al.
1987).

Nunes et al. (1993) tease out several different levels of transfer, such as:

1   application to problems *with unfamiliar parameters*, for example to non-
    standard ratios or scales on drawings for buildings
2   *reversibility* or use of a procedure in the opposite direction from its usual use,
    such as calculating a unit price, given the cost of n items
3   *transfer across situation(s)*, for example asking fishermen to solve unfamiliar
    ratio problems concerning the relationship between unprocessed and processed

seafood that were isomorphic, in the researchers' view, to familiar problems about weight–price relationships.

Thus in one set of experiments Nunes et al. (1993) found that fishermen were able to use their everyday mathematics thinking in a conceptual, rather than just a procedural, way, to solve a range of problems, demonstrating (2) reversibility and (3) transfer across situations.

Similarly, Schliemann and Carraher (1992) conclude that

> learners can develop proportional reasoning first in a limited range of contexts. . . . *Given the proper conditions*, similarities of relations can be detected and transfer and generalisation become possible. This recognition may then act as a bridge for transfer of procedures to the unknown contexts.
> (Schliemann and Carraher 1992: 61, emphasis added)

Schliemann (1995) concludes more generally

> mathematical knowledge developed in everyday contexts is flexible and general. Strategies developed to solve problems in a specific context can be applied to other contexts, *provided that the relations between the quantities in the target context are known by the subject as being related in the same manner as the quantities in the initial context are.*
> (Schliemann 1995: 49, emphasis added)

This is an important conclusion, which focuses on *similarity* in relations between *quantities*. I shall seek to build on it, in the next section.

Nunes et al.'s research has made a very substantial contribution to efforts to describe cognition in context, through their seeking out and description of numerate thinking in a wide range of work contexts. Further, the experimental phase of many of their studies has focused on, and isolated out, the *situation* within which they assume the subject is thinking, rather than the *context*. Here the situation is understood as that part of the context that provides the overt background to problems, for example as computation exercises, word problems or a 'store simulation'. The context additionally includes the setting and social relations, such as a clinical interview, written test, or market sales transaction. This separation allows an ingenious attempt to bring the crucial bases for thinking under the researcher's control: the experimenter allocates the *situation* for a given task in a controlled way, while the *context* of setting and social relations is assumed to be held constant over different situations (Nunes et al. 1993: Ch. 3).

However, this move deals with the complexity of the context of problem-solving methodologically, but not theoretically. It has the effect of limiting the capacity of the research to study cognition in context, fully understood: it limits its 'ecological validity' or generalisability, since being tested in the simulated store situation is not the same as functioning in the market context. Furthermore, it diminishes the importance of the context which is kept in the background of the analysis, since its multiple facets (language, goals, social relations; see earlier) cannot be captured by the situation. The situation is foregrounded, and is seen as given by the wording and format of the problem, by any physical object available, and by any background information given by the interviewer, or taken-as-given by all subjects. However, focusing on the situation in this way, at the expense of the context (at least in some of Nunes et al.'s experiments) runs the risk of taking us back towards the traditional approaches discussed earlier.

Moreover, as I argued in connection with Lave's work, the context is not simply 'given': the subject is involved in 'constructing', or construing, it! There may thus be significant variation in subjects' experience of the context. In particular, schooled subjects may call up school, as well as everyday, practices, as the basis for the context for their problem-solving. Nunes et al. acknowledge this possibility by recording the number of years of schooling for their subjects, and by comparing school students' and working-people's performances in some of their designs. But this may not adequately capture the likely variation across subjects in construing and experiencing the context.

Nunes et al.'s use of the distinction between oral and written 'practices' or 'representations' in calculation appears practical, since the basis of the distinction is overt, but it can also be deceptive.[12] Further, as suggested by Schliemann and Acioly's (1989) study of bookies, and Saxe's illustration earlier, some examples of problem-solving may combine or mix informal procedures with modified taught school procedures.[13] In any case, Nunes et al. make it clear that the distinction is based in the context where the methods were learned – in school or in the street – rather than simply in the overt character of the procedure (Nunes et al. 1993: 74).

Nunes et al. sum up their work by reflecting on the street maths – school maths distinctions (ibid.: Ch. 7). They refuse to characterise the couple as particular versus general, and indeed refuse the polarisation of forms of knowledge along these lines. They thus reject the idea of people using syntactic, domain-independent logical rules in reasoning, as envisioned by traditional approaches to transfer, as well as the image of the subject using narrower rules tied to particular domains in which s/he has actual experience, as in situated cognition. Instead they are attracted to the idea of 'pragmatic schemas' (Cheng and Holyoak 1985), or 'logico-mathematical concepts that can be used in a general way even though they may not involve context-free reasoning'. An example in street mathematics would be the 'additive composition of money': counting and calculating with money involve representing money totals as constant, even if coins of different values may be used to compose the total (Nunes et al. 1993: 144).

To summarise, Nunes, Schliemann and Carraher's studies of non-academic 'street mathematics' used by a broad range of occupational communities have contributed to the growing corpus of accounts of practical mathematics. They have clarified the conditions under which transfer from school to outside might be effected, and have also considered the harnessing of street maths to aid school learning (for example Schliemann 1995). In a number of studies, they have used an ingenious way of controlling for differences in situation, although this has limited the scope of their research to describe fully the context. Also important for my purposes here is their stress on the use of representations: in street mathematics, the representations are especially suitable in their power to 'evoke the situation', or to help the reasoner keep it in mind more clearly (Nunes et al. 1993: 144).

Richard Noss, Celia Hoyles and their colleagues at the London Institute of Education have studied a 'mathematical orientation' among bankers (Noss and Hoyles 1996a), nurses (Pozzi et al. 1998) and airline pilots; overall, see Noss et al. (1998). Their studies also generally begin with a phase describing the work practices of the relevant community – combining workplace observation, interviews with senior practitioners and analysis of textbooks – in order to describe what they call 'routine visible mathematics'. This is followed by a phase of simulation interviews and questionnaires, which investigate how practitioners would handle what Noss et al. call 'breakdowns' of workplace routines. Finally, they have added a third phase of 'teaching experiments', where they have encouraged practitioners to use mathematical modelling in unfamiliar settings.

In their general discussion of context, Noss et al. emphasise mathematical representations, particularly computer-based mathematical models or 'computer microworlds' (Noss and Hoyles 1996b). At the same time, they problematise the dichotomies formal versus informal, concrete versus abstract, and contextualised versus decontextualised, and hence the characterisation of mathematics as involving decontextualisation and abstraction, thereby questioning further most traditional and constructivist positions (Noss et al 1998, Noss and Hoyles 1996b). Their work has developed the idea of 'situated abstraction', a way 'to describe how mathematical meanings are shaped and constrained by the tools and language of settings, yet simultaneously capture salient mathematical relationships' (Noss et al. 1998: 2). This concept clearly overlaps with the concerns of Nunes et al.'s use of 'pragmatic schemas' (see earlier).

Key points of their work contribute to the argument here. As an illustration, I refer briefly to their discussion of 'banking mathematics' (BM), as used by employees of a major investment bank. First, following situated cognition, there is a recognition that work practices and school mathematics discourses are distinct. Noss and Hoyles found that the inhabitants of the bank 'spoke a different language' (Noss and Hoyles 1996a: 7). Further, in banking maths, *standards of accuracy* are distinctive (for example, the tolerance of $25 allowed on large-scale transfers in the bank). Certain *well known results* (cf. Lawler 1981) can be used to avoid the need for calculations, such as the fact that a Treasury Bill yielding £100 after one year and discounted at 8 per cent over the year, will have a lower purchase price than a simple interest 'instrument' that will yield the same amount after one year at 8 per cent interest.[14] Further, familiar representations, such as graphs, are 'read' differently: in BM, graphs tend to be considered as *displays of data* whereas, in academic mathematics (AM), they are read as a 'medium for expressing relationships' (Noss and Hoyles 1996a: 13–15).

Noss and Hoyles contribute ideas about facilitating the 'transfer' of learning, in several ways. First, they show how to identify areas where work practices might usefully 'overlap' with academic mathematics. For example, they used the idea of a function as a 'bridging concept' between BM and AM, and computer programming as a way of building models, so that their students would learn

what it means to construct a mathematical relationship, and how and why the

language of mathematics assists in conferring expressive power to the
description of relationships . . . programming is a way by which learners can
express the state of their current understandings symbolically while holding on
to the meanings which can all-too-easily become lost in the passage to conventional mathematical discourse.

(Noss and Hoyles 1996a: 8)

Second, they show how to develop new ideas on 'building bridges' across
practices, by developing a deeper mathematisation, by posing 'provocative'
problems that appear 'innocent' in BM, but are deeply significant in AM. For
example, they use their knowledge of the meanings within the practical context
(BM) to seize on the idea of 'continuous compounding', where the periods over
which interest is calculated shrink continually (from yearly to monthly, to daily
and so on) and which is normally considered only an 'exotic' topic in financial
mathematics texts (Noss and Hoyles 1996a).

Noss and Hoyles's notion of situated abstraction elucidates ways in which
mathematical meanings are shaped and constrained by the tools and language of
settings. For example, care is required in discussing interest calculations (using
percentages), where the conceptual priority (in AM terms) of simple interest
(prior to compound interest) conflicts with its relative rarity in BM practice
(ibid.). They have extended the programme of workplace studies described in this
section to professional groups, in particular, those making substantial use of
information technology. Their work thus shows how to build on research on
workplace mathematics, so as to enhance curriculum development, and to facilitate 'transfer'.

The contribution of those working under a broad 'sociocultural' banner to my
development of a fuller concept of the context of mathematical thinking will be
summarised in the conclusion to this chapter. However, several aspects of context
have not been adequately discussed yet. First, although language, 'sign systems',
and so on are mentioned by many of these researchers, no systematic way seems
available to characterise the effects of *language* or *discourse* in different
contexts. Similarly, there is little systematic attention to *social difference* such as
gender and social class, although social class is referred to by Scribner, Nunes at
al. and Saxe (1994) in a US school study.

These two aspects – language, and social difference – are discussed in the next
section.

The Turn to Language: Structuralist and Poststructuralist Approaches

For an indication of how a concern with language might help clarify the idea of
boundaries between practices, we can turn to structuralist approaches, and their
characterisation of context. Muller and Taylor (1995) draw on Basil Bernstein's
(for example 1996) general sociological discussion of types of knowledge and
boundaries between knowledges, where different knowledges are seen as
different discourses, based on different 'codes' of language.

Discourses and Boundaries: Basil Bernstein

According to this perspective, school knowledge is reinterpreted or transformed from academic knowledge in the universities, by processes of 'recontextualisation', which produce a new discourse with distinctive principles of selection, ordering and focusing. Bernstein (1971) used the concept of 'classification' to describe the strength of boundaries between curricular subject contents in educational knowledge. These subject contents are socially recognised, and Bernstein's use of the term is based on structural differences between the language codes used in the different contexts.

In Bernstein, the structural issues of the strength of a discourse's *classification* (and framing), are related to an individual's having (competent knowledge of) recognition (and realisation) rules. When different discourses/knowledges are strongly classified, 'recognition rules' can be explicit and unambiguous. The 'able' subject can recognise a context as school or everyday (vertical or horizontal discourse, respectively), and can use 'realisation rules', sometimes multiple ones, to complete a task; for example, to produce one or more classifications of objects, according either to 'general' or context-bounded criteria (e.g. Holland 1981, reported in Bernstein 1996, Cooper and Dunne 1998).

Bernstein also suggests that the strength of classification in the educational system has profound effects on one's views of knowledge:

> Any collection code [i.e. system with strong classification] involves an hierarchical organisation of knowledge, such that . . . only the few *experience* in their bones the notion that knowledge is permeable, that its orderings are provisional, that the dialectic of knowledge is closure and openness. For the many, socialization into knowledge is socialization into order, the existing order, into the experience that the world's educational knowledge is impermeable.
>
> (Bernstein 1971: 57, his emphasis)

Bernstein's account owes much to Durkheim, and the latter's distinction between the 'sacred' and the 'profane'. In this view 'boundary-maintenance' is very important, requiring energy, and transgressing boundaries is exceedingly dangerous. Thus, Bernstein tends to be an insulator, for whom 'curricular knowledge is part of that large class of esoteric discourses, separated from everyday knowledge by a hard boundary that we weaken at our peril' (Muller and Taylor 1995: 262–3).

Bernstein has not himself addressed much explicit attention to the problem of 'transfer', but in his approach, it would relate to the structure of the discourse – namely, whether vertical or horizontal/everyday. *Vertical discourses*, for instance academic disciplines such as mathematics or sociology, are specialised symbolic systems. They are explicitly assembled via recontextualisation(s) for teaching purposes (Bernstein 1996), and acquired through general principles. For a learner to transfer or apply ideas within a vertical discourse, therefore, s/he needs to grasp the principles of recontextualisation. *Horizontal discourses* are organised in segments, related to organised activities, such as work practices, shopping, playing football; they are acquired through exemplars (cf. Schon 1983). Transfer within horizontal discourses may well be accomplished through intuitive recognition of analogies (Bernstein 1996: 172). Therefore, the problematic of transfer is understood in terms of recontextualisation.

Paul Dowling (1998) develops these ideas by showing that the mathematical text-books prescribed for 'lower ability' students and which incorporate numerous examples intending to model everyday situations, have the consequence of excluding their readers from the 'esoteric' discourse of school mathematics proper. For Dowling, the recontextualisation of everyday life material into the curriculum ends up by being neither 'real maths' nor 'real life'. Besides distorting the everyday setting – in which the 'lower ability' learners are meant to feel 'at home' – it also inculcates an anodyne view of mathematics as a series of algorithmic solutions, very different from the view of mathematics as a connected set of generalisable principles, into which only the 'higher ability' students are inducted.

Cooper and Dunne (1998, 2000) use interviews to present items that they have previously categorised as 'esoteric' (school-type) or 'realistic' (practical). In addition, they also attempt to assess systematically whether the students' are 'calling up' esoteric or realistic responses to particular items. This work is drawn on further in Chapter 9.

This structuralist work points to ways of analysing practices and the boundaries between them. However, there is still the threat of arriving at a similar *cul-de-sac* as with the strongly situated approach: 'Dowling's strong position would seem to imply that school mathematics should incorporate no "real world" examples' (Muller and Taylor: 268). This raises the question as to whether clear-cut boundaries between different knowledge discourses, and between knowledge and social discourses can be *guaranteed* to exist, and can be maintained.

Put another way, we might ask whether the structuralist approach as so far outlined allows sufficiently for *intertextuality*, the capacity of a term (or other element) in one discourse (or text) to recall, or to reverberate with, a similar element in another text (Fairclough 1992). This brings us on to the work of poststructuralists, particularly that of Valerie Walkerdine.

Discursive Practice and Relations of Signification: Valerie Walkerdine

Valerie Walkerdine's work is sensitive to the need to avoid the pitfalls of recontex-tualisation revealed by Dowling. Like the constructivists, she is committed to bridging the space between everyday and school knowledge, but, unlike them, rather than assuming it away, she sees the importance of *theorising* the boundary (Muller and Taylor 1995). As with situated cognition, she recognises different practices as in principle distinct, but sees this distinction as requiring analysis, rather than leading to hopelessness.

Walkerdine's work has ranged across child development, mathematics education, and cultural studies more broadly. In particular, she has studied the role of language and 'discourse' in learning; social difference and oppression, especially related to gender and social class; and the pain, anxiety and anger that form part of the 'lived experience' of these differences (for example Walkerdine 1990a, 1990b). She also brings a systematic socio-historical dimension to her analyses of cognition and reason.

Her work in mathematics education comprises a series of empirical studies focusing on gender differences in the learning of mathematics at primary and secondary school (such as Walden and Walkerdine 1982, 1985, Walkerdine and Girls and Mathematics Unit 1989), and several theoretical statements (e.g. Henriques et al. 1984, Walkerdine 1988, Walkerdine 1997). Walkerdine's position can be labelled *poststructuralist*; its main tenets are outlined below.

Walkerdine's early work criticised the notion of context used in Piagetian and post-Piagetian work (for example Donaldson 1978) as something which '*is external to*, and *exists in an additive relation to*, thinking' (Walkerdine 1982: 131; emphasis added). She urged that language, cognition, and context not be seen as separate systems, nor be understood in a narrow, a-social way. For example, to understand the relation of the actions and vocalisations of a new-born baby to the actions and decisions of the parents, we have to understand

> what sense they make of its cries, and what this sense suggests as courses of action . . . to understand their actions and their 'discourse' we do have to look at action, at gesture, at sound, at word; but we . . . must also include current thinking and writing, fashions etc. about feeding, mothering and so on. There is a historical and social dimension which we must include . . . It is the positioning of their discourse in relation to a number of other discourses and practices which enables us to make sense of its functioning in the process of signification. These discourses and practices are not the context but actually have a constitutive effect.
> (Walkerdine 1982: 132)

Thus, different contexts are characterised by – indeed, are 'constituted' by – different practices and discourses, related sets of terms and meanings: these practices can be called *discursive practices* since they are based in, and regulated by, discourse. Sometimes, instead of 'discursive practice', for ease of expression, I shall use the terms 'practice' or 'discourse' as broadly equivalent.

Here 'regulation' means subjecting someone, or something, to rules and standards of evaluation. How this may work and the differences of meanings across home and school contexts is illustrated by the following:

> Mathematical meanings – indeed, the development of language and word meanings in general – cannot be separated from the practices in which the girls grow up. The mother is positioned as regulative in these practices, in which desires, fears, and fantasies are deeply involved. So 'mathematical meanings' are not simply intellectual, nor are they comprehensible outside the practices of their production. Yet in school . . . children have to learn that there are special meanings to these terms, which are not necessarily those used at home.
> (Walkerdine and Girls and Mathematics Unit 1989: 52–3)

In both quotations, the particular practice or mix of practices in which subjects are engaged, *positions* the latter within that practice (or mix). Thus, the parents in the first quotation are positioned as 'the carers' in child-care discourses; in the second, the mother is positioned as the one who must regulate the child's eating at home. To take a different example, relevant to the interviews analysed in

Chapters 9 and 10, in the practice of 'eating out', being the one who pays is determined in many contexts in a complex interplay of cultural conventions, gender and age positioning, as well as the use of certain ploys and so on. Clearly, power is implicated in the positioning of subjects in social relations, as are oppression and resistance. However, power is worked out in relations at the micro level, not conferred by positions in a predetermined way (Henriques et al. 1984: 115–18, Foucault 1982).

These illustrations show further that positioning may depend on social differences such as social class or gender. For example, there may be strong social class differences in terms of how much money is available, and how it is handled: many working class families need to regulate the spending of money and consumption generally, while middle class families may be freer to allow choice in consumption, and to make calculations around money into a game, for the children at least. Walkerdine has called these relationships with calculation ones of 'material necessity' and 'symbolic control' respectively (Walkerdine 1990b: 52).[15] Thus children from different social class backgrounds may be positioned very differently in practices which include calculation tasks (see also Chapter 7).

As for gender, the discourses of primary school mathematics teaching and 'child-centred pedagogy' (Walkerdine 1984) tend to view as laudable 'active learning', 'breaking set' and so on, and to view as pathological 'rote-learning' and rule-following. But these ideas are 'gendered'. Girls tend to be positioned in the social interactions of the classroom as neat, helpful, hard-working and well-behaved, and then their production of behaviour consistent with such 'characteristics' tends to be read as evidence of their passivity. In contrast, boys' naughtiness and restlessness in the classroom is seen as testifying to their 'potential', 'mathematical flair' and so on (Walkerdine et al. 1989). We can note that many of the adjectives used in these gender 'stereotypes' purport to be descriptive, but they can be seen as producing meaning, and hence performance! Thus discourses can be seen not simply as 'representative' of reality, but rather as *productive* of it.

Walkerdine also puts these ideas into historical perspective, drawing on the work of Foucault (1977, 1979) on the description of discourses and of the 'subject-positions' within them. Though the ideas above are part of relatively recent discourses, they can also be seen to relate to ideas from the last century – and earlier – that held women to be excessively swayed by emotions and therefore lacking in capacity for rational judgement. These ideas live on in today's 'common sense' that 'women's minds' are not fertile ground for mathematics and the 'hard sciences' (Walkerdine 1985, Walkerdine et al. 1989); see also the next chapter.

Thus the way a person is positioned in discourse will determine and delimit, to a great extent, his/her *subjectivity*. We can understand subjectivity as including thinking and emotions, and what traditional psychological discourses call 'abilities', 'attitudes', 'personality and 'identity'.[16]

My reservation about Walkerdine's position, as so far described, is that it is somewhat determinist. While people are 'positioned' as described above, they nevertheless appear to be free within limits to interpret a particular task/situation in a variety of ways. For example, Winter (1992) reports a study which used systematic observation of his daughter Jessie's experiences with numbers and so forth at home. In two cases of sharing by Jessie (aged 2 years, 7 months) of dates (to eat) with her father, she first recalls or 'calls up' counting out dominoes, and the next day, she calls up taking turns playing with a toy with a young friend. Winter concluded that problem-solving in mathematics is a form of *metaphoric* thinking, in young children at least. That is, the problem is made sense of by substituting for it another problem selected from those previously encountered by the child, and meaningful to her/him.

To summarise, Walkerdine sees cognition as inseparable from its context. The context of any social action is constituted, or 'highlighted', by the practices in play, and the related discourses. These are the practices which 'position' the subjects, and which are the basis for the subjects' making sense of what is happening, of formulating problems and thinking about them, of expectations, for example as to what they ought to do. Social differences such as gender and social class are related to the positioning of a subject within a particular practice.

The fact that the particular discourse(s) called up provide(s) the basis for the subject's examining a problem and thinking about it, means that cognition will be *'specific' to the discourse* called up, as is also argued by situated cognition researchers. Here, however, as shown by the examples, the specific meanings of a word, a gesture – or any other 'signifier' – depend on the specific discourse through which the signifier is read. I show in the next section how the discourses in use are systems of meaning which can be analysed by considering 'relations of signification', and devices such as metaphor and metonymy. This brings a *systematic* quality to our discussion of discourse, practice and context.

Conceptualising Contexts, Practices, Boundaries and Bridges

The discussion so far shows that several issues need to be addressed, so as to formulate the problems of context, and of 'transfer', satisfactorily. They are:

1 How to define and delineate the contexts of thinking, activity and learning, and the related practices at play in them.
2 How to describe the relations between practices, e.g. what the boundaries between them might be like, and how they might be bridged.
3 How to acknowledge the importance of affect, and emotion, so as to avoid separating thought, feeling and value.

Issues (1) and (2) will be addressed here; (3) will be taken up in the next chapter.

In attempting to elucidate these issues, my position is built on the contributions of other positions outlined above, especially those of Valerie Walkerdine. Other approaches drawing on poststructuralist insights have also been helpful here (for example Taylor 1989, 1990, Muller and Taylor 1995, Evans and Tsatsaroni 1994, 1996, 1998).

Contexts and the Practices at Play

The approach I am advocating focuses on *practices:* examples would be school mathematics, academic (research) mathematics, work practices such as nursing

(Pozzi et al. 1998) and banking (Noss and Hoyles 1996a), apprenticeship into fields such as tailoring (Lave and Wenger 1991), and everyday practices such as shopping (Lave 1988). Each context is constituted by one or more practices, and by related *discourses*. Discourses are systems of ideas expressed in terms of *signs;* they give meaning to the practice by expressing the goals and values of the practice, and regulate it in a systematic way, by setting down standards of performance (for example precision).

However, it must be stressed that practices 'are both material and discursive; they are not simply created in language' (Walkerdine 1997: 63). Practices tend to be institutionalised, in several ways. Relevant material resources may be developed, promoted (for some) and/or their use discouraged (for others); a good example is the contemporary debate about the use of calculators for mathematics in British primary schools. Also, practices tend to be associated with a community of practice, a subculture of individuals with (some) shared *goals*, and a set of *social relations* (power, difference). Different people take up different *subject-positions*. For example, the basic positions available in mathematics in school are normally 'teacher' and 'pupil'; in shopping or street-selling, they would be 'seller' and 'buyer'. In a particular setting, we can analyse the practices at play, that would be involved in the positioning of participants.[17] 'Situated cognition' for Walkerdine is 'not people thinking in different contexts, but subjects produced differently in different practices' (Walkerdine 1997: 65).

My approach, like situated cognition, recognises different practices as in principle distinct, as discontinuous: for example, school mathematics and calculation in everyday practices like street selling. However, using the approach recommended here, we can go further, to *analyse* the differences between practices, and also their similarities.

My analysis focuses on 'relations of signification', especially relations of similarity and difference between 'signifiers' and 'signifieds', and devices such as metaphor and metonymy. This terminology comes from linguistics. A linguistic sign is considered as the unification of two elements: a *signifier*, which may be thought of as the word, sound, symbol, gesture, or, say, part of a diagram; and the *signified*, which may be thought of as the concept or mental image (but not the 'thing itself') to which the signifier relates (de Saussure 1974, Hawkes 1977).[18]

Further, the process by which linguistic utterances are formed is based on two dimensions: the *combination* of words in a chain in a 'horizontal' movement, and the *selection* of a word, from those available, for a particular position in the chain in a 'vertical' movement. This two-fold process is underpinned by two ways of relating words as 'equivalent': metonymy and metaphor. *Metonymy* is based on relations of contiguity, and is the mode of the combinative dimension of language. *Metaphor* is based on relations of substitution or analogy, and is the mode of the dimension of selection (Hawkes 1977: 76–9). Examples can be given of the way these two rhetorical figures convey meaning. The metonymic phrase 'Ten Downing Street considers . . .' proposes an equivalence between a specific building and the Prime Minister of the United Kingdom; the metaphor 'My boss steamrollered me' proposes that the boss has an equivalent effect to that of a particular machine.

So far this draws on Saussure's structural linguistics. Going further, various writers have shown how to use poststructuralist ideas about the inevitable tendency of the signifier to slip into other contexts, thereby making links with other discourses, and producing a play of multiple meanings, so as to provide insight into meaning-making in mathematics; see Winter's (1992) example earlier, and that given later of the different possible meanings of 'more'; also Walkerdine (1988: Ch. 2) on children's use of language to indicate relations of size, Brown (1994) and Evans and Tsatsaroni (1994). Thus, rather than attempting to specify the context of a school mathematics problem by looking only at its wording (and format) – or by naming the context as if straightforwardly based in 'natural' settings – we can describe it as *socially constructed* in discourse through attention to particular signifiers and their relations in texts, such as interview transcripts (see the analysis of the case studies in Chapter 10).

Different practices may also be characterised in terms of their 'well-known results', and their familiar methods (see 'The Turn to the Social', this chapter). An example can be given by contrasting a street seller's calculation (using repeated addition) of the cost of 10 coconuts (@ 35cr. each) – as 105 (three 35's, a 'well-known result'), plus 105 (making six), plus 105 (making nine), plus a tenth 35 (Nunes et al. 1993) – with a pupil's doing 35 x 10 using school methods.

Relations Between Practices: Boundaries and Bridges

To build bridges between practices, one must try to identify areas where out-of-school practices might usefully 'overlap' or 'interrelate' with school mathematics. This requires first of all that, consequent on the type of analysis outlined in the previous subsection, distinctions are made between those relations of signification in the learner's everyday practices that can provide fruitful 'points of articulation' with school mathematics, and those that may be misleading (cf. Muller and Taylor 1995).

I shall clarify what I mean by giving examples of both fruitful and misleading interrelations of two practices. The first three examples involve attempts to 'harness' out-of-school practices for pedagogic purposes, to help with the learning of school mathematics.

An example of a misleading interrelation would be an attempt to harness young children's everyday understanding of 'more' to teach the comparison of quantity at school. The problem is that, in school discourses, 'more' is meant to form an oppositional couple with *less* – whereas in home discourses, the opposite of 'more' is *no more* – as in 'no more ice cream for you' (Walkerdine and Girls and Mathematics Unit 1989: 52–3). Here the signifier 'more' signifies differently in home and school practices.

Besides identifying fruitful (non-misleading) points of interrelation, the pedagogic task is to structure the school discourse so as to work systematically through a process of *translation* from the everyday discourse (Muller and Taylor 1995). This translation is done through the construction of 'semiotic chains' or chains of meaning, where a sequence of new signs is formed. A very simple example is that of a mother who uses a discussion with her child on the number of drinks needed for a party of the child's friends to teach the child to count by following transformations from one step to another; see Figure 6.1.

At the first step, the mother–teacher, encourages the child to form a sign linking the name of each child (signifier) with the 'idea' of that child (signified). At each subsequent step, the signifier from the previous stage becomes the new signified, which is in turn linked to a new signifier (gesture, spoken numeral, written numeral). Here, the different steps do not represent different discourses in any straightforward sense, but the overall chain nevertheless shows how a series of care-fully-constructed 'discursive shifts' could provide the basis for transforming the relations of signification from those related to home practices to those of school mathematics. Such a series of shifts, and the 'semiotic chain' thereby generated (see Figure 6.1), provide the basis for crossing boundaries, or transfer across practices (cf. Walkerdine 1997). [19]

Another example is provided by a primary teacher aiming to harness the children's prior knowledge from outside school about counting objects and so on, to lead to learning about addition in school mathematics (Walkerdine 1988: Ch. 6). Again, she shows how the process of 'translation' or 'transformation' of discourses must be accomplished through careful attention to the relating of signifiers and signifieds in particular chains of meaning. Thus,

> teachers manage in very subtle ways to move the children . . . by a process in which the metonymic form of the statement remains the same while the rela-tions on the metaphoric axis are successfully transformed, until the children are left with a written metonymic statement.
>
> (Walkerdine 1982: 153–4)

Other examples from school show that it is possible for pupils to become confused when there is misleading 'discursive overlap', for example when a task appears to be part of an everyday discourse, but its purpose is pedagogic. One example is given by the pupil in the quotation at the head of this chapter. Another comes from a primary school 'shopping game' observed by Walkerdine (1988: Ch. 7). There a boy made 'errors' in his sums because he did not realise that, in the

Step				
1	Child *(signified)*			
	Name *(signifier)*			
2		Name *(signified)*		
		Finger *(iconic signifier)*		
3			Finger *(signified)*	
			Spoken numeral *(symbolic signifier)*	
4				Spoken numeral *(signified)*
				Written numeral *(symbolic signifier)*

Figure 6.1 Steps in the Construction of a Semiotic Chain
Source: Muller and Taylor (1995: 271), based on Walkerdine (1998: 128ff.)

game, one was allowed – indeed, one was *required by* the rules, made to ensure the game's pedagogic effectiveness – to start afresh with a new 10p after each purchase. Though the child called up – that is, identified the task as – practical shopping, through which he 'made sense' of the apparent demands of the task, he nonetheless made errors because he was positioned in, and *regulated by*, the pedagogic shopping game.[20]

While some aspects of everyday shopping practice might also be useful in the game – say, remembering (if a 'well-known result') that 'when you have 10p and buy something worth 9p, you will have 1p left', other aspects of shopping – for example, the knowledge of the requirement of giving up money to obtain a purchase – were not 'included' in the discourse of the school shopping game. Also, importantly, the goals or 'products' – a subtraction calculation in the 'game', and a purchased item in actual shopping (Walkerdine 1997) – were quite different in the two practices.

In my discussion of street mathematics and Nunes, Schliemann and Carraher's position on transfer earlier, I quoted Schliemann's conclusion that

> [mathematical] strategies developed to solve problems in a specific context can be applied to other contexts, *provided that the relations between the quantities in the target context are known by the subject as being related in the same manner as the quantities in the initial context are.*
>
> (Schliemann 1995: 49, my emphasis)

It will be noted that this conclusion focuses on similarity in relations between quantities, where the latter are considered to be 'known' by the subject. Thus, for example, a knowledge of counting will transfer to playing cards, but only up to a point: to play most card games with the standard deck, you must know the ranking of numbers up to ten – but also that the ace, though signified by a single heart (or whatever) will beat all others in that suit.

Similarly, Walkerdine (1988) argued that activity within one discourse – say, playing the card game whist – can be harnessed to help with school mathematics in those, and only those, aspects of the game which are both contained in school maths and which enter into *similar relations of signification*. As with Schliemann, similarity matters, but here it is in the relations of signification (which can be read from the rules, or from everyday, card-playing, knowledge).

However, it is important to note that the analysis of relations of signification described here focuses both on similarity and *on difference* between signifiers and signifieds. Thus we need an approach to transfer, or translation, which attends to both similarities and differences. The consideration of values in standard card games alongside counting in school mathematics shows how a careful analysis of differences can enhance translation, indeed, that such an analysis is essential to it. The discussion here of the example of the multiple meanings of 'more', and the analysis of the shopping game (earlier) also show how a discussion of differences in relations of signification is basic to any process of harnessing or transfer that is fruitful, that is meaningful and not misleading.[21]

Thus it is important to conclude that there is nothing *intrinsically* 'transferable' or non-transferable about particular signifiers, ideas, or subject contents like mathematics.[22] What is important is how the relations of signification are set up between two practices. And crucial in this process is the presence of a 'knowledgeable' subject who desires to look with a 'mathematical gaze' (cf. Noss and Hoyles 1996a) at another area for application of his/her knowledge. In teaching, the issue is how we structure the pedagogy so as to work systematically through a process of transfer or translation. Examples given in this section show how this can be done.

Conclusions

My aims in this chapter have been to develop my conception of the contexts of mathematical thinking, and to examine the related ideas about transfer of learning, in several alternative views.

The discussion here highlights several aspects that are crucial in describing the context of activity or thinking more fully:

* the goals and values of the practice
* the social relations in the setting and around the practice
* the material and institutional resources available
* 'macro-cultural' aspects
* the 'activity structure' of the practice
* the discourse or language underlying practice(s).

The *goals* of human activity are emphasised strongly by sociocultural theorists (see 'The Turn to the Social', this chapter), and include both 'higher-order' goals, and goals at the more specific level of actions. The *values* of the practice include the level of precision required in calculations (Maier 1980), and the need for 'flexibility' and 'efficiency' in problem-solving (Scribner and Lave).

Several authors emphasise *social relations*, though these are understood in several ways. Saxe emphasises face-to-face social interaction as a resource in learning. Lave's later work focuses on 'communities of practice' (though it is not clear that this notion is relevant to all practices). Walkerdine discusses the ways in which subjects are regulated in particular practices, thus pointing to differences in power, as well as to social differences, related to gender and social class. Social class is focused on by some researchers (Nunes et al., Saxe, Dowling, Cooper and Dunne), but not all.

Several authors refer to *material and institutional resources* as an important part of the context. These include: computation technology (Noss and Hoyles); the physical layout of supermarkets (as 'arenas', Lave) and schools (Walkerdine 1984); and the physical characteristics of packaged goods (Lave and Saxe). Saxe and Lave also refer to 'macro-cultural' aspects, such as currency and number systems.

Saxe has produced a useful account of the 'activity structure' of candy-selling among adolescents in his studies in Brazil, in terms of the tasks or actions which they had to complete in a determinate cycle; see also Scribner (1984) on the activity of dairy employees, and Dowling (1998) on distinguishing the structure of shopping and school mathematics activity. This is a promising additional dimension of the analysis of situated activity which can be developed in further research.

Overall, however, although the importance of language is alluded to by several other researchers, my approach, based on that of Walkerdine, aims to be more systematic, by using ideas about discourse and also poststructuralist ideas about relations of signification, so as to analyse the ways that meaning is conveyed and interpreted.

Turning to the transfer of mathematical learning, based on the discussion so far, I take a sceptical position on several widely-held views, yet am optimistic about the reformulation of the problem, as begun here. The main points in my position are:

1   Continuity between practices (e.g. school and out-of-school activities) is not as straightforward as utilitarian and constructivist views assume, and hence claims that transfer is in principle straightforward are misguided.

2   Indeed, responding to arguments from situated cognition and others, we can agree that there is a *distinction* – but not a total *disjunction* – between doing mathematics problems in school, and numerate problems in everyday life. But the distinctions can be analysed (see point 4), and hence we can be more optimistic than situated approaches suggest.

3   There are problems in practice with transfer. It is not dependable: although people do sometimes seem to accomplish it, often the ideas, feelings and so on. that they transfer are not what we as educators expect, because of the (sometimes unexpected) fluidities of signification, and also of emotion (see Chapter 7). *The ability of a signifier to form different signs, to take different meanings, within different practices, constitutes a severe limitation on the possibilities of transfer.*

4   *Yet the fluidity of meaning also provides the basis for any such possibilities.* Though the successful building of bridges cannot be guaranteed 'risk-free', we can sketch some steps it is *necessary* to follow. For anything like transfer to occur, a 'translation', a making of meaning, across discourses would have to be accomplished through careful attention to the relating of signifiers and signifieds, representations and other linguistic devices that are used in each discourse, so as to find those crucial ones that function *differently* (though in a specifiable way) – as well as those that function *in the same way* – in each discourse. This translation is not straightforward, but it often will be possible. If possible, it will be built on, first, analysis of the practices, and the related discourses, involved (in the transfer relationship), as systems of signs; and second, analysis of the similarities and differences between discourses (such as school versus everyday maths), so as to identify fruitful 'points of interrelation' between school mathematics and outside ('target') activities.

I shall continue discussion of the issues around transfer/translation in the next chapter.

The ideas reviewed in this chapter support a shift in my way of thinking about the practical character of everyday mathematical thinking or numeracy. The issue is not one of defining (and finding a valid measure for) a 'practical mathematics', as I was attempting to do in the earlier chapters, since the relevant mathematics would depend on which practices we are considering. Further, a specified activity cannot be prejudged as *essentially* 'mathematics in practice', since it can be described from multiple points of view – not only by mathematics education researchers, but also by those researching in other fields, and also by those practising the activity. Nor is the issue to be resolved by describing a number of different 'situated mathematics', since that threatens to lead us to the 'dead-end' of the strong form of situated cognition described earlier (see also Noss and Hoyles 1996b).

In order to describe numerate thinking in context, it is necessary:

- to describe the numerate aspects of a practice, through attention to signs and relations of signification, as illustrated in the last two sections, while aiming
- to describe the ways that different subjects are constru(ct)ing the context, as noted earlier.

I shall develop this approach in the rest of the book. For the moment, it is important to note that the discussion in this chapter has not treated *affective aspects* fully. Indeed, Walkerdine has been virtually alone among the researchers discussed here[23] in emphasising the importance of the relations between thinking and emotion; for her, ' "mathematical meanings" are not simply intellectual'. Her approach to affect, and that of other researchers, will be discussed in the next chapter.

In Chapter 8, I outline the type of interview which I developed on the basis of ideas discussed here, and which was used in this study. The general ideas of *discursive practice* and *positioning* are central. In studying numerate thinking in problem-solving episodes in an interview, a number of specific issues arise:

- What are the practices at play or 'available' in a particular situation?
- What determines the subject's positioning at any point?
- What indicators are there for the subject's positioning?

These issues will be discussed further in the next four chapters.

7 Rethinking Mathematical Affect as Emotion

When we approach the problem of the interrelation between thought and language and other aspects of the mind, the first question that arises is that of intellect and affect. Their separation as subjects of study is a major weakness of traditional psychology.

(Vygotsky 1962: 10)

In this chapter, I broaden the earlier discussion of affect and mathematics. In Chapter 4, affect in general, and mathematics anxiety in particular, were considered as relatively stable characteristics of an individual, able to have an ongoing effect on mathematical thinking, performance, and participation in mathematics courses. My own model, and Fennema's (1989) 'individual-differential' model provide examples of this approach.

This approach focuses on the causal links assumed between affective and cognitive. Further, the social aspects of experience are separated from the individual: the social, the cultural, socialises the individual, so that values and affect are 'internalised'; affect in turn influences differences in cognitive outcomes across individuals. We have seen that this process of socialisation is, to a greater or lesser extent, bracketed as being largely outside the individual, and influences the affective via 'cultural transmission' of beliefs, values and so on. Affect is usually conceptualised as attitudes, one of the three types of affect in McLeod's (1992) typology[1] of 'beliefs', 'attitudes', and 'emotion', and measured by scores on attitude scales or, in my work, by 'trait' anxiety scales such as the Mathematics Anxiety Rating Scale (MARS). Cognitive outcomes are measured by the number of mathematics exams passed, scores on standard(ised) tests and so on. I call this 'individual differences' approach Model A (Evans and Tsatsaroni 1996).

In this chapter, I draw on the reflections in Chapter 5 on the quantitative phase of the study, and also on the rethinking of context and thinking in context presented in Chapter 6. I discuss several other theories of affect, which focus more directly on *emotion*, and its relationship with mathematical thinking. I review a selection of 'micro' studies, which describe the process of mathematical problem solving, focusing on the role of emotion in it, and preferring 'qualitative' research methods; these studies are taken as examples of what I call Model B. Next I discuss Model C, which is informed by psychoanalytic ideas that much thought and activity takes place outside of conscious awareness; thus it begins to question the idea of a rational subject. Finally, the integration of psychoanalytic ideas into poststructuralist approaches (such as Walkerdine's) in 'Model D', proposes the fluidity of language as a basis for understanding unexpected flows of meaning and emotion; it thereby further calls into question the idea of a rational, unitary subject, as well as the closed nature of mathematical discourse. This model also allows an understanding of how discourse in its historical development positions social groups differently in relation to 'rationality' and emotionality.

First though, I reconsider several of the major studies of (mostly adult) cognition in context discussed in Chapter 6, with respect to the attention paid to affect and emotion in them.

'No Emotion, Please! We're Researching Mathematics'

In much of the research on the use of mathematics by adults, the situation has not changed much from that described in psychology by Vygotsky three generations ago (see the quotation at the beginning of the chapter). There is still little or no explicit acknowledgement of the importance of the affective – feelings of anxiety, frustration, pleasure, and/or satisfaction which attend the learning of mathematics and the solution of numerate problems. As Nick Taylor puts it:

> Cognitive theories of learning tend to view the subject as making completely rational choices from alternatives provided by her environment; affective aspects, if taken into account at all, are considered as having a purely additive influence.
>
> (Taylor 1989: 162)

For example, Carraher, Carraher and Schliemann, in a rare mention of affect in their work, play down the importance of anxiety in the problem-situation as 'some peculiarity' which is to be distinguished from 'the meaning that problems have for children' (Carraher et al. 1987: 95). Both of these points imply that the problem-solver is basically a cognitive subject, and that the meaning of problems for him or her can only be fundamentally cognitive.

More than any of the research reviewed in the previous chapter (except for Walkerdine's), Cole and Traupmann's (1979) report of findings from the after-school cooking club appears to engage with affect. For example, they describe several situations where Archie, a child considered by his teacher as 'learning disabled', is being tested by a psychologist, or being asked a question in a team quiz. Yet here the distress evinced by Archie, and the supposedly related distress produced in the tester and in his fellow students, are seen as simply part of the set of outcomes of this action in context, rather than as having a distinct *emotional* quality. In addition, there is no description of the process whereby the distress of the tester, for example, might be a response to Archie's. This focusing on action with a predominantly cognitive lens is typical of the US work based on activity theory, and in general of the studies of everyday mathematical thinking discussed in Chapter 6.

That state of affairs is somewhat surprising, given an earlier article by Sylvia Scribner and Michael Cole arguing the need for studies of 'informal learning'. They noted that informal education tends to fuse the intellectual and the emotional: because of the 'affective charge . . . associated with almost everything that is learned within that context . . . the content of learning, especially for children, is often inseparable from the identity of their teacher' (Scribner and Cole 1973: 555).

Again, D'Andrade (1981), cited approvingly by Lave, argues for the general inseparability of the intellectual and the affective, on the cultural grounds given by Scribner and Cole, and also on the grounds of what might be termed the indivisibility of meaning. For example, the sentences 'The stove is hot!' and 'Joe is a cheat!' convey both ideational and affective meaning. The latter encodes not only a representation of the speaker's feelings, but also directs how the listener should act by virtue of an assumption of intersubjectivity. Thus, the distinction affective versus ideational is analytic only. Again, the strength and effectiveness of the affective component of 'cultural representations' like those above is due to its being communicated 'through face and voice by the important people in one's life' (D'Andrade 1981:193). Yet, even when we come to the more recent discussions of 'situated learning' as compared with school learning, for example Lave and Wenger (1991), we again find basically a void concerning the affective.

This brief discussion illustrates the surprisingly low level of attention given to the affective in discussions of mathematical activity, among adults in particular, and also more generally. However, a few researchers have argued for the inseparability of affect and cognition in solving problems. For example, Ginsburg and Asmussen (1988) argue that the tremendous difficulties associated with the learning of mathematics could not possibly arise only from deficiencies in knowledge or in the use of cognitive procedures and strategies; the difficulties must also be bound up with what the authors see as non-cognitive factors such as beliefs, cognitive style, motivation, confidence, anxiety and identity.

For these reasons Ginsburg and Asmussen stress that 'mathematical thinking is clearly 'hot'! (Ginsburg and Asmussen 1988: 107) In this chapter, we consider the work of those relatively few researchers who consider the crucial interactions of emotional factors and cognitive factors in mathematical activity.

Model B: Process Conceptions of Affect and Anxiety in Mathematics

We can distinguish between 'macro' approaches, such as individual-difference frameworks of Model A, and 'micro' approaches (Mandler 1989a). One type of micro approach focuses on the process of an individual attempting a particular task or problem, including the emotions experienced.

For example, a model termed 'cognitive-constructivist' (Mandler 1989a; McLeod 1989a) describes the process of emotional experience as follows:

1   A *discrepancy* (or interruption) between the individual's *expectations* and the demands of ongoing activity leads to visceral arousal.
2   The *physiological arousal*, on the one hand, and the person's *evaluation* of the situation, on the other, lead to the 'construction' of emotion.
3   Experiencing emotion may lead to a *reduction in conscious capacity* available

for problem-solving (because the process of emotional construction itself, on this view, requires conscious capacity).

This makes the experiencing of emotion seem to be generally somewhat debilitating (capacity-reducing), even if the emotion is 'positive', but see the different views of Buxton and Nimier later. In any case, emotion is here more 'hot', more intense, than affect in Model A earlier.

Although questionnaire measures are perhaps appropriate for the measurement of repeated emotional reactions to a category of (say mathematical) tasks, more process-sensitive methods are seen as necessary for describing reactions which are not yet so automatised (McLeod 1989b). Thus, the methods of research used here tend to be:

* description of particular episodes of mathematical problem-solving
* 'cross-subject' comparisons; for example, McLeod et al. (1989) compare 'experts' and 'novices' in mathematical problem-solving.

Micro models can focus on the processes based on interactions in face-to-face social settings like a classroom, not only on individuals. Cobb, Yackel and Wood (1989) focus on the role of beliefs, norms (including expectations and obligations), and 'emotional acts' in mathematics learning. They characterise an *emotional act* as conveying an appraisal or construal of a physiological arousal, or *emotional state*. That is, they take a basically cognitive-constructivist position, and emphasise the mutual dependence of emotional acts, norms, and beliefs, all of which they see as basically cognitive. Yet they also see beliefs as spanning the individual and the social domains. In school, the teacher can show the pupils norms for construing their own (or others') actions, as a basis for emotional acts; for example, if blockages in problem-solving are seen as to be expected, then the expression of frustration, rather than anger, can be seen as normal. Thus, Cobb et al. discuss how the rationality required for learning mathematics or other subjects might be negotiated intersubjectively through emotional acts based on specific beliefs and norms.

Carter and Yackel (1989) deploy these ideas in a study of adults participating in an eight-week programme to combat mathematics anxiety at a large research university in the USA. This aimed 'to encourage participants to become aware of their beliefs about mathematics and their emotional responses, and . . . of their contexts of mathematical activity' (Carter and Yackel 1989: 4). To begin describing the relevant beliefs, they draw on the distinction between 'instrumental' and 'relational' understanding (Skemp 1976): 'instrumental' learners view mathematics as rules and algorithms, where the goal is to produce 'the right answer'; 'relational' learners view mathematics as a group of related concepts and ideas, where the goal is to construct new mathematical relationships, or understanding. They consider different interpretations of the emotion aroused by an interruption to the expected solution path of a problem: for an instrumental solver, anxiety, they claim, is an 'appropriate' interpretation; for a relational solver, anxiety would not be 'appropriate', though challenge and excitement perhaps would be. They conclude that 'use of the term "mathematics anxiety" can obscure the cognitive basis of these emotional acts' (Carter and Yackel 1989: 32).

This argument, and its striking conclusion, show some of the strengths and the limitations of the cognitive-constructivist approach. The fact that emotional acts are explained through having a logical rationale with respect to the local (for example classroom) social and moral order of beliefs and norms means these researchers avoid the limitations of assuming a generally *one-way* causality, of the form 'mathematics anxiety is debilitating for thinking'. Thus Ginsburg and Asmussen (1988) argue that cognitive and non-cognitive factors interact in a complex dynamic, where the direction of causality may be in both directions. For example, for one of their interviewees (a woman, aged 30+), anxiety seemed to influence effort and learning, and these in turn influenced anxiety.

In addition, the idea that emotions are constructed in relation to the social order is important, and parallels ideas on the social nature of cognition developed in Chapter 6. However, the claim that particular emotions are *appropriate* for holders of certain beliefs – as in Carter and Yackel's discussion of mathematics anxiety – is highly problematical. First, a concern with norms may turn into an inappropriate normativeness; for example, Carter and Yackel imply that feeling anxiety (or frustration) may not always be the emotional act or response that is 'socially appropriate' for the situation (Carter and Yackel 1989: 23)! Second, this seems to suggest that, if one's beliefs and expectations (themselves related to the 'operant social norms') are right, one need not feel anxiety at all, a view which seems rather optimistic (and much at variance with the psychoanalytic approaches discussed in the next section)!

In any case, it is important to allow, as indeed Carter and Yackel do, not only that beliefs, norms and emotions are context-dependent, but also that, even in situations using an open, liberatory pedagogy, learners may bring in beliefs from other (or earlier) contexts of activity, or simply from prevailing beliefs in the culture at large. These may lead to conflicts, such as to how to interpret an emotional arousal. This raises the now familiar question of how to describe the context – this time, of emotion – to which I shall return in discussing Model D later in this chapter.

The widespread distress from mathematics anxiety is not doubted by Laurie Buxton, who has made an important contribution to developing the idea of mathematics anxiety in *Do You Panic About Maths? Coping with Maths Anxiety* (1981). He bases his work on a 'cybernetic' theory of learning as a goal-directed activity (Skemp 1979), which has resemblances to the cognitive-constructivist theory discussed above; for example, the origins of anxiety for both are in the interruption of a goal-directed plan.

However, in arguing for the inseparability of cognition and emotion, Buxton allows priority to the affective:

> It is inappropriate to believe that there is such a thing as the cognitive power of a particular person. We operate well or badly in learning, and more especially in problem-solving, according to the drive provided by our emotions. Reason is powered by emotion, or, more often, hampered by it.
>
> (Buxton 1991: 3)

He distinguishes three levels of rejection of mathematics which correspond to three different emotions, only one of which is closely related to anxiety.[2]

1   simple boredom or lack of 'affinity'
2   a feeling that mathematics offends commonsense (for example in the idea that the product of two negative numbers is positive), and a consequent lack of 'emotional acceptance' or belief
3   *panic*, a strong form of anxiety, the only level of rejection which is 'pathological' (ibid.).

Buxton (1981) characterises panic as being preceded by tension mounting to a critical point, issuing as a sudden discontinuity of behaviour, leading either to a type of frenzy, a 'mind in chaos', or else to a freezing of the mind and a sense of paralysis.

Buxton studied a series of group meetings (four women, three men) over one year, where the aims were to explore negative feelings about mathematics, and to learn some mathematics. Since he gave members of the group mathematics problems to do, Buxton was able to observe panic and anxiety at first-hand, as well as discussing them.

His account of the origins of mathematics anxiety includes the following factors:

•   unpleasurable or distressing feedback
•   the apparently arbitrary rules
•   the ambiguity of jargon (for example 'x is unknown')
•   the moral connotations of 'right' and 'wrong' answers.

He particularly emphasises the importance of time pressure in the classroom, and early unhappy encounters around mathematics with 'authority figures' – teachers or parents – linked with the threat of disapproval.

Drawing on the metaphor used by two subjects, of mathematics as a 'secret garden' into which they could not enter (itself a metaphor in psychoanalytic approaches for sex) Buxton (1981) speculates whether gender differences in feelings about sex might explain gender differences in emotions about mathematics. He also briefly explores what 'right' and 'wrong' answers might mean in superego terms.

Thus Buxton argues clearly for the inseparability of emotion from mathematical learning and problem-solving. His work also points to the need to engage with psychoanalytic ideas.

Overall, the studies reported here suggest that research with adults might use interviews and direct observation to produce a fuller description of problem-solving experience including its emotional aspects. However, in the cognitive-constructivist account, the affective is described as threatening to block or 'interfere with' clear thinking; this tends to present the affective as generally debilitating (of cognitive capacity), and as subordinated to the cognitive.

Related to this, Zajonc (1984) addresses the claim (Lazarus 1982) that affective arousal *always* requires *prior* cognitive appraisal – a claim that appears to be a strong form of one made by the cognitive-constructivist approach (Mandler, (1989a 1989b) – and criticises it on a number of empirical grounds. In addition Zajonc usefully analyses emotional states as having three 'aspects' or 'manifestations':

- somatic or bodily processes
- overt behavioural expression
- subjective experience or feeling.

This same trilogy can also be found in Burkitt's (1997) sociological discussion of social relationships and emotion. Thus, in a 'fight-flight' reaction to danger, the somatic components of a racing heartbeat and raised levels of adrenaline prepare the body for sudden action and produce the sharp behavioural reaction to sudden movement. However, in human beings, there is also a feeling component, claimed to be entirely learned, so that a fight-flight reaction can be felt and expressed as anger or rage, as fear, dread or anxiety, or as a combination of these depending on the situation. 'The somatic, behavioural and feeling components are all aspects of the same thing – the emotion itself'. (Burkitt 1997: 45)

Burkitt insists on this triple aspect of emotion so that behaviour does not become a secondary effect, a mere 'expression' of the person's feeling. He cites Elias's (1987) conclusion that the function of emotional expression for human beings is not as an outer signal of 'inner' feelings, but as *'signs in the network of social relations'* (Burkitt 1997: 45, author's emphasis). This suggests that the poststructuralist insights into the role of language discussed in Chapter 6 can be used in our study of emotions around mathematics. This is developed in 'Model D' later, but first we consider insights from psychoanalysis.

Model C: Approaches Informed by Psychoanalysis

A number of researchers have studied affect around mathematics, using psychoanalytic approaches. These approaches start from the Freudian position that affect can be thought of as *a 'charge' attached to particular ideas*. Ideas which have strong negative charges, such as anxiety, or which mobilise intrapsychic conflict, tend to meet defences, and to be pushed into *the unconscious,* through the operation of repression, one of the defence mechanisms.

Therefore, much thought and activity takes place outside of conscious awareness: everyday life is mediated by unconscious images, thoughts and fantasies (Hunt 1989). If repressed contents 'return' to consciousness, they retain their charge but tend to be found in a disguised or distorted form, for example as jokes, or 'slips of the tongue', or in dreams. This unconscious material is linked to complex webs of meaning. The affective charge can move from one idea to another along chains of associations by *displacement*, and can build up on one particular idea through *condensation*.

Basic Ideas: Freud and Lacan

Freud's views on anxiety were introduced in Chapter 4. He characterised anxiety as unpleasant feelings linked with motor discharges which are 'perceived' by the subject. As indicated earlier, there is a tendency for anxiety, or for the ideas associated with it, to be repressed into the unconscious: 'defence can be directed not only against instinctual claims, but also against everything which is liable to give rise to anxiety: emotions, situations, superego demands, etc.' (Laplanche and Pontalis 1973: 110). Therefore anxiety cannot be assumed to be observable in any straightforward way, let alone susceptible to dependable self-report. Because of defences, anxiety may appear *in distorted form*: as 'no feeling at all', or indeed as the opposite of anxiety, such as over-confidence. What may also be distorted is the focus or *the object* of the anxiety. For example, in agoraphobia, the fear of the feelings one may have in a crowd is displaced into an apparent fear of the street. We could expect a similar displacement in a mathematics phobia.

In his later work ([1926] 1979, [1933] 1973), Freud made a distinction between anxiety as an automatic reaction to trauma, of which the prototype is the birth trauma, and anxiety as a warning of the approach of such trauma. Examples of the latter, and therefore possible focuses of anxiety are:

- loss of the mother as object
- loss of the object's love
- loss of the penis
- loss of the superego's love (that is, guilt).

Freud and his successors postulated *mechanisms of defence* against anxiety (and other psychic threats). These include: repression, regression, projection, sublimation, reversal into the opposite; and they might use a range of processes including fantasy, dreams, intellectualisation (Laplanche and Pontalis 1973: 103–11).

Melanie Klein extended Freud's work in her account of defences in pre-Oedipal children, especially 'splitting of the object' and projection. *Projection* involves the rejection from the self, and the location in another person or thing (for example mathematics) of qualities, feelings, wishes which the subject refuses to recognise in him/herself (Laplanche and Pontalis 1973: 349–56). *Splitting* occurs, for example, when the mother is absent: the child does not just produce a positive fantasy of the mother, but also 'splits' the mother (or the breast) into a 'good' and a 'bad' part. Thus the child can want both to have the breast, and to destroy the breast. (For more on Kleinian ideas, see Hinshelwood 1991.)

The work of Jacques Lacan is relevant here (see for example Henriques et al. 1984: 212 ff., Urwin 1984, Tahta 1991). First, Lacan used a theory of signification in the form of structural linguistics (see Chapter 6) to establish links between words and ideas (see later), whereas Freud's more biologically-based approach referred to 'memory traces', associations, and so on. Lacan thus allowed a space for the social, by giving priority to the 'symbolic order', or language, which predates the infant's birth, and into which (s)he must enter in order to become a full member of the community, 'within the terms set by pre-existing social relations and cultural laws' (Henriques et al. 1984: 213).

In Lacan's work, *desire* permeates the working of language. Unlike a need which can be met in principle, desire, because of the fundamental 'loss' involved in its production, cannot be totally satisfied, and must be fulfilled by a fantasy or dream.[3]

> Like Freud, Lacan regards the mother as providing the infant with his or her first experience of satisfaction. But the infant must come to terms with the loss

of satisfaction, or the absence of its source, the mother. In Lacan's account, the child uses his or her first words to establish, in fantasy, control over the loss of the object which first gave satisfaction. *As words displace the original object, we see the first step in the process of repression which forms the unconscious; entry into language inaugurates the production of subjectivity. . . .*

(Henriques et al. 1984: 215, emphasis added)

Fantasies and dreams involve 'the restoration of signs which are bound to the earliest experiences of satisfaction' (Laplanche and Pontalis 1973: 481–3), along with transformations and transpositions through a series of condensations and displacements (see later). Though full satisfaction is never possible, the infant needs to embark on 'filling the gap', mastering the loss.[4] In this section and the next, I show how mathematics can fill the gap, for some.

Anxiety is created by the absence of satisfaction itself, or by the fear of loss of the source of satisfaction. For Lacan, like Freud, the attempt to master this anxiety and ultimately to control desire is the impetus to acquire language. In entering into language, the child's own thought is inevitably regulated through cultural laws.

For Freud, repression was turning something away, and keeping it at a distance, from the conscious. Repressed material is to be contrasted with ideas and memories which are 'forgotten' in the preconscious, but which can be made conscious relatively easily. It is repulsed by the preconscious, and attracted by a (repressed) chain already existing in the unconscious. For Lacan, this may be seen as a *chain of signification*, rather than as simply a chain of associations.

Therefore, for Lacan, since language is by definition social, the social (the 'signifying order') enters into the formation of the unconscious. This leads to the idea that the unconscious is 'structured like a language' (Thom 1981, Henriques et al. 1984: 213). Lacan inverts Saussure's idea of the relation between the signifier and the signified, to produce the relation on the left-hand side of Figure 7.1, for several reasons. First, the signified becomes less and less important, as it is always receding, eluding us. Meaning therefore springs from metonymic and metaphoric relations between signifiers. Also, Lacan sees repression as metaphor, that is as the process of a new signifier, S', replacing the original signifier, S, which now 'falls to the level of the signified' (see Figure 7.1).

Lacan argued that there were isomorphic relationships between the semiotic processes of metaphor and metonymy (described in Chapter 6), and the two key mechanisms of the unconscious (such as in dreams), condensation and displacement, respectively (e.g. Thom 1981). *Condensation* occurs when meanings

Figure 7.1 Lacan's View of Repression: The Signifier Falling to the Level of the Signified

from multiple elements in the 'latent content' of the dream 'pile up' on a single signifier in the 'manifest dream-text'. *Displacement* was for Freud a process by which energy is channelled from one object to another object; in dream analysis, it appears as a form of 'distortion', in that elements that are central to the manifest content may be peripheral to the latent dream-thoughts, and vice versa. This distortion is made necessary by the existence of 'censorship' between the conscious (and preconscious) on the one hand, and the unconscious, on the other.

Thus the operation of metaphor and metonymy are interdependent. Metaphor creates a superimposition of signifiers; metonymy effects a continual 'sliding under' of signifiers, as depicted in Figure 7.1[5]. For an illustration see Martin Thom's (1981) 'reading of Freud through Lacanian spectacles'.

The emphasis on signification in Lacan's work provides a way of elucidating how unconscious processes might work, and also how the social might be implicated. However, his emphasis on a universal and timeless symbolic order seems to limit the extent to which the influence of the social can be *specific* to a given culture and period, as was argued in Chapter 6.

This analysis drawing on psychoanalysis will be extended in the next section's description of Model D, which takes on 'poststructuralist' insights. However, first I consider some relevant empirical work.

Studies Done Within a Psychoanalytic Perspective

Nimier (1977, 1978, 1993) reports on a series of studies which used both clinical and statistical methods. His early work in France used interviews with about sixty students, posing questions about:

* progress in mathematical studies
* feelings while doing mathematics
* whether they had discussed mathematics with parents.

He then developed questionnaires including semantic differential, 'closed' (Likert-type), and verb-choice items, based on the themes arising in the interviews, and addressed to 600 students in final year *lycée*, including males and females, human-ities and science specialists. Later phases of the study included a further 800 students in Belgium and Quebec, and a final 810 students in the USA, Britain and Canada (Nimier 1993).

In the interviews, Nimier (1978) found a certain amount of anxiety (*angoisse*) about mathematics. This took different forms:

* anxiety about the loss of one's own personality
* anxiety about destruction
* anxiety about separation or solitude
* anxiety about castration.

These anxieties are similar to the set of anxieties about loss set out by Freud (see p. 114). Nimier (1977) considers they are not only matters of conscious fears, but there is something deeper, under the surface, as in an iceberg. Metaphors (indicating condensations) and associations of ideas (suggesting displacements) are two ways of reaching the unconscious. These anxieties may be displaced onto mathematics, but they may be recovered by:

1   associative chains produced spontaneously by the interviewee, or
2   the 'slipping of meaning' through a signifier (Nimier 1978).

The use of associative chains is illustrated by the following quotation. Several symbols are used regularly in quoting from interviews throughout the book:

> ...   indicates a pause by the speaker
> [...]   indicates an omission, effected by my editing of the transcript

Subject: Right from the start I set myself against learning algebra ... Why algebra? I've been asking myself that question [for three years]. [...] the teacher [...] that poor woman had a *voice to put one to sleep*. [...]

Researcher: A voice ... a voice that reminded you of what?

Subject   What did that remind me of? Yes, definitely, it must have recalled something [...] particularly disagreeable certainly ... [...] something or someone that struck me or displeased me ....

Researcher   Someone?

Subject   [...] It was like a *purring*. [...] very soft. Oh, I think it was in connection with something else, but I don't want to tell myself that it's possible that it was that ...

Researcher   What are you thinking of? Even if it's not that, it doesn't matter ...

Subject   All right! Because there was [...] a very big disagreement between my mother and father [...] all these nights, I didn't sleep because I heard the arguments [...] in a half-*sleep* and each time I arrived in maths class I heard that voice [...] that made about the same noise, the same *purring*. So that got on my nerves ....

                    (Nimier 1978: 169; author's emphasis, my translation)

Here we see that the student presents her experiences in the algebra classroom in a way that suggests they are associated with a time when she was trying to sleep next to where her parents were arguing. The link is through the 'purring' sounds of the voice(s) – the signifiers – that were features of both situations. The 'disagreeable' affective charge has been displaced from the memory of the bedroom/falling asleep situation to the current experience of the algebra classroom, which 'gets on her nerves' and against which she 'set herself'. We can see the process of the original signifier, the parents' voices (linked with parental strife as signified), dropping to the level of the signified, where the purring voice of the teacher becomes the new signifier (cf. Figure 7.1).

An example of the 'slipping of meaning through a signifier' comes from the following:

R: What does mathematics bring to you?
S: I think it's above all the rigour which is important in that. Mathematical rigour
     is something fundamental.
R: Yes. Why?
S: [. . .] There is a solution: it's that or it isn't that. [. . .] In mathematics it's
     something pure [. . .] And rigour, I've always liked rigour: when I was little, I
     recall, I asked to be whipped (*fouetté*).

                                        (Nimier 1978: 169, my translation)

Here the signifier 'rigour' allows the subject to effect a sliding between the mathematical and the physical. This illustrates another form of displacement.
    Nimier(1993) offers a typology of defences against anxiety around mathematics:

> Either the anxiety and its supporting fantasies are displaced onto mathematics,
> and defences are directed against mathematics, so indirectly containing the
> anxiety; or the anxiety is contained in some other way, and defences can be seen
> to be mounted against this anxiety, mathematics serving as an instrument of this
> defence. Mathematics, then, through the fantasies that it calls forth, can be
> either that which you defend yourself against, or – on the other hand – that
> which participates in a defence against anxiety. It can even sometimes, by
> splitting, serve as both.

                                                          (Nimier 1993: 30)

Each type of defence corresponds to a grouping of the statements used as Likert-
type items in the questionnaire; an example from Nimier's questionnaire is given for
each. The first category of 'phobic' defences *against mathematics* includes:

1   Phobic avoidance, which may bring 'a sense of peace' (ibid.: 33), such as
    Question 3: At the start of a mathematical problem, I feel as if I'm in front
    of a black hole.
2   Repression, or denial of reality; for example claiming mathematics is
    meaningless, e.g. Question 9: Doing maths, it represents nothing, it's absurd.
3   Projection, or the rejection of 'unacceptable' feelings, wishes and so on from
    the subject onto mathematics, e.g. Question 18: Doing mathematics sometimes
    risks bringing destruction, you only have to think of the atom bomb.

The second category of 'manic' defences *by* (or using) *mathematics* include:

4   Reparation (against destruction anxiety) where mathematics is felt as useful,
    constructive, an object of value, e.g. Question 19: Mathematics brings you the
    pleasure of creating something.
5   Introjection, of order and stability, e.g. Question 17: Mathematics is a way of
    getting a strong character.
6   Reversal into the opposite, seeking to neutralise a disagreeable feeling; for
    example whenever a solution to a mathematical problem is found, e.g. Question
    5: When I work something out, I feel like a void is being filled.

Success or failure in mathematics tends to be played out in the balance between the anxieties produced by mathematics, and the defences using mathematics, or not, to combat these anxieties. The first set of defences (likely to be used by Arts students) is likely not to be favourable to success in mathematics – whereas the second set of defences (likely to be used by Science students) is likely to be favourable to such success (Nimier 1978). In addition, positive or negative *attitudes* may be linked with these defence mechanisms (Nimier 1993).

Nimier's work shows the importance of the defences against anxiety, and the operation of unconscious processes such as displacement in the appropriation of knowledge, especially mathematics. However, there is a danger that this sort of analysis on its own may reduce mathematics to a 'good' or a 'bad' object – as some of these interviewees do – without allowing for mathematics to be characterised in richer, and multiple, ways in discourse.

Legault (1987) also attempted to study the cognitive and the affective together, by using Piagetian tests and projective tests (the Rorschach and the Thematic Apperception Test, see Chapter 4) using two small groups (ten in each) of schoolgirls, one group judged as good in mathematics and one as having specific difficulties.

Legault drew two important conclusions. First, success or failure is inscribed in a whole dynamic *particular to each student*. Therefore, it is not possible to tie success in a general fashion to a specific affective factor. This suggests the analysis of clinical interviews, such as those discussed in Chapters 8 to 10, should focus on the whole range of affective factors, and not just anxiety. Second, in order to understand anxiety, it is necessary to look for *evidence of defences*, rather than attempting to measure the level of anxiety itself. This helps to make possible empirical work in this area, in the light of the Freudian view that anxiety may often not be observable. Thus, the problem would become one of attempting to find sayings, gestures, etc. that might indicate the operation of defences against anxiety; for example making jokes or 'slips', speaking or behaving in a manner that is not customary. That is, we must consider the *exhibiting* of anxiety which is not admitted by the subject, and not only assertions or *expressions* of it.

Thus the methods used for Model C feature clinical or semi-structured interviews, but also sometimes questionnaires, as in Nimier's work. The interviews can be treated as case studies, or used in between-subject comparisons, as in Legault's study.

The advantage of Model C is the depth possible in the treatment of affect. However, as already hinted at, opening up the problematic of affect has had its effects. For one thing, as Freud himself once remarked, the 'discovery' of the qualities of the affective was the last blow, namely to the 'psychological' subject – as conscious, rational, unified – that scientific research dealt to 'the universal narcissism of men', the other two being the 'cosmological', associated with Copernicus; and the 'biological', associated with Darwin (Freud [1917] 1953–). Also however, within mathematics and mathematics education, the effects have included the conceptualisation of the field of mathematics knowledge as an open system: open to its context, to the social. This has therefore presented researchers with the difficulty of distinguishing neatly between the structure of mathematics and its social context. Model D addresses this problem.

Model D: A Psychoanalytic Approach, Informed by Poststructuralism

The work of some poststructuralists radicalises Model C by questioning some of its assumptions. Walkerdine (1988, 1990b, 1997) has shown the need to understand thinking in terms of social difference and deprivation (see Chapter 6), *as well as* in terms of pleasure, anxiety and defences, which tend be seen in Model C as related to early family dynamics. Taylor (1989) argues for the importance of both socially available discourses and fantasies, in understanding 'mathematical' problem solving, and motivations such as career hopes (see later).

In Chapter 6 I argued that cognition is 'specific' to the context, and hence to the discourse(s) called up. That is, the meaning of a word, gesture, diagram, problem-text – or any other 'signifier' – depends on the specific discourse through which the signifier is read. Therefore, the same problem could serve as a 'puzzle' in a Sunday newspaper or as a test-item in school mathematics.

Thus, one of the ways that poststructuralist researchers develop the basic psycho-analytic approach is by emphasising signification, the active production of meaning through the play of signifiers, rather than seeing language as simply a representation of pre-established meanings from an inner or outer social reality. They also analyse the elements and relations of discursive practices in a culturally and historically based way, which provides a basis for moving beyond the constraints of assuming a timeless social order, as portrayed by Lacan (Henriques et al. 1984, Section 3).

As signposted in Chapter 6, the cognitive and the affective are viewed in Walkerdine's work as inseparable constituents of one ensemble of meanings acquired through learning – since learning takes place within discursive practices, and these are emotionally charged (see also Scribner and Cole (1973) and D'Andrade (1981), discussed in the first section of this chapter). This inseparability is illustrated by the following:

> Mathematical meanings – indeed, the development of language and word meanings in general – cannot be separated from the practices in which the girls grow up. The mother is positioned as regulative in these practices, in which desires, fears, and fantasies are deeply involved. So 'mathematical meanings' are not simply intellectual, nor are they comprehensible outside the practices of their production. . . . at home . . . these terms . . . carry strong emotional . . . content and act as signifiers in very different [ways] from the word pair 'more'/'less' as used in school mathematics.
>
> (Walkerdine and Girls and Mathematics Unit 1989: 52–3)

Thus, 'mathematical meanings' are both cognitive and affective, and they depend on the child's positioning in specific social practices, some of which may be sited solely (or largely) in the home, with others sited in the school.[6]

However, the affective qualities of these meanings cannot be read off simply from the name of the site of the particular practice. The inadequacy of simple naming as a way of providing a 'natural' definition of the context was discussed in Chapter 6. For example, Josette Adda, in a passage criticising the artificiality of many word problems in school, points out that the sign of the affective charge (that is positive or negative) cannot be predicted from the familiarity or 'everyday' qualities of the site:

> [For] problems of the type 'Mummy goes shopping, she buys . . . ' [. . .] this variable 'Mummy' (each pupil supposedly feeling involved) introduces an emotional factor that is not necessarily positive: for example, when the mother has financial difficulties, has little time to do the shopping, is sick, far away or deceased.
>
> (Adda 1986: 59)

This suggests that the affective meanings cannot be read off from the site in any simple way such as:

home, out of school	. . .	familiar	. . .	'positive'
school, mathematics	. . .	unfamiliar	. . .	'negative'

That is, we expect that the affective response to 'the same site' may be different for different subjects; see the analysis of the case studies in Chapters 9 and 10. Further it is not sufficient to reduce the affective meanings of mathematics to its being a 'good' or a 'bad' (or 'protective' versus 'threatening') object for that particular subject, as Nimier seems to do, in his classification of defences. Nor can we expect to predict a subject's affective response from their social structural location, for example their gender or social class. In order more fully to appreciate the particularity of each person's feelings towards 'mathematics', we need to consider the specific discourses in which both the person – and mathematics – are positioned, as well as the person's particular 'history of desire'.

The next two subsections elaborate on this.

Cultural and Emotional Differences in Mathematical Activity

Nick Taylor's research (1989, 1990a, 1990b) aimed to evaluate educational television programmes as aids to mathematics learning, for a class of black thirteen-year-olds in Soweto, South Africa. Both pre- and post-tests included a question about sharing out evenly, among five or six children, a whole loaf of bread or a cake, respectively.

Taylor's argument follows a similar path to that of the last two chapters. He started from the 'undeconstructed cognitivist notion that the situational referent of a party and the cake would somehow provide the handle for better understanding the mathematical ideas, without examining the specificity of such referents' (Taylor 1990a: 280). The 'handle' presumably involved the greater *familiarity* of the party/cake context (compared with an abstract one), or perhaps its more *positive affective charge*.

In addition to testing the whole class, Taylor interviewed four of the pupils on several occasions, presenting some of the same problems. He also sought to elicit other associations, such as by asking who cut the bread and the cake in specific contexts. Taylor found that Paul, one interviewee, dealt with the sharing of a loaf of bread among five children by visualising the slicing of the loaf into ten slices, and then giving each child 'two-tenths'. In contrast, to share a cake amongst six children, Paul proposed the formal solution of one-sixth. Taylor argues that the difference in strategy used in the two contexts shows that mathematical activity is not merely 'rational but is somehow tied in to profound emotional and/or cultural forces.' (Taylor 1989: 162)

Now, these forces are *not only cultural* since Taylor observed differences in approach to the two questions among children from the same social milieu. For the bread problem, about half the pupils in the class appeared to call up 'culturally embodied' practices involving bread, the other half calling up an 'abstract' discourse, school mathematics. For the cake problem, only about one-third of the students appeared to call up a culturally embodied response (Taylor 1990a).

Taylor sees the bread and the cake as different metaphors for the same 'metonymic' principle, that is, 1 divided by n = 1/n. He argues that the different metaphors tend to call up different discursive practices, with different associated positions, for each particular subject. The subject-position in turn is instrumental in determining the kind of strategy adopted in making sense of the problem. For example, for Paul, the 'bread metaphor' tends to call up practices such as having lunch with his grandmother, and thereby familiar methods of thinking (and hence calculation) grounded in the procedures of slicing a loaf.

The importance of emotional forces is shown by the positive associations for Paul (and another young male, Camel) of bread with his participation in sport and with the care of his body supported by science teaching at school, and also by lifelong memories. When asked why he likes bread more than cake, Paul replies, 'Because, meneer [sir], I have liked bread since I was young', (Taylor 1990a: 259). In contrast, cake is seen as sweet and unhealthy, as signifying birthday parties. Thus the cake metaphor is less familiar, and less positive in affective terms. For the problem based on cutting the cake, Paul and Camel apparently need to fall back on algorithms, which, being less familiar, tend in Paul's case to decay into forgetfulness and 'garbled numerese'.

Taylor draws on Lacan's ideas (see the previous section) to interpret these associations:

> Bread is part of the web of his life. It nourishes him and provides energy for feats of physical prowess. It is associated with the routine lunch provided by his grandmother. Bread as signified and signifier is tied to earliest memory and thus ... closely linked to originary loss and the primal phase of signification. ... the subject, through the production of word signifiers, attempts to satisfy the unfulfillable desire and to exert fantasy control over the lost world. The bread metaphor calls up a process embedded in everyday practices: [Paul] moves with great assurance toward the identical argument ... on each of two [interview] occasions separated by seven months.
>
> (Taylor 1990a: 260)

Thus 'bread', as a key signifier in the life of Paul (and Camel), is culturally specific, and is implicated in wish-fulfilment, through processes in the unconscious attempting to satisfy (unfulfillable) desire. Taylor's interpretation suggests that these processes involve the *condensation* of multiple meanings on the signifier 'bread', and also that desire is *displaced* from the satisfactions of eating with his family, to bread.

Therefore, Taylor argues that any attempt to understand the pupils' thinking about these apparently 'mathematical' problems in terms of cognitive and cultural processes only, will be unsettled because the thinking is also infused and invested with affect, which he sees as relating to unconscious desire and wish-fulfilment. Wish-fulfilment is a 'primary element in orienting any individual toward a particular subject-position' (Taylor 1990a: 278), as are emotional charges generally.

We must be cautious about Taylor's data and its interpretation. First, any conclusions based on two boys may well not be generalisable. Even a series of three or four semi-structured interviews per child is limited for producing the sort of psychoanalytic interpretation offered here, especially if we allow for the boys' youth, their inarticulateness, and the oppression of blacks under *apartheid*. These interpretations are plausible and fascinating, but very tentative!

Taylor makes several contributions to the argument here. He deals with the problematical notion that posing problems in familiar or emotionally positive contexts, such as cutting bread or cake will straightforwardly facilitate problem-solving, as follows. He considers instead the *specificity of meanings* within different practices, such as those in which 'bread' and 'cake' respectively are central terms. At the same time, he emphasises the *particularity* of the meanings within these practices for each subject. That is, he considers both cultural and emotional aspects. As for what facilitates problem-solving, Taylor explains as follows: the 'roles, rules and procedures within the "bread" practice are *clearer* to [Paul] than those of the "cake" discourse' (Taylor 1990a: 257; emphasis added). This suggests that some of the important aspects of difference between the practices are to do with familiarity and the strategies deployed.

Taylor uses interviews to study problem solving under an alternative set of discursive conditions to those of the test. Thus we can say that the pupils are 'positioned' differently because of different social relations in the two contexts. In addition, the interviews clarify the relationship of the 'metaphoric content' of the problem with the discourse (and subject-position) called up, and hence with the strategy used to produce a response. In the examples given, this is based on a distinction between an 'embodied' response from within an everyday practice or an 'abstract' response based on school mathematics. This process of 'calling up' depends both on cultural factors – such as the existence of bread and birthday parties, family formations and child-rearing practices – and on emotional processes, including their unconscious aspects. Insights into these emotional processes are provided by the life-history element of the interviews.

Taylor's work also confirms findings of Adda and others that the affective meanings of everyday practices are not uniformly positive. Nor are the affective dynamics of school mathematics uniformly negative. Indeed, Walkerdine's (1988) classroom research describes two sorts of pleasure, relating to two very different relations to mathematics. One type of pleasure is experienced by two boys, working on place value in a 'faster' group of six and seven year olds. These boys derived pleasure from the apparent power of the methods they were using. In learning the discourse of mathematics, these boys were also enjoying the fantasy of mastery and control (Walkerdine 1988: 199 and Ch. 8).

However another type of pleasure was evident in a 'shopping game' played in the same classroom (see Chapter 6). Here the pupils (several girls and one boy) had continually renewed resources (10p), and paid unrealistically low prices such as 2p for a yacht! Hence, most seemed to call up a practice of shopping in which they were positioned as powerful customers, which they were not accustomed to since they came from working class families. This allowed them to derive a great deal of pleasure from the fantasies invested in their positions.

It was not clear how much mathematics they learned, because of their involvement in the fantasies, which they did not 'suppress', but which provided the basis for their enjoyment. Walkerdine suggests therefore that this group's pleasure was double-edged, a double-bind: 'While they fantasise about being rich, they cannot "master" subtraction'. Further,

> Their inscription as subjects within everyday practices is . . . cross-cut, in the very relations of signification themselves, by desire. Absence, lack, loss, prohibition are present. And the subject's experience . . . therefore of the practices in which 'numeracy' is produced, must be relations of desire. They are not formal systems, but lived relations of power and powerlessness, of wanting, having, being; they are continually open and shifting, not closed axiomatic systems like mathematics.
>
> (Walkerdine 1988: 198)

As this quotation suggests, the practices within which subjects are positioned, including their numerate aspects, are generally charged with desire, and the workings of desire are influenced by power relations (see also Urwin 1984). Thus, to appreciate the particularity of each subject's affect towards mathematics, we need to understand its production within a positioning – a sometimes *contradictory* one – within a web of interlinked discursive practices (Henriques et al. 1984: 218-26).

The quotation also emphasises an important feature of the discourses forming the basis for the everyday practices in which numeracy is produced: they are open and shifting, not closed and timeless, as mathematics *appears* to be (Tsatsaroni and Evans 1994). What it means, say, to be performing a calculation is to a great extent open and indeterminate. This is why every time a mathematics teacher reaches for an everyday example, the mathematics is 'at risk', in the sense that meanings from the everyday practice threaten to invade and subvert the 'purity' of what may first appear to be 'mathematical' signifiers; see the discussion of Adda (1986) earlier.

In addition, we do not need to be restricted by the idea of a universal and timeless social order, attributed to Lacan in the previous section. If discursive practices and relations are culturally and historically specific, so too will be the content of unconscious processes; see Taylor's (1989) suggestive evidence (discussed earlier) and the case studies (such as Ellen's) in Chapter 10.

Wendy Hollway (1984, 1989) provides a clear poststructuralist response to these theoretical challenges. Along with the other authors of Henriques et al. (1984), she aims to build on Lacan's emphasis on language and signification, while moving beyond his ideas of a universal social order and timeless sets of meanings, to analyse discourses as historically specific, drawing on Foucault.

Hollway (1984) investigated gender identity in heterosexual relationships in a group of people aged 30+ in early-1980s London, using interviews and transcripts from an ongoing 'consciousness-raising' group (Hollway 1989). She considered several discourses concerning sexuality in heterosexual relationships to co-exist in the contemporary epoch:

- the 'male sexual drive' (MSD) discourse, which emphasises men's (supposed) sexual needs
- the 'have/hold' (HH) discourse, which stresses that sex should take place within a lasting relationship with spouse/partner
- the 'permissive' discourse, based on the assumption that sexuality is entirely natural and should not be repressed
- 'feminist' discourses.

These discourses offer alternative bases for action and for readings of actions. None is hegemonic, and they may often be contradictory. For further description and evidence about their salience, in the form of quotations from influential magazines and other 'authorities', see Hollway (1984, 1989).

Hollway differs from some poststructuralists in seeing discourses as *making available* subject-positions for subjects to take up, a less determinist position than Foucault. The first two discourses are gender-differentiated, in that the taking up particular positions is 'differentially available' (Hollway 1984: 236) to men and to women. Practices or actions have different meanings, if read through different discourses; for example, 'having sex' signifies differently within the MSD discourse – as a male need, and within the HH discourse – as intimacy; and differently again in the permissive discourse.

In seeking to avoid discourse determinism, Hollway argues that the social availability of a position in discourse must be accompanied by an *investment* (*Besetzung* in Freud) for the person to take up the position. This means there will be some 'satisfaction' in taking up a particular subject-position, though this satisfaction may contradict other resultant feelings (for example guilt), and the investment may not necessarily be a 'rational', nor even a conscious 'choice', though 'there is a reason'. Satisfaction comes, say, from one's ongoing 'identity' being confirmed, or from the conferment of power: for example, a man who defines the woman as subject in the HH discourse, thereby suppresses his own wishes for intimacy and avoids being let down or hurt, therefore remaining powerful (Hollway 1984).

Taylor's study (earlier) found 'bread' a key signifier in the lives of at least two boys in Soweto. A similar finding for 'oranges' – though rather more particular – is based on one of Hollway's episodes. A woman, during a phase when she was not paying her male partner much attention, noticed that he was 'getting at her' in little ways. When they tried to discuss it, he first came up with a blank – then, 'oranges', as if from nowhere. After some reflection, he said it had something to do with his relations with women. A woman's peeling an orange showed she cared for him. Then he said that his mother used to do it for him, even when he could do it for himself.

Here, 'peeling oranges' has meanings for the man which are not rationally or culturally accessible through, say, the definitions of oranges or of peeling. In

Lacanian terms, the signifier 'oranges', fallen to the level of the signified – that is, repressed – is part of the metaphoric axis whose links are formed by desire. Through condensation and displacement, it connects to a wider set of meanings around proof of loving and of caring, through women doing things for him. It is connected with a suppressed signifier established early in his history, through its links with his desire for the unconditional love of his mother. It is also part of the signifying chain from mother to 'Other', which, according to Freudian theory, is 'historically [that is biographically] unbroken for men, though savagely repressed' (Hollway 1984: 250). Thus, the significations occupied by desire may be idiosyncratic, but they are not arbitrary, for they are a product of a person's history. This shows how Lacanian ideas can be used to make *links between discourse and subjectivity* (Hollway 1989).

Hollway (1989) presents case studies of decisions made by several couples, such as by Will and Beverley early in their relationship on whether to have an abortion. Will's accounts of their discussions indicate for himself two contradictory 'positions', which Hollway calls 'the woman's right to choose', and 'I want a child'. The first we can see as related to feminist discourse, but the second is more difficult to locate. It could be related to Will's putting himself as subject in the have/hold discourse, or perhaps to a type of 'paternal' discourse, likely to have been learned early in life, through *identification* with his father.

This gives a basis for distinguishing analytically two aspects or stages in the process of positioning in discourses. First, Will is *positioned* in the feminist and humanistic discourses since they hold sway in the contexts where the research is conducted, namely consciousness-raising groups and interviews: it is 'taken for granted that he will speak and act within the meanings' of these discourses (Hollway 1989: 81).[7] At the same time, he is able to *call up* – as part of the mix of discourses in play – others such as the 'have/hold' or 'paternal' discourses in a way that flows from his particular subjectivity. Subjectivity is thus the product of the person's 'history' of positioning in discourses, and of the way this constructs their 'investments' in taking up specific positions.

Hollway's work shows how to move beyond an abstract notion of language and signification, to a conception of discourses as historically and culturally specific, and also potentially in conflict in their positioning of persons. Yet simply acknowledging the importance of the discourses risks losing a psychodynamic – and hence affective – account of the subject. However, it is not sufficient simply to refer to what is 'other' than the subject; what is needed, according to Hollway, is to study 'the continuous [that is, fluid], everyday, defensive negotiation of intersubjective relationships within the effects of power-knowledge relations' (Hollway 1989: 84).

Melanie Klein's emphasis on defence mechanisms which work between people, rather than within a person, is relevant here. For example, with splitting, gender-differentiated characteristics are located in one partner, for example the expressing of feelings in the woman. This involves a repression by the man of his feelings through projection and a consequent position as powerful, rational, supportive. This position is 'invested' in that it protects the vulnerability of the man – conditional on positioning the woman as having the feelings (perhaps through her own introjection). Here, splitting works not as a permanent accomplishment of socialisation of one individual, but as a dynamic and inter-subjective process. The possibility of intersubjective, and institutional, splitting will be illustrated next.

Poststructuralist Views on the Gendering of Mathematics

Much discussion of gender and mathematics draws on the argument that, in school mathematics discourses, many of the main terms are understood as 'gendered', rather than gender-neutral. For example, active learning, rule-breaking, naughtiness, aggressiveness and (potential) rationality tend to be seen as 'masculine', while passivity, rule-following, helpfulness, nurturing and emotionality tend to be linked with 'femininity'.

Walkerdine (1985, 1988; Walkerdine and Girls and Mathematics Unit 1989) seeks to provide an explanation why these links have been made so persistently made, and a description of the consequences, including the discounting of girls' good performance in the classroom. Her argument has three parts. First, she argues historically that the development of science and mathematics from the seventeenth century has been closely connected to the control of nature by man. Since this time, 'reason has been a capacity invested within the body, and later the mind, of the man, from which the female was, by definition, excluded'. Women were excluded – for example, from higher education and the professions in the nineteenth century – on the grounds that they were swayed by emotion, and not, therefore, capable of rational judgements (Walkerdine 1985). Thus we can say that rationality and mathematical understanding are gendered. Nowadays evidence of the 'lack' on the part of women in mathematical understanding, and also can be produced by the sciences of psychology and education (Lee 1992).

The second part of the argument suggests deep-seated emotional reasons, *at the level of cultural groups and institutions*, for the tenacity of these ideas. Walkerdine suggests parallels with Homi Bhabha's (1983) discussion of the 'fear of the Other' inscribed in the stereotypes of the colonised people that form part of the coloniser's discourse. Much of the material of such views is fantasy, based on the projection of characteristics feared or disapproved in oneself onto 'the Other'; for example laziness, dishonesty, excessive sexuality and so on; these processes are largely unconscious.[8] In the case of gender relations and stereotypes, the unconscious fear of irrationality may lead to its being projected onto the woman: 'the Other of mathematics is uncertainty, irrationality, out of control' (Walkerdine 1988: 199).

The third part of the argument considers what might be the emotional 'invest ments' in involvement in mathematics for particular individuals. 'Reason's dream' is described as follows:

> The desire's object is a pure, timeless unchanging discourse, where assertions proved stay proved forever (and must somehow always have been true), and where all the questions are determinate, and all the answers totally certain. In terms of the world, the desire is for a discourse that proxies the manipulation of physical reality achieving a perfect and total control of 'things', where no realizable process falls outside mathematics' reach.
>
> (Rotman 1980: 219)

'Reason's dream' is embraced by some, generally boys. It may also be sought by girls, but to the extent that reason and mathematics are made to signify in a gendered way, the girl cannot 'have it'.[9] The dream is of an all-comprehending, unchanging, infallible discourse, based on an omnipotent fantasy of mastery through the use of intellectual reason over a universe which is thereby ordered and controllable. Walkerdine (1988) traces the basis of the dream of control to the apparently universal applicability of mathematics, based on its seeming 'decontextualisation'.

These features, however, mean that the learner has to 'suppress' the metaphoric content of the statements made in mathematics to leave simply a metonymic string – though the signifiers within it remain linked by semiotic chains to other discourses; see the example of the mother teaching her child to count the friends invited for a party in Chapter 6 (p. 102). Nevertheless, these links including those to aspects of value, emotionality and desire, need to be 'forgotten' by the successful learner of mathematics. To the extent that this involves repression, it needs to be coped with by 'wish-fulfilment' through unconscious processes such as dreams and fantasies; see Model C and Walkerdine (1988). Thus Reason's dream and the mastery of mathematics in particular provide one way for a fantasy of omnipotence and control to be 'lived'. Whether this way is taken up by a particular subject will depend on a particular subject's 'investments'.

This account of why women have been so considered persistently as unsuited to reason and to mathematics raises many further issues, including the dependence on the three parts of the argument on different types of evidence.[10] However I can focus here on only a few issues.

Some support for the idea of collective defences against anxiety, relevant to the second part of the argument, is provided by Ellen Gottheil (no date). She reconsiders the sort of evidence on gender differences in mathematics performance and mathematics anxiety produced by psychology and educational studies, and referred to by Walkerdine as providing the basis for the 'expected', the 'normal'. (Some of this evidence is discussed in Chapters 3 and 4.) Gottheil cites a large study of men and women postgraduates in science, engineering and medicine at Stanford (Zappert and Stansbury 1984; see also Becker 1990), which showed no gender differences in ability or undergraduate performance. However women assessed themselves as less competent in mathematics and sciences, and reported greater anxiety and stress-related symptoms, and both genders expected men to outperform women in mathematics. Since the sample was just the sort of group where the 'truth' of such expectations should be most in doubt, Gottheil argues that these students were exhibiting defences of *denial, repression, and rationalisation.*

Gottheil explains the link between gender and mathematics, from a Kleinian perspective, as related to attempts at a cultural level to defend against the most basic anxieties:

- anxiety about the destruction of self, or ego, or personality
- anxiety about the loss of the loved object (mother)
- anxiety about dependency and helplessness.

(This list recalls Nimier's; see p. 116.)

Gottheil draws on Eliot Jaques's studies of organisations (for example Jaques 1977, Menzies 1960) to support the idea of *collective defences against anxiety*. For example, the primitive Kleinian defences of splitting and projection may combine to produce rigidly differentiated gender characterisations, such as active versus passive, rational versus emotional, and so on – just as in the gender stereotypes discussed earlier. Splitting, along with the manic defence against helplessness and dependence, may provide boys/men with payoffs resulting from the derogation of girls/women as passive, 'merely' emotional, and so on, and from fantasies of control. These arguments about the bases of gender stereotypes, like Bhaba's about colonial discourses, are plausible, though it is not possible to consider further evidence for them here.

However, descriptions of how particular subjects might have investments in a fantasy of omnipotence and control 'lived' through mathematics, could be produced by clinical interviews like those reported by Nimier earlier; see also the next three chapters.

Thus we have the basis of an argument as to why it should be more 'expected', more 'normal' – in general and in UK schools in particular – for girls to perform less well, or at least in ways lacking in 'flair', and for boys, at least some, to excel. Put another way, we can see how girls' 'failure' in mathematics is socially produced, in (discursive) practice. However it is still not clear why girls should be any more anxious about this 'normal' state of affairs, or about mathematics – or indeed if they are!

Ellen Gottheil reminds us that anxiety for Freud was a signal of unconscious intrapsychic conflict that leads to the mobilisation of defences. She therefore argues that mathematics anxiety is such an indicator of intrapsychic conflict for some women, especially 'high performers' in mathematics.

This argument is supported by Marina Horner's (1968, 1972) positing of the 'fear of success', or motive to avoid success.[11] Horner considers women to be more anxious in testing or achievement-oriented situations than men, because there are for them negative consequences, and hence anxiety, associated not only with failure, but also with *success*. These negative consequences for women include loss of self-esteem, doubts about their femininity, and fear of rejection. They arise for women, because intense intellectual striving can be viewed as 'competitively aggressive behaviour' (Mead 1949); this is reinforced if one accepts Freud's ([1933a] 1973a) claim that the essence of femininity lies in repressing aggressiveness. Thus, Horner's work provides an explanation for higher female levels of expressed test anxiety, especially on items which specify the situation, but not what *consequences* one is anxious about.

Horner also notes, however, that the psychoanalytical literature suggests that, besides anxiety, defensive reactions against anxiety should be considered. An example of the latter would be the defensive projection by women of achievement motivation into less conflictual situations, such as women engaged in activities in the home, and men engaged in more intellectual and achievement-oriented types of pursuits (Horner 1968: 17).

I round off this discussion by considering *social class differences* in anxiety. I recall Walkerdine's (1990b) discussion of possible differences in the sort of relations the working classes and the middle classes tend to have with calculation in money-based practices – labelled 'material necessity' and 'symbolic control', respectively (see Chapter 6). Here the general arguments above suggest that it may be more difficult to mathematise, or to intellectualise, a problem – that is, to suppress the metaphoric links in the discourse called up to solve it – if there is much anxiety, or other 'negative' emotion, around it. Thus, when people for whom contexts involving money, and consumer spending, have in the past generally involved the pain of deprivation and strict regulation, are put in learning contexts ostensibly based on such practices, they may tend to fantasise about wish-fulfilment rather than going for mastery; see the example of the working class children in the shopping game. Of course, middle-class children may be anxious about other matters such as possible academic failure.

Conclusions

Much previous research on adults and mathematical thinking which has taken the context seriously, has tended to ignore, or to play down, the importance of affect generally; this can be seen in most studies reviewed in Chapter 6. The work reviewed in this chapter focuses generally on *emotions*, which are 'hotter' (more intense) and less stable types of affect than attitudes. The latter are the focus of much of Chapter 4, where 'Model A', an 'individual differences' approach to relating affect and cognition, is used as a basis for analysing the survey results.

Model B sees emotion as based on the evaluation of the physiological arousal resulting from the interruption of a plan. Research here is normally based on observations of problem-solving episodes. Some of this work has theoretical affinities with recent psychological and sociological analyses of emotion as having three aspects: *physiological*, *behavioural* and *feeling*. Some also argue that emotional (behavioural) expression does not function as outer signals of 'inner' feelings, but as 'signs in the network of social relations' (Burkitt 1997). This points to the use of insights concerning language and discourse developed in this and the previous chapter. The tripartite division of emotion allows us to understand the 'charge' of the affective as relating especially to the *physiological* and *feeling* components of the emotion.

Model C develops the Freudian notion of affect as charges attached to particular ideas into the Lacanian view of affect as *charges attached to particular signifiers*. This allows us to see the unconscious as 'structured like a language', as a repository of repressed chains of meaning, and to analyse repression as metaphor, and displacement as the movement of charges of feeling along a chain of signification. Because defences can operate to distort (or occlude) the expression of emotion, anxiety in particular is not necessarily observable (in any straightforward way), let alone available for self-report. Hence we may need to look for evidence of the operation of defences – in jokes or 'slips', or in fantasies or 'free associations' – and to allow for the possibility that anxiety may be presented as 'no feeling at all', or even as a different feeling.

Model D develops the basic psychoanalytic model, by infusing it with post-structuralist ideas – that is by emphasising *signification*, the active production of meaning through the play of signifiers, rather than seeing language as simply a representation of pre-established meanings. It also emphasises the social relations and power aspects of discourse. This approach allows the analysis of meanings, both generally and for particular subjects, thus providing a basis for embracing both culture and emotion in the study of mathematical thinking.

This approach also provides the basis for seeing affect or emotion as related specifically to discourse and practices. The emotional meanings of 'mathematics' thus depend on how the particular subject has been positioned in discourses which involve 'mathematics'. Understanding these is challenging because the salience of mathematical – or more broadly, of quantitative or spatial – ideas in this culture means that the signifier 'mathematics' may connect frequently with chains of signification in the unconscious – that at first may seem unrelated to mathematics. (See Chapters 9 and 10 for illustrations.)

For example, Taylor (1989) describes how the performance of two of his subjects seems to depend on the 'clarity', or familiarity, of the practice called up in response to different problems: the boys appear to reason better when calling up a culturally embodied strategy based on everyday practices, than with an abstract strategy based on school mathematics. Taylor is not saying simply that 'positive affect leads to better performance', the sort of 'purely additive influence' that he has elsewhere criticised (see the first section of this chapter). Rather, I read his work as suggesting a two-stage explanation: first, having a positive affective charge means that a practice is more likely to be called up more frequently, and thus to be more familiar; and second, familiarity (with the practice) normally means clearer thinking and more effective problem-solving. All these processes – developing affect, becoming familiar, and thinking – depend on specific practice(s) and the particular subject(s) involved. This is a more complex and specific account of the relationship between cognition and affect, and needs to be assessed in further research.

Summing up the relationship between the cognitive and affective, many of the conclusions drawn from using Models A and B describe the relationship in terms of affect 'interfering with' cognition, or sometimes 'supporting' it. This would be in line with discourses which define subjectivity as the identity of a rational, unique and unified self, and which mark the affective as different from, as 'other' to, the cognitive. Now, in the psychoanalytically informed Model C, the affective is rather more privileged: it provides the charge attached to (or infusing) ideas, and is thus related to the cognitive, and not entirely 'other' to it.

Moreover, as we have seen, affect can be displaced onto ideas different from those to which it was originally attached. This means that, although affect is not entirely 'other' to cognition, neither is it completely 'at one with' cognition. One implication of this is that anxiety apparently relating to mathematics may have been displaced onto it, from another setting (see Nimier's first example earlier, and Chapters 9 and 10). Model D draws on theories of discourse and signification, to analyse the elements of discursive practices, including devices such as metaphor and metonymy, linked with condensation and displacement, respectively.

I can sum up the ideas on *context*, and *positioning in practices*, developed in the last two chapters. In Chapter 6 I argue that problems can be recognised, and thought about, only within a specific context, which is constituted by one or more discursive practices. In this chapter, I have further argued that the person acting in a setting generally will have a 'positioning' particular to him/herself, because of the possibility of 'contradictory' (conflicting) positions in multiple practices, and especially because of his/her peculiar history of emotional 'investments'. Thus, in order to understand mathematical 'performance', we must describe the subject's 'positioning' in practices when confronting problems, in an interview or other setting.

However, describing such positionings is challenging. It requires relating to very complex ideas about subjectivity, where the terminology is slippery. It requires going beyond both the work of some poststructuralists and that of psychoanalysis. Some poststructuralists have described the positioning of subjects in social practices, but still lack a satisfactory theory of subjectivity; many psychoanalysts have addressed subjectivity, but tend to ignore its social/historical dimensions. For all these reasons Model D is developed here to underpin my idea of *positioning in practices*.

There are two 'aspects' or 'stages' of the process. In any particular situation, certain discourses hold sway generally; we say that they *position* any person acting in that setting, because the discourse(s) are used to 'read' – interpret, evaluate, regulate – these persons' activities, and power is exercised in these processes. Much of Foucault's work (for example 1977, 1979) was oriented to describing such subject-positions. Also Hollway has argued that certain positions in gender-differentiated discourses, for example, tend to be more available to males or to females. So far, positioning seems rather determinist: any particular white middle class female, say, should be positioned in basically the same way in a given discourse.

However, this is where the other aspect comes into play, because of subjectivity. In any situation, the play of signifiers in language, and the (related) flow of emotional charge along them, is such that a particular subject may *call up* (recognise, recall, select) one or more discourses – which may *or may not* be from among those considered 'at play' for subjects in general – which are used to examine, understand and resolve any problem.

Even deciding how to call the process is complex. 'Being positioned' emphasises the determined aspects of the process, and 'calling up' the 'voluntary' aspects. Talking about them together, as aspects or stages of a process, is an attempt to avoid either tendency towards one-sidedness. 'Being positioned' emphasises that the person may have no choice about being 'subjected' to a reading (by another) of his/her performance through the particular discourse, nor may he/she be able to offer resistance to the power exercised in social relations by others. For example, Walkerdine (1988) describes her powerful position as an experimenter reading the performance of cognitive tasks by some pupils who had been told they were playing a 'game'. On the other hand, 'calling up' suggests that there can be variation across persons, and what may be experienced as 'choice'. Even those apparently 'less powerful' may attempt to call up a different discourse, and thereby to vary their positioning in certain contexts; an extreme example comes from the two little boy pupils who challenge their teacher by using sexist terminology (Walkerdine and Girls and Mathematics Unit 1989).

Thus, the subject's positioning in a specific situation, such as mathematical problem-solving, results from a range of features and processes. First we can say that the subject is positioned in ways that are:

- influenced by the subject's *place in a structure* according to class or gender, but not determined by it (which would be' social structural determinism')
- moulded by the *discursive features of the situation*, for example the overt social relations of the setting, the wording and representation of the questions, but not determined by them ('discourse determinism').

Then, we can see that the discourses (and positions) called up by the subject are:

- subject to the *unpredictable 'play of the signifier'*
- dependent on the subject's *complex* of *'emotional investments'*, related to his/her 'history of desire', and subject to the operation of the unconscious, including (possibly inter-subjective) defence mechanisms, rather than being necessarily 'rationally', or even consciously, chosen.

One's positioning thus results both from the general social availability of positions in discourse, and from the investment for the particular person to take up a specific position. It is not at all 'freely' chosen, nor is it fully determined, but there are 'reasons' for it (Hollway 1989).

The subject will generally have an 'inter-discursive' positioning in *multiple* discourses (Henriques et al. 1984), which makes the task of description more complex. Further, the positioning develops in a *fluid*, changing way, and hence one must describe this positioning at crucial points of an interview (or other social interaction), rather than attempting to describe some 'overall' positioning for each subject.

Thus, the agency of subjects is created within the situations and statuses or positions that are conferred on them. This recalls Marx's famous dictum: 'Men make their own history, but they do not make it just as they please . . . under circumstances chosen by themselves' (Marx [1852] 1968: 96). In parallel, we can say:

> Subjects have their own agency but not through discourses and positioning of their own choosing.

This chapter has implications for formulating the 'problem' of transfer. The connections and discontinuities between practices (described in Chapter 6) involve not only ideas, strategies, and so on, but also values and feelings, carried by chains of signification. Therefore the quality and intensity of affective charges may often be a *major influence in the success or failure of many attempts at transfer*, an influence that has so far been largely ignored in the mathematics education literature.

I argued in Chapter 6 that one reason that a particular set of relations of signification may not provide fruitful interrelations with the discourse of school mathematics, say, was that these relations might be *misleading*. The arguments in this chapter, and illustrations such as that from Adda (1986), suggest that another reason may be the relations between the discourses might be *distracting* or *distressing*.

In this chapter, I have begun to sketch an alternative view of mathematics anxiety, and emotion around mathematics generally. Emotion is specific to specific discursive practices in the sense that it invests or infuses particular signifiers. These signifiers may be linked in chains in unpredictable ways to other signifiers, and indeed, may be linked with unconscious ideas. The last two chapters provide the basis of the themes investigated, and the methodology, of the interview phase of the study. I take these up in the next three chapters, where I shall consider issues such as what practices can be said to be at play generally in a given setting, and what indicators there are for the subject's positioning in a particular episode in that setting.

8 Developing A Complementary Qualitative Methodology

Every night before going to bed, he would write down the number of stones he had added to the wall that day. The figures themselves were unimportant to him, but . . . he began to take pleasure in the simple accumulation, studying the results . . . At first, he imagined it was a purely statistical pleasure, but after a while he sensed that it was fulfilling some inner need, some compulsion to keep track of himself and not lose sight of where he was.

(Auster 1991: 203)

At this point it is appropriate to produce a list of themes or 'foreshadowed problems' that will guide the production and analysis of the interview material in the second phase of this research. I then discuss the aims, methodology and execution of the interviews. Finally, I provide an overview of the analyses to be reported in Chapters 9 and 10.

Themes or Foreshadowed Problems

Ten themes or foreshadowed problems have been formulated on the basis described in the previous two chapters.

Theme (1) Contexts of Thinking Ascribed in Terms of Positioning in Practices

In the last two chapters, I have developed my alternative way of understanding the context of mathematical thinking, namely, that the context is *constituted by the discursive practice(s) in which subjects have their positioning.* Developing earlier work, I explore how, in a specific situation such as problem-solving interviews, it is possible to make judgements about which practice(s) provide the positioning within which the subject responds to a particular problem.

Using the interviews, I also illustrate the *multiple positioning* of some subjects, and the resultant conflicts and contradictions for them. This leads to the crucial notion of the subject's *predominant positioning* in a particular episode of the interview; this idea is used throughout the 'cross-subject' analyses of Chapter 9. In

Chapter 10, the emotional investments of particular subjects, and their effects on positioning are considered.

Theme (2) Inseparability of Task and Context

In Chapter 2, I distinguish between 'proficiency' and 'functional' views, on the basis of differences in their notions of context and practical mathematics. The proficiency approach tends to see abstract questions formulated in the school mathematics context as 'the same as' others formulated in practical or everyday terms – as long as the mathematics which can be abstracted from both is seen as being the same. In the functional approach, the wording and format of the problem are normally assumed to be sufficient to position it in a school mathematics context, or alternatively in a 'practical mathematics' context.

In Chapter 6, both of these approaches were considered together, as 'utilitarian'. I contrasted them with others, including my own, on context and transfer. My view takes on board the ideas that cognition is situated – though not the strong form of that view (see Chapter 6) – and that there are discontinuities between doing school mathematics problems and doing numerate problems in practical contexts, that relate to more than wording, for example to social relations. This means that task, thinking, and context should be seen as a whole, and therefore we might explore these three approaches empirically, as follows:

(a) by comparing each subject's positioning for two items where the 'mathematical content' appears to be the same, but where the wording (and format) of the questions is different
(b) by comparing each subject's positioning for two items where the mathematical content and the situation described seem the same, but where other aspects of the context differ – here, the social relations of the survey/test, or the interview setting.

If differences in positioning for a substantial proportion of subjects are observed for each comparison, we can conclude that the subjects' experience of the contexts as indicated by their positioning differs for both pairs of questions. This will confirm the 'situated' view that task (and cognition) depend on the context – understood to include both the language of the task and the social relations of the setting – and in this sense are 'inseparable' from it.

Theme (3) Gender and Social Class Differences in Performance Related to Positioning

In my questionnaire results, the initial gender differences in performance in school mathematics and in practical maths were substantially reduced when controls for qualification in mathematics, age and social class were introduced. Because of the small numbers in the interview sample, it is normally possible to control only for one alternative explanatory variable at a time. Therefore performance differences on selected questions will be examined as follows. I consider whether there are gender differences in the correctness of the answers given and then whether they are related to differences in mathematics qualification (as in the questionnaire analysis). Similar analyses will be done for social class.

Following the discussion of positioning within practices in the last two chapters, I then explore whether any observed gender differences in performance are related to ways that men and women might be positioned differently in the relevant interview episode.

This analysis will allow us to confirm (or not) whether, *within* the groups of women, and of men, there are differences in positioning for a particular problem-solving episode. This will be repeated for social class group.

Theme (4) Numerate Thinking as Specific to the Subject's Positioning

The context-specificity of numerate thinking will be examined further by considering differences in thinking such as the strategies used to solve a problem, and methods used to evaluate answers. Such differences will be related to subjects' positioning in practice.

Theme (5) Emotion Pervades Mathematical Thinking

This pervasiveness of emotion can be observed in the way that interviewees speak of their current and previous experiences of mathematical activity. We can look first for *expressions* of feelings, in particular those of anxiety or fear (see Chapter 4), confidence, pleasure and anger (Frankenstein 1989: Ch. 2). However, these first impressions need to be interpreted carefully.

Theme (6) Gender Differences in Expressing Anxiety

Earlier research has often produced gender differences in anxiety expressed in self-report questionnaires; the survey results for this study produced confirmation of a gender difference for maths test and course anxiety (related to school mathematics), but only borderline support for a numerical anxiety gender difference (related to everyday mathematics) (see Chapter 4). However, for these interviews, it will not be easy to try to replicate these findings, by categorising each interview problem as school mathematics or practical, since my approach aims to take account of the fact that particular subjects may have different perceptions of the same problem (see Chapter 6).

Recent work on 'gender identity' has been vast, and includes Walkerdine's work (for example Walkerdine 1990b) and Hollway's (1984, 1989) discussed earlier, and also studies aiming to develop insights into the construction of a 'masculine identity' (e.g. Metcalfe and Humphries 1985, Connell 1987, Segal 1990). I consider here the points that are relevant to mathematics anxiety. First, these studies confirm the view that men are often less willing – or less able – to express their feelings openly than women (e.g. Tolson 1977). Second and more generally, given that a man, or a woman, can be positioned in other ways, for example in social class terms, they may be subject to contradictory positionings; this itself may tend to generate anxiety, because of simultaneous positionings of power and powerlessness (Henriques et al. 1984: 225), or intrapsychic conflict more generally (Hunt 1989); see later.

I explore the idea that men have more difficulty with expressing their anxiety than women in the interview analysis. However, as argued earlier, anxiety is not necessarily fully conscious, and hence we may have to attend to 'symptoms', or indicators of the functioning of defences.

Therefore, I shall explore the following hypotheses:

1   Fewer men tend to *express* anxiety than women.
2   Rather than expressing anxiety, men tend instead to *exhibit* anxiety (or defences against it), for example, by being impatient to know whether he got the answer 'right' (to the problem set).

This raises the question of indicators, which will be discussed later in this chapter.

Theme (7) Anxiety as Specific to the Subject's Positioning

Just as utilitarian researchers have chosen to view many numerate tasks as 'in essence' mathematical, so too researchers interested in affect have considered that the items they have labelled as 'mathematical affect' concern essentially this area. However, if the context of a particular action is seen not as 'mathematics' (or not as *only* mathematics), then any anxiety expressed is not (necessarily, or only) 'mathematics anxiety'. Therefore we need to consider cases of what may appear to be mathematics anxiety in terms of subjects' positioning in practice(s).

Theme (8) Emotion that is not Expressed may be Exhibited, Because of Defences

Further, we have seen that, anxiety may on occasion be expressed as 'no feeling at all', or else expressed as its opposite, that is over-confidence, because of the operation of psychological defences. The previous chapter emphasised that anxiety is not necessarily fully conscious, and hence it may be necessary to attend to indicators of the functioning of defences.

An illustration of the problem is given by the growing realisation expressed by an interviewee in Hollway's study, that:

> The very signs I took to signify confidence were, for him . . . actually the signs of his lack of confidence, like – talking too much . . . being opinionated and things that I couldn't bear. And when I read it back as lack of confidence, I could see.
>
> (Hollway 1984: 248)

In this research, I adopted Carl Rogers' (1971) distinction between *expressing* a feeling, and *exhibiting* it without expressing it. That is, in a specific situation, a subject may not express a feeling, or indeed may not even be aware of it. I developed the distinction to explain why using the simple expression of anxiety in the process of problem solving might not be a fully valid indicator for measuring anxiety. (Note that expressing anxiety in such a 'live' setting is different from producing self-report responses estimating one's feelings generally in a series of situations described briefly as 'mathematical' in a questionnaire: see Chapter 5).

The idea of exhibiting anxiety, or other feelings, goes some way towards taking on board Freudian ideas about defences distorting or occluding the expression, and indeed the conscious awareness, of feelings such as anxiety (see the discussion of indicators later). It thereby keeps open the possibility of empirical study of the affective area. Thus it is important to examine episodes where the subject appears to be expressing other feelings (or no feelings) – whereas s/he may nevertheless be *exhibiting anxiety*.

Each interview may provide other opportunities to apply psychoanalytic insights, concerning, for example, the *displacement* of emotion onto mathematics; *transference* reactions to teachers (or to researchers); and *fantasy*.

Theme (9) The Relationship Between Thinking and Emotion as Particular to the Subject, and Specific to the Positioning

In the statistical modelling of the survey results, the relationship between anxiety (measured by Rounds and Hendel's 'maths test/course anxiety' scale) and performance (measured by the school mathematics items) was reported as approximating the shape of an inverted U, and was considered as a general relationship. However, in Chapter 5, I noted that both the generality of the proposed relationship, and the direction of influence might be problematical, the former because of the model's limited fit with the data. Hence there was a need to consider the relationship between cognition and emotion for particular subjects and for particular episodes of action. In these analyses the examination of 'slips' or surprising errors will be useful.

We also need to understand the positioning of the subject. The discussion in the previous chapter cautions against simple conjectures about the relationship between cognition and affect: familiarity with an object or site does not *guarantee* a positive affective charge, nor does positive affect *necessarily* lead to better thinking or performance. Instead, I expect that first, a practice with a positive affective charge is more likely to be called up more often, and hence to become familiar, other things being equal; and second, that familiarity with a practice should lead to clearer thinking and more effective problem-solving. Thus I shall examine familiarity as an important concept straddling the cognitive and the affective.

Theme (10) The Possibilities of Transfer, as Dependent on Similarities and Differences in Signification, and the Role of Affect

Though the interviews were not set up as transfer experiments in any sense, we can assess whether particular subjects were able to exhibit the careful attention to the relating of signifiers that would support the kind of meaningful interrelations between practices on which transfer/translation of learning and thinking is based (see Chapter 6). Other subjects may illustrate how mathematics may be connected with other *apparently unrelated* discourses, by the unexpected flow of affective charge along particular chains of meaning.

Because the argument in the last two chapters has emphasised both structural similarities (that is, gender and social class) of categories of subjects and their positioning, and also the particularity of thinking and emotional processes for each person, the methods of analysis for these themes will be based both on 'cross-subject' analysis and on 'within-subject' case studies. At a the end of the chapter, the methods to be used for analysing each theme will be summarised.

Focus and Methodology of the Interview

Many of the themes developed in the previous section suggest an overall aim for the interviews of describing mathematical thinking and emotion in context. This can be analysed into the following objectives:

1   describing the context(s) in which the interview takes place, in terms of the setting and particularly in terms of the subject's positioning in practices during particular episodes
2   describing numerate thinking and problem-solving performance, both as produced in the interview, and as recalled from other contexts by the subjects themselves
3   describing emotion and especially anxiety, both as produced in the interview and as reported by the subject in connection with earlier experiences
4   describing the relationship of thinking and emotion.

These emphases suggested a 'qualitative' or ethnographic approach to the interviews (Hammersley and Atkinson 1983).

My conception of the positioning of a subject in any context includes crucially the subject's particular perceptions of the task(s) or problem(s) facing them. Thus objective (1), including in particular the description of these positionings, is well served by ethnographic approaches, whose central methodological concerns include:

- *understanding*: of the goals and purposes of subjects, and of the meanings they perceive in events, institutional arrangements, and so on, hence the need to describe the perceptions of (various groups of) the subjects themselves
- *holism*: the actions of individuals are motivated by events within the larger (subcultural or organisational) whole, hence the need to describe social activity within its context
- *reflexivity*: social researchers are part of the social world studied, hence the need to document the social interaction or 'relational dynamics' involved in the production of the data (Atkinson 1979, Hammersley and Atkinson 1983).

I considered several types of interview. Brigid Sewell's (1981) interviews for the Cockcroft Committee asked respondents to attempt some everyday numerical problems and noted not only their answers, but also their emotional reactions. This approach has some affinities with Piagetian 'clinical interviews' generally (Ginsburg et al. 1983; see also Ginsburg and Asmussen 1988), and seemed an effective way to pursue objectives (2) and (3). Similarly Laurie Buxton chose an interview strategy where he would present his adult subjects with mathematical problems so as to 'induce some of the feelings, though hopefully not at too stressful a level, that I was concerned to study' (Buxton 1981: 131). This sort of strategy seemed useful for studying objective (4).

In order to elicit information about the individual's life experiences and ongoing motivations towards mathematics, it also was appropriate to draw on elements of a 'life history' approach (Hammersley 1979), despite the fact that I was unlikely to have more than one semi-structured interview with each subject.

Thus I constructed an interview strategy based on a combination of the life history approach and a problem-solving approach, with the crucial addition of contexting questions (see later) meant to help synthesise the two approaches, and to produce information which could be used specifically to provide information about the subject's positioning during each problem-solving episode in the interview.[1]

Psychoanalytic Insights in the Interview Process

Hunt (1989) has contrasted the psychoanalytic perspective in social research with classical ethnographic views in terms of the distinctive assumptions of the former:

1   Much thought and activity takes place outside of conscious awareness; thus, everyday life is mediated by unconscious images, fantasies, and thoughts, which sometimes appear as jokes, slips, dreams, or subtly disguised as rational instrumental action.
2   The unconscious meanings which mediate everyday life are linked to complex webs of significance (or signification) which are ultimately traceable to childhood experiences.
3   Intrapsychic conflict among id, ego, superego – or among desires, reason, ideals, norms – is routinely mobilised *vis-à-vis* external events, especially if they arouse anxiety or link to unresolved issues from childhood.

A very general implication of (1) and (3) is that any product of mental activity – including interview talk – may, upon deeper investigation, reveal hidden aggression, forbidden desire, and defences against these wishes. I would add suppressed anxiety to this list of possible revelations.

An important implication of (2) is that *transference*, the imposition of 'archaic', childhood images on to everyday objects (people and situations), is a routine feature of most relationships.[2] Thus, unconscious aspects of communication may affect empathy and rapport in the fieldwork interview; for example, transference may facilitate, or blind, the researcher's understanding of some dimensions of the subject's world. Transference by subjects of feelings onto the researcher is especially likely where a close, long-term relationship between researcher and subject develops (Hunt 1989).

As possible indications of transference, she recommends attention to:

1   the expression by the subject of strong emotions, such as anger, love, anxiety, shame, annoyance, boredom
2   reactions that seem 'inappropriate or peculiar in social context', suggesting defences against transference-generated anxiety (Hunt 1989: 62)

3   'unusual' data, for example, about fantasies, dreams, jokes and slips
4   reports of dreams by the subject, where the researcher appears clearly, or thinly disguised, and descriptions of other persons or experiences which resemble some aspect of the researcher, or the research setting
5   introspection of the researcher's own emotional responses to subjects, and scrutiny of the possible significance of his/her own fantasies or dreams.

Therefore, to allow for the possibilities of transference, *reflexive accounts* need another, deeper dimension, especially where a close relationship between researcher and subject has developed. The researcher also needs to be prepared to conduct unstructured, 'free associative' interviews, following stressful events or, instead of them, if the planned interview schedule itself becomes too stressful.

Examples of the use of these recommendations in the conduct of my interviews will be given in the discussion of the interview results.

Areas for Exploration and Developing Indicators

My dual-purpose method for the interview used two main types of question: life history questions, and problems to be solved, plus contexting questions.

Life history questions focused on the following areas:

- gender, age, social class position[3] and degree specialisation (including any changes since enrolment)
- qualifications passed (or attempted) in mathematics
- recent academic studies and the use of mathematics in them
- recent paid work and the use of numbers involved
- other salient aspects of the respondents' use of and experience with mathematics and with numbers in education, work, and everyday life.

Even the apparently 'factual' life history questions gave the respondent scope to tell stories about experiences with, and feelings about, mathematics and numbers in a range of contexts.

The questions related to *problem-solving* included:

- how the subject was thinking about each problem
- his/her 'answer'.

Nine questions were used in the interviews, six adapted from those of Sewell (1981), and three constructed by myself. On the basis both of my experience in teaching entrants to the social science degree, and on views within mathematics education (for example Sewell 1981), I considered *percentages* to be the most fruitful mathematics topic to focus on in the interview. Thus four questions on percentages were offered:

- an 'abstract' calculation of 10 per cent (Question 2)
- calculation of a 10 per cent tip (Question 4)

- estimation, then exact calculation, of a 9 per cent rise for a wage slip (Question 5)
- calculation of the effect of a 10 per cent cut in course tutorial times (Question 8).

There were two further questions based on *graphical information:*

- reading a pie chart (Question 1)
- reading a level and comparing rates of change on a line-graph (Question 3).

One question can be seen as being about *ratio*:

- choosing a 'best buy' from two bottles of ketchup (Question 6).

Much of the interest here centres on comparing my results with those of Lave (1988).

Finally, two questions involved *fractions or decimals:*

- calculating the cost of baking a cake from a recipe (Question 7)
- calculating one person's share of the costs of the uniforms for a team (Question 9).

I first aimed to use only problems that were 'balanced' in their interest and appeal to men and women. However, Question 7 on the cost of baking a cake was considered possibly biased towards traditional 'feminine' interests. Therefore I constructed Question 9 on the cost of sports kit to re-establish a rough 'balance' in gender terms.

The problems to be solved were arranged in a specific order. Question 1 on the pie chart was considered an 'ice-breaker' (because it seemed easy). Question 2 was the only problem which I considered 'abstract' in the design of the interview. Question 3 and subsequent problems were considered practical. (However, in the actual interviews, Question 3 seemed to call up school mathematics for many respondents; see the next chapter.)

Finally, for every problem, two *contexting questions* were posed. When the problem was first presented, I asked:

Does this remind you of any of your current activities?

After the subject had thought about the problem and given their answer, I asked:

What sorts of earlier experiences with numbers it reminds you of, or feelings it brings up?

The use of contexting questions was an important innovation in my interviews, based on the earlier theoretical discussions about positioning. The contexting questions were aimed to provide information on what practices the subject was 'bringing to mind' in response to each problem posed, as the interview unrolled. Their responses provided the data supporting my judgements about which practices were positioning the subject. I aimed to create a situation, where the respondent was not 'led', either to call up school mathematics (or another academic subject) or to call up an everyday practice. In fact the first contexting question was designed to allow the subject to indicate the practice(s) called up, by the 'first impression' of the diagram or facsimile introducing most of the problems, before any numerical or 'mathematical' question was posed.

For the problems used in the interviews, see Appendix 2 and, for further details, Evans (1993).

I now turn to the indicators used to describe thinking and problem-solving 'performance'. These varied from question to question, but I generally considered some or all of the following:

- the subject's manner of describing her/his own thinking, including any expression of emotion
- the broad strategies used, for example unit price or other approach to comparisons in Question 6
- the *methods* used, for example written versus 'oral' procedures, ways of approximation in Question 5
- the correctness or cogency of the response, within the context apparently called up.

As for emotion, I considered the indicators for *expressing* anxiety, say, to be relatively straightforward. They would include statements by the subject of the form:

'I feel anxious, scared, unsettled . . . ' (at this moment or generally these days)

or

'I felt anxious, scared, unsettled . . . ' (at that time, or generally in that period).

The indicators for *exhibiting* anxiety were less straightforward. However, following Legault (1987) and Hunt (see earlier), I looked for indications of defences against anxiety, or other intrapsychic conflicts. Several types of indicators were considered promising:

1   'Freudian slips' (parapraxes) or jokes made by the subject: for example a slight (but possibly 'motivated') mispronouncing of words or forgetting of a name (Freud [1901] 1975); a 'surprising' error or memory failure on one of the interview problems, given the student's previous performance or experience; striking symbolic gestures
2   denial of anxiety: for example 'protesting too much' or making an assertive 'statement' that the subject feels exceedingly confident about mathematics, or alternatively, feels nothing about mathematics
3   relating of dreams and fantasies
4   behaving 'strangely' or unusually: for example laughing a lot, especially 'nervously', talking unusually quietly, or unusually loudly
5   impatience to know the 'right answers' for one or more interview problems.

(The last two may be indicators of anxiety itself, rather than of defences against it.)

Illustrations of these will be given in the accounts of the case study interviews (see Chapter 10). Of course the set of all conceivable indicators is not specifiable in advance.

Doing the Interviews

The interviews were done at a the end of the first year with students from Cohorts 2 and 3 of the study.[4] For practical reasons, they were all conducted at a the site of the institution where I had my office. That meant I interviewed students from only one of the two courses studied in the survey, the BA Social Science.

Sampling Methods and Recruitment

I aimed for the interview to be long enough to produce sufficient life history material and to allow most subjects to attempt six or seven problems, but not so long as to deter participation. Therefore half an hour seemed about right at a the outset, although, in the event, some students agreed to stay for three quarters of an hour.

Cohort 2 interviewees were chosen by a combination of random and volunteer methods; this produced nine subjects. In Cohort 3, interviewees were chosen by a process of stratified random sampling, in an attempt to enhance the representativeness of the results. The set of completed questionnaires from the previous autumn was used as a sampling frame, stratified according to the three social structural variables found to be important in the statistical modelling of the survey results: gender, age and parental social class (see Chapters 3 and 4). Eventually sixteen students in Cohort 3 were interviewed. Thus a total of twenty-five students were interviewed in both cohorts. For further details of the sample breakdown, see Evans (1993: App. I4).[5]

The Conduct of the Interview

At a the beginning of the interview, I attempted to produce a relaxed atmosphere, by offering coffee or tea. I asked the student's agreement to record the interview: all but one (number 25) agreed. I began by describing my work as

> doing research on people's experience with numbers, and on what sorts of things help people feel comfortable with numbers, and what stands in their way. . . . So what I would like to do in this interview is to give you some space to talk about your experience with numbers, and your feelings about them.
>
> (Evans 1993: App. I1)

After this information, I emphasised to the student that he/she did not have to answer any question if they did not want to. This was part of an attempt to position both of us in a research, rather than a 'college maths', discourse. It was also based on my commitment to treat interviewees according to the principles of 'informed consent' (International Statistical Institute 1985). I began with the 'life history' questions, and then moved on to the problems to be solved, each preceded by the first contexting question, and followed by the second (see earlier).

The student was given at most only neutral feedback while attempting the problems. Since I was careful about time-keeping, I stopped at the problem being discussed about ten minutes before the agreed time. This meant that a few students attempted all nine problems, though the average number tried was about six. Some did as few as four.

Towards the end of the interview, I reverted in most cases (in some, explicitly) to my position as teacher in college mathematics. Several times, the student asked to discuss 'the answers' to the problems; this was always done. Further, in one or two cases where the student's thinking had indicated basic misconceptions, I myself offered a tutorial to help clear them up. For further details on the 'script' or the conduct of the interview, see Evans (1993).

The General Reflexive Account

Hunt's (1989) work on psychoanalytic issues in interviewing recommends assessing the possibilities of transference in several ways, including producing 'reflexive accounts' concerning the researcher's relations with key subjects in the setting. Here I produced two sorts of reflexive account: a general one, given next for the interview phase as a whole, and a particular one for each student (see Chapter 10 for examples of the latter).

At the beginning of the study, I was already an experienced lecturer in statistics at the polytechnic, though because most of my teaching was with BA Social Science students, I was strongly identified with that area. In the years when the interviews were conducted, I was very much involved with the first year 'Maths' course, not only in giving some of the lectures, but also as the coordinator. In these lectures the team of staff attempted to present mathematics as a 'social and historical product' (that is, consistent with what was done in the other strands of the 'Methods and Models' course, namely philosophy of science and computing), and in a way that was exciting and reassuring to those who had had unpleasant earlier experiences with mathematics.

Thus, by June when the interviews were conducted, almost all of the cohort of some 150 students would know me by name, and would have seen me in lectures at a least. Further, about a third would already have had me as a tutor (for Maths or Social Policy, or as personal tutor). In both of these years I would have been known to most students, and would myself have known up to half of them by name. Because of the quality of the Methods and Models Maths course, my 'mathematician' colleagues and I were on good terms with most of the students, considering that most of them would have arrived with unpleasant associations with mathematics! I personally had good relations with most of the students in those two years, especially those in my Maths seminar groups. Thus it was possible to expect that many, if not all, of the students would accept an invitation from me to an interview as described, and that in general they would basically trust my interest in their experiences with mathematics and numbers.

Nevertheless, there was some variation in the extent to which I knew each student who came for an interview.[6] For further discussion of this issue, see Chapter 10.

Overview of the Analysis of the Interviews

The interview data included:

- the subject's stories about their experiences with mathematics at a school or numbers at a work
- their descriptions of current uses of mathematics and numbers
- my observations of their thinking, performance and emotional expression in context
- my description of the context.

For each completed interview, an initial account was prepared, and then linked with the questionnaire for that subject. Then, both 'cross-subject' and case study analyses were done.

The *cross-subject* analyses aimed to produce summaries, frequencies and relationships for indicators on which the sample of twenty-five subjects could be compared. These analyses were inspired by Miles and Huberman's (1994) approach, which is more systematic than most for analysing qualitative data.[7] The indicators included the subject's structural positions, experiences, positioning in 'school mathematics' or everyday practices, the 'correctness' of answers, the expressing or 'exhibiting' of anxiety, and others discussed in 'Focus and Methodology of the Interview', this chapter.

The single-subject analyses produced detailed *case studies* and shorter illustrations, based on the life histories, problem-solving episodes and accounts given by selected interviewees. Here the webs of meaning linking practices, positioning, thinking, 'performance', anxiety, other emotions, and earlier experiences, could be explored for a particular subject.

Although all twenty-five interviews were used in the cross-subject analyses, here I can present extended case studies only for a subsample. The choice was made on the basis of which interviews appeared to be most fruitful in terms of the themes given priority in this study, and by a desire to balance the subsample on gender, age, and social class.

The actual selection began with three cases which seemed especially fruitful:

1 Donald (assumed name, as are all others): male; age 40+; working class (WC) parents, middle class (MC) himself (having worked in the London money markets). The initial analysis of the interview shows, in particular, a sensitivity to differences in language used between practices, and a range of feelings expressed.
2 Ellen: female; age 18–20 at entry; MC parents. The interview shows much expression of confidence, and a 'surprising slip'.
3 Fiona: female; age 25+; MC parents, MC herself (several jobs). The interview shows 'mock-anxiety', turning into a range of strong feelings about her father during discussion of Question 3.

Eventually, I decided I needed to include accounts of interviews with two further women and two more men:

4   Harriet: female; age 25+; WC parents, MC herself (as an unqualified social worker). The interview shows her diffidence with mathematics turning into confidence through her work, and pleasure in using formulae.

5   Alan: male; age 20; MC parents. The interview displays his claims of 'no

Table 8.1 Summary of Themes for the Qualitative Phase of the Study, and Strategies Used for the Analysis of Each

Theme	Cross-subject analysis	Case-study analysis	Results
1 Contexts of thinking ascribed on the basis of positioning in practices	√		Chapter 9 (section on Theme 1)
2 The inseparability of task and context	√		Chapter 9 (section on Theme 2)
3 Gender and social class differences in performance related to positioning	√	√	Chapter 9 (section on Theme 3) / Chapter 10
4 Numerate thinking as specific to the subject's positioning	√	√	Chapter 9 (section on Theme 4) / Chapter 10
5 Emotion pervades mathematical thinking	√	√	Chapter 9 (section on Theme 5) / Chapter 10
6 Gender differences in expressing anxiety	√	√	Chapter 9 (section on Theme 6) / Chapter 10
7 Anxiety as specific to the context of the subject's positioning		√ Ellen, Jean	Chapter 10
8 Emotion that is not expressed may be exhibited, because of defences		√ Alan	Chapter 10
9 The relationship between thinking and emotion as particular to the subject, and specific to the positioning		√ (all of subsample)	Chapter 10
10 The possibilities of transfer, as dependent on the similarities and differences in signification, and the role of affect		√ Donald	Chapter 10

feelings about maths', and the importance of positioning *vis-à-vis* money.

6   Peter: male; age 20; MC parents. The interview shows his diffidence about his understanding of school mathematics, related to his father's (and older brothers') attempting to 'help' with homework.

7   Jean: female; age 18–20; WC parents (see note 3). The interview shows the multiple bases of 'mathematics anxiety', and the importance of positioning *vis-à-vis* money.

Summary

The 'qualitative' phase of the study uses interviews designed to focus on a set of themes, or 'foreshadowed problems'. My methodology of semi-structured interviewing draws on the ethnographic tradition (for example Hammersley and Atkinson 1983), and on insights from the psychoanalytic critique of traditional fieldwork (Hunt 1989) and from poststructuralism.

My interviews have several distinctive features. In broad terms, I attempted to combine the qualities of life history and problem-solving interviews. In particular, through the use of 'contexting questions' related to each problem, I aimed to study the student's thinking and emotion in the interview context, as well as in earlier experiences with mathematics and numbers, in relation to his/her positioning in specific discursive practices. In addition I deployed two strategies of qualitative data analysis.

Twenty-five interviews were conducted at a the end of the first year for Cohorts 2 and 3. The sample was based on a combination of probability and volunteer methods. The interviews, which took place in my office, appear to have been largely successful in establishing a relaxed and productive atmosphere. In order to try to assess possible influences on the results, from the interview arrangements, as well as from my established relationships with the students, I produced a general reflexive account, as well as particular reflexive accounts for each interview.

For problem-solving performance, I considered not only correctness of answers, but also methods and strategies of thinking. For emotion, in particular for anxiety, I considered not only its expression, but also situations where it might be 'exhibited', though not expressed.

Two strategies for analysing the interviews were used. The cross-subject analyses aimed to produce summaries of results considered comparable across the sample of subjects. The single-subject analyses aimed to produce detailed case studies (and shorter illustrations), based on the problem-solving episodes, life histories and accounts given by selected interviewees emphasising the webs of meaning linking practices, thinking and affect. In Table 8.1, I summarise the strategies to be used for the analysis of each of the themes or foreshadowed problems set out.

In Chapter 9, I present the results of the cross-subject analyses. The case study analyses are found in Chapter 10. Illustrations are produced from all twenty-five interviews as relevant.

9 Reconsidering Mathematical Thinking and Emotion in Practice

The approach . . .] must refer to the specificities of the different practices in order to describe the different subject positions and the different power relations played out in them. It cannot simply speak of a subject's behaviour and attitudes or ascribe in advance the subject's position according to class or gender.

(Henriques et al. 1984: 117)

This chapter considers cross-subject analyses relevant to themes formulated in the previous chapter for the qualitative phase of the study. These themes include:

1 developing my way of understanding the context of activity as positioning in discursive practices, drawing on discussions in Chapters 6 and 7
2 assessing the idea of the situatedness of thinking, that is, the inseparability of task and context
3a reconsidering gender differences in performance, understood not as resulting from differences in biology or 'essence' between males and females, but from differences in positioning within practices
3b reconsidering social class differences using a similar analysis
4 reconsidering the idea that numerate cognition is specific to the context – in the light of understanding the context as in (1)
5 exploring the idea that mathematical thinking is 'hot', that is, infused with emotion
6 reconsidering the claim that there are gender differences in experiencing anxiety.

Theme 1: Contexts of Thinking Ascribed in Terms of Positioning in Practices

This theme generates subsidiary questions. First, given the importance of the specific discourses available to position subjects in any particular setting, what are the *available discursive practices* and subject-positions in my interview situation? Then, given that different subjects in the same situation may call up different practices, what *indicators* are there *for the subject's positioning*?

We can analyse the interview situation, to ascertain the available practices. First, each subject was positioned as a student on the BA Social Science course at the Polytechnic. The 'general reflexive account' (see Chapter 8) indicates that each student knew me in contexts which positioned me as a teacher, and as an authority, especially in mathematics, and in the institution.

At the same time, a particular student attended the interview itself because s/he had been 'chosen' (in most cases) in the random sampling exercises described earlier. The letter of invitation, and my interview script, were designed to talk about 'research', 'interview' and 'numbers ', rather than 'mathematics' or 'test', and to position the student as interviewee; see Chapter 8, and Evans (1993) for further details.

Thus, I considered that two discourses provided the overt possibilities for the subject to be positioned in the interview setting:

- *college mathematics (CM)*, or *school mathematics (SM)* with subject-positions teacher and student
- *research interviewing (RI)*, with subject-positions researcher and interviewee.[1]

Throughout this chapter and the next, the terms 'college mathematics' (CM) and 'school mathematics' (SM) are used interchangeably (and sometimes, to emphasise the point, thus: 'college mathematics (SM)'). These terms denote the cluster of practices of formal teaching and learning in mathematics that provide the basis for students' positionings. The mathematical content studied in school and college was recognised as overlapping by most students, and there are clearly emotional reverberations for many between the two activities. The main point is to distinguish these discourses from other kinds of everyday and work discourses.

Now, the subject's positioning at any moment in the setting depends on the discourse(s) s/he calls up – which may be from among the available discourses and positions 'at play' in the setting, or may be another, 'brought forth' by this particular subject, acting in this setting. Whatever practice is called up, the subject will have access to ideas, methods of reasoning, 'skills', and emotions from that practice.

Here I considered the RI discourse would allow space for the subject to call up out-of-college, everyday practices. Therefore, through the way I defined the interview, in the invitation and in the script, I attempted to shift the discourse and the positioning from those relating to college mathematics, to those of a research interview. I hoped the interview would thus tend to create space for the subject to call up one or more practices from his/her 'everyday life', and that the *ecological validity* of the interview findings would be enhanced.[2]

I specified some indicators for the subject's positioning, elaborating on Walkerdine (1988: 53ff):

1 The explicit 'discursive features' of the tasks: for example how the interview itself was described (in the 'invitation', see Chapter 8); how each task or problem was introduced (in the interview script), the terms and constructions used.
2 The unscripted aspects of interaction between researcher and subjects: including the (possibly unconscious) emission of different verbal or vocal signs for 'correct' and 'incorrect' answers,

3   The student's talk, both during the problem-solving phase of the interview, and afterwards: for example the language used in discussing the problems; 'confessions' produced at the end, as to the subject's expectations (or fantasies) about the interview's requirements; see the quotations here and in Chapter 10, and also transcripts for selected interviews in Evans (1993).

4   'Reflexive' accounts of the 'history' of my ongoing relationships with subjects: the ways I was in the position of 'maths teacher', or 'researcher', or otherwise related, to a cohort of students in general (see Chapter 8), or to particular subjects (see Chapter 10),

5   'Messages' given off by the setting for the interview: reading of the meanings of using my office at college, the arrangement of furniture, the use of a tape-recorder, and so on.

The researcher normally has some control over features (1), (2) and (5). For example, I aimed to minimise the potential problems from (2) unscripted talk, on my part, by having standard prompts and probes; for example, when the subject gave an 'answer', my 'neutral' response – 'fine, thank you' – aimed to avoid the interview's drifting into discussing 'answers', and hence into college mathematics discourse.

It was usually possible, but not always straightforward, to make a judgement as to the subject's positioning at a particular point in the interview. Sometimes the subject was considered to call up more than one practice – that is to have a *multiple positioning*, in some 'mix' of college mathematics, research interviewing and perhaps some other 'everyday' practice. Examples will be given here and in the next chapter. In addition, such multiple positionings often appeared 'contradictory'. This is because the contexting questions, a distinctive feature of these interviews, often brought up much life history material, and would tend to position the subject in the 'research interview' (RI) discourse. However, the problem itself also included a mathematical aspect, a 'pseudo-question', so-called because I 'knew the answer' (or at least, I thought I did, at the start of the interviewing!). This would tend to position the subject in the college mathematics (CM) discourse. This dilemma will need to be assessed in relation to particular problems for particular subjects.

In general, we would expect a particular subject, confronting a particular problem in the interview, to have a multiple or interdiscursive positioning – rather than a positioning in a single practice. However, for most of the cross-subject analyses, I aimed to record the practice which *predominated* in each subject's positioning, so as to present the key relationships clearly.

I illustrate the ideas of positioning in practices, by showing how I made the relevant judgements for the first problem in the interview. This question presented a 'pie chart showing water consumption' and I asked the following three questions, with time for discussion between them:

(Contexting question C)   Does this remind you of any of your current activities?
(Question 1)   Looking at this 'pie' chart, which do you think uses more water; households or industry with meters?
(Contexting question R)   Now, could you tell me about any sorts of earlier experiences it reminds you of, or feelings it brings up?

(See Appendix 2 for the interview problems in full.) This problem was intended as an 'ice-breaker', and indeed it seemed easy for most subjects. However, for a researcher, judging the subjects' positioning was more difficult than for other problems, since the subject was likely to decide on the answer without producing much talk that could be used as an indicator (see above).

My judgements as to the subjects' predominant positionings, that is the main practices called up, are summarised in Table 9.1(a).

For Question 1, college mathematics appeared to be called up in most cases (at least 13 out of the 22 where I could make a judgement), with five more subjects considered to have called up the numerate aspects of work or consumer practices. In some cases this seemed straightforward: for example, Jean's (interviewee number 3) response to one of the contexting questions suggests she has called up school mathematics and school practices generally.[3]

S: Well, it reminds us of when I first went into the comprehensive school, and you were given certain tasks to see which Set they wanted to allocate you to – and concerning maths, there was a lot of these kind of charts.

(Jean's transcript)

However simply mentioning a practice does not necessarily mean that the subject has *called up* that practice as the basis for addressing the problem. For example, number 21 (previously a manager) mentioned both statistical analysis in psychology (her chosen specialism), and data presentation and training exercises in management. However, the way she dwelled on the *meaning* of the division of the pic led me to conclude that she was addressing the problem from within business/financial discourses, rather than college mathematics or statistics. Therefore her positioning was classed as work practices.

The two cases classed as multiple positioning were more complex. Harriet (number 16), an intending social worker, mentioned three activities:

1   an essay for the Methods and Models course, which included mathematics, since the pie chart reminded her of a 'chart' of types of poverty, in a relevant article
2   reading dials on electricity meters, each marked from 0 to 9

Table 9.1 (a) Positionings for Interview Question 1 (Pie-Chart): Predominant Practice Called up by Gender

Predominant positioning	Men	Women
College maths, school maths	6	7
Work maths, 'consumer maths'	3	2
Multiple positioning	—	2
'Common sense'	1	1
Not possible to categorise	1	1
Problem not attempted	1	—
Total	12	13

3   (in response to the contexting question about reminders of 'your earlier experi-
     ences') being 'back at school again, using a similar chart like that . . . for a
     similar question' (Harriet's transcript: 6).

Here, her mentioning school mathematics, especially only *after* giving her
answer to the problem, does not mean that she had called it up. Similarly, it seems
unlikely that the ideas, relations and variations associated with reading an electricity meter – for example, the equal distances between marks on each dial would
form helpful interrelations with (that is 'transfer' to) reading a pie chart representing *unequal* shares, as in Question 1. But, despite its being unhelpful for
solving the problem, I judge that she called up reading electricity meters, because
of an *emotional charge*:

JE: What comes up when you think of the electricity meter?
S: The bill! [laughs].

(Harriet's transcript: 6)

I categorise Harriet's positioning as multiple – both in college mathematics, and in
what we might call 'consumer maths', a set of practices that includes the reading of
quantitative information in everyday documents, such as electricity bills. This
conclusion is provisional, given limited information.

   Another example of multiple positioning was interviewee number 6. The first
thing she called up was exams in CSE Mathematics, which was 'hard', because
'you've got to use a protractor'. Then,

JE: does anything else come up around that question?
S: . . . no, apart from – this may sound silly, but – a cake divided into bits . . . .
JE: [. . .] is that something you do very often?
S: No.
JE: Uh huh [three lines] . . . was that a big thing about dividing the cake up
      equally?
S: Yes, yeah. Still is . . . . [laughs]
JE: [. . .] and who usually does the dividing up?
S: Oh, that depends – 'cause my brother doesn't like me cutting it, 'cause he
      reckons I don't do them equal . . . used to get Mum to do it . . . .

(Number 6's transcript: 3–4)

   Thus she also called up home practices, to do with sharing out food treats, on rare
occasions. This recalls Nick Taylor's use of test and interview questions based on a
bread (or cake) 'metaphor'. Here, the cake metaphor was not part of the pie-chart
problem, but was called up by the subject herself. Its emotional charge was at best
mixed, because of friction with her brother about whether she was 'doing them
equal'. Nevertheless, she got the question right. However, we might wonder whether
the 'difficulty' and unpleasantness around sharing 'equally' at home might have
been displaced onto the need to use a protractor to cut out 'equal' parts of pie charts
at school.

Two subjects called up positionings which prevented their answering correctly. Number 18 called up school geography, and Fiona (number 5) called up what was purportedly 'everyday knowledge'. These discourses seemed to them to allow the importing of 'outside information', rather than *reading* the pie chart! Thus each appears to 'refuse the terms' of the question. For Fiona, however, bringing in ideas about defences from psychoanalysis allows a deeper interpretation of her case study (see next chapter). This underlines the provisional character of many of the judgements made about positioning, and of the interview analysis overall.

Overall, twenty-one of the twenty-four subjects produced the correct answer.[4] The response of Peter (number 19) illustrates how answers may occasionally be difficult to classify as 'correct' or 'wrong'. He switches his response from 'households' (correct) to 'industry, because of the metered and unmetered' (not answering the question), and back to 'households' when I reread the question to him. I decided to code this as 'correct'.

Thus, a seemingly straightforward problem elicited a range of positionings, and a variety of responses. Further, some practices seem to bring up an emotional charge, such as that of anxiety related to household bills, or of unpleasantness associated with sibling rivalry.

Problem 3 illustrates some different positionings. It referred to a graph showing how the price of gold varied in one day's trading in London, and asked:

(Contexting question C)   Does this remind you of any of your current activities?
(Question 3A)   Which part of the graph shows where the price was rising fastest?
(Question 3B)   What was the lowest price that day?
(Contexting question R)   Could you tell me about any sorts of earlier experiences it reminds you of, or feelings it brings up?

(For the full interview problems, see Appendix 2.)

Here eighteen of the twenty-two subjects were classified as calling up school mathematics or college mathematics, sometimes along with another academic subject. In contrast, the predominant positioning of interviewee number 21, previously a manager, was classed as business maths, based on her response to contexting question (R):

JE: Does that remind you of earlier experiences?
S: Well, [. . .], when I was working, we would use these graphs, but not so much: we just had wall charts to show levels of business [. . .] from the previous financial year and that was about it . . . .

(interview transcript: 7–8)

Therefore, number 21 was classed as calling up 'everyday (or practical) maths' (PM). In reporting the interview results, I shall use everyday maths (PM) as shorthand for work, business or other non-school, practices which have numerate elements. For these problems, it includes consumer mathematics and 'maths for citizenship'.

Two of the others also called up PM: Donald (previously in the money markets) and number 9 (previously a stockbroker). All three had extensive work experience with graphs such as these, as well as recognising them from school or college mathematics. The fourth subject calling up business maths, Fiona, seemed to 'refuse the terms' of the question again (see the discussion of Question 1 earlier), purporting to respond using knowledge of her father's work as a stockbroker. She also appeared to experience much emotion in confronting the question; see her case study in Chapter 10.

Table 9.1(b) summarises my judgements of subjects' predominant positionings for the first six questions. (Later questions received too few responses.) For Question 1, the two respondents classed as 'multiply positioned' in Table 9.1(a) have been reclassified here as having a predominant positioning in CM/SM, since the PM activity that each also called up concerned equal 'shares' or 'markings', whereas the question represented unequal shares.

Table 9.1(b) Predominant Positionings for Interview Questions 1 to 6: Number of Subjects Calling up College Mathematics (SM) and Everyday Maths (PM) Practices

Predominant positioning		College maths, school maths (SM)	Everyday maths, work maths, etc. (PM)
Question 1	(reading pie-chart)	15	7
Question 2	(abstract 10%)	17	6
Question 3	(reading graph)	18	4
Question 4	(10% tip)	5	18
Question 5A	(approx. 9% increase)	2	15
Question 5B	(exact 9% increase)	11	6
Question 6	(best buy comparison)	2	12

We can note two things in Table 9.1(b). First, the majority of subjects have their predominant positioning in college maths/school maths (SM) for the first three problems, and in what I call PM (everyday maths, work maths and so on) for Questions 4, 5A, and 6. The result was surprising for Question 3, which I had classed *a priori* as a 'PM-type'. Second, the majority of students switch from PM to SM between Question 5A and 5B. This suggests flexibility of positioning, and therefore of thinking, among some students at least (see also 'Theme 4', this chapter).

Theme 2: Inseparability of Task and Context

In Chapter 8, I discussed contrasting proficiency, functional, and my own 'moderately situated' views, on the relationship of thinking and context – by examining two carefully chosen comparisons. To study the differences, I selected three problems as follows:

1   an 'abstract' (school mathematics type) 10 per cent calculation, from the interview (Question 2)

2   a 'practical' 10 percent tip, from the interview (Question 4)
3   a 'practical' 10 percent tip, from the survey (Question 18).

These questions were worded as shown in Table 9.2(a).

For comparison (a), my view (and the functional view) will be supported over the proficiency view, if the two problems – where the 'mathematical content' appears the same, but the wording and format of the questions is different – call up different practices, for many subjects, at least. For interview problems 2 and 4, the 'mathematical content' is a 10 per cent calculation for both, but the formats are abstract and 'tipping-related', respectively.

For comparison (b), my situated view will be supported over both the others, if the tipping problems from the survey and the interview settings – where the mathematical content, format and situation described are seemingly the same, but the social relations of the settings differ – call up different practices, in many cases, at least. If differences are observed in both, this will allow us to confirm the importance of considering the task (and cognition) as situated, or specific to the context, and therefore as not separable from it, in any simple way.

For these problems, indicators for a predominant college maths or school maths (SM) positioning were considered to be:

- for all questions, the use of written calculations; and/or
- for Question 2, expressed confusion as to where to put the decimal point; or
- for Questions 4 and 18, the giving of an answer which involved a fraction of 1p.[5]

Indicators for a predominantly practical maths (PM) positioning were:

- for all questions, the use of mental calculation; and/or
- the formulation of an answer in terms of practical (for example money) units,

Table 9.2(a) Interview and Survey Questions involving a Calculation of 10%

Interview Problem 2: [Show the question.]
(C)   Does this remind you of any of your current activities?
      What is 10% of 6.65?
(R)   Does this remind you of any earlier experiences?

Interview Problem 4: [Show the facsimile menu.]
(CA) Do you ever go to a restaurant with a menu anything like this?
(CB) Would you please choose a dish from this menu?
(A)   Suppose the amount of 'service' that you leave is up to the customer: what would you do?
(B)   Could you tell me what a 10% service charge would be?
(R)   Does this remind you of any earlier experiences?

Survey Question 18: Suppose you go to a restaurant and the bill comes to a total of £3.72p. If you wanted to leave a 10% tip, how much would the tip be?

Answer...

Sources: Appendix 1 for survey question; Appendix 2 for interview problems

especially for Question 2, which had been posed in abstract terms.

Now, specifying indicators for positioning like this may not be valid for every subject. Thus, although doing mental calculations would normally indicate a positioning predominantly in practical maths, a subject such as Peter (interviewee number 19), even when (I would judge) his positioning was in SM, does as many calculations as possible *mentally* – for personal reasons (see Chapter 10).

Another difficult subject to classify as to positioning was number 9 (previously a stockbroker) who used a method of decomposition:

$$10\% \text{ of } 6.65 = (10\% \text{ of } 6) + (10\% \text{ of } 0.6) + (10\% \text{ of } 0.05)$$

This is often associated with non-school contexts (Nunes et al. 1993) since it is appropriate for mental calculation, and can be used when a school algorithm is not known or forgotten. However, after completing his calculation, in responding to contexting question (C), on whether 'this reminds you of any current activities', he responded:

> although I'm practising at home constantly, *in this environment*, I'd like to have my calculator and my notes. That's why I divided it up [i.e. decomposed] like that.
>
> (Number 9's transcript: 6; my emphasis)

The key to the practice called up is his perception of what 'this environment' involves: he would like to have his notes – presumably notes from school (or from tutoring) – and hence his predominant positioning was coded as 'SM'.

For the first comparison, to see whether the 'abstract' 10 per cent problem tends to call up the same practice as the 10 per cent tipping problem, see Table 9.2(b). First, note that for Question 2, the abstract 10 per cent question, we see that the majority (seventeen of twenty-three), but not all, of the students, have a positioning predominantly in school mathematics (SM) discourses. For Question 4, the 10 per cent tip, the majority called up out-of-school discourses (PM), notably 'eating out', but there were still at least five subjects whose positioning was 'predominantly' in school/college mathematics. For example, interviewee number 21, a former manager who was unenthusiastic about the menu shown (since she was vegetarian) gave an answer that would have been 'inappropriate' in the eating out discourse – namely '25.3p' (for a 10 per cent tip on a meal costing £2.53). But she 'corrected' it to '26p' when prompted. Thus we can see that, both for Question 2 and for Question 4, the students in the sample address the problem in varying ways, since they call up different practices.

For the two problems considered together, the bare majority – twelve of twenty-three – called up a different practice. All of these are coded as being positioned in SM for Question 2, and in PM, eating out and tipping practices, for Question 4. Of the five coded in SM for both questions, Jean is a straightforward illustration, since she writes down both calculations as a formula. Of the six classed as having 'PM' positioning for both questions, Harriet and number 11 (a working class young

Table 9.2(b) Predominant Positionings for Interview Question 2 (Abstract 10%) and
Interview Question 4 (10% Tip): Cross-Tabulation of Numbers of Subjects

| Positioning for Question 2 | Positioning for Question 4 | | |
	School maths (SM)	Tipping practices (PM)	Total
SM	5	12	17
PM	0	6	6
Total	5	18	23

Note: This table excludes 2 of the 25 subjects for whom either response is not available

man) gave quick mental answers in money terms, to both questions!

Overall, the 'disparity' in positioning for the two questions for many subjects provides contrary evidence to traditional proficiency views that, because both Question 2 and Question 4 'are' (essentially) 10 per cent questions, they both will be (or should be) thought about in the same way, using the same positioning.

For the second comparison, to see whether a 10 per cent tipping problem on the questionnaire tends to call up the same practice as the 10 per cent tipping problem on the interview, see Table 9.2(c).

Now, if the context of the task were completely specified by the situation described by the wording of the problem, then we would expect all, or almost all, cases to have the same positioning for both problems, that is, to be on the main diagonal of the table (top left corner to lower right corner). However, almost half (nine of twenty) of the cases coded are 'off' the main diagonal. Thus there is support here for the idea that the context, as constituted by the discursive practices called up, depends on the social relations in the respective settings, and not simply on the wording of the problem.

Taken together, the two analyses in this section confirm the importance of understanding the task, and the thinking done on it, as situated, or specific to the context. Put another way, it confirms the importance for positioning of discursive features of the situation, including both the language and representation used, and the social relations in the setting. (See the conclusions of Chapter 7.)

Table 9.2(c) Predominant Positionings for Survey Question 18 (10% Tip) and Interview
Question 4 (10% Tip): Cross-Tabulation of Numbers of Subjects

| Positioning for Survey Question 18 | Positioning for Interview Question 4 | | |
	School maths (SM)	Tipping practices (PM)	Total
SM	1	5	6
PM	4	10	14
Total	5	15	20

Note: This table excludes 5 of the 25 subjects for whom either questionnaire or interview response is not available

Theme 3: Gender and Social Class Differences in Performance Related to Positioning

Differences in performance related to gender or social class might be expected for a 'school maths' type problem like Question 2 ('abstract' 10 per cent calculation) – and also for problems like Question 3 (reading a graph) and Question 5B (exact 9 per cent calculation), which, despite my classifying them *a priori* as 'practical maths' problems relevant to everyday practices had a majority of the subjects call up a school/college mathematics positioning in response; see Table 9.1(b).[6] It will also be useful to compare these results with those for Question 4 (calculation of a 10 per cent tip), a problem considered as 'everyday', both by myself and by the students interviewed.

Gender and social class differences in performance for Questions 2, 4 and 5B are examined in the following subsections. No further results for Question 3 are presented here, because of several anomalies in performance noticed during the analysis.[7]

The same social class indicator, based on parents' occupations, is used here, as was used in the survey. However, because of the small number of cases, it is collapsed into two groups, middle class and working class, rather than the three used earlier.[8]

Abstract Calculation of 10 per cent

The full set of questions posed for this problem, a 10 per cent calculation, is given in Table 9.2(a). Table 9.3(a) shows subjects' performances for Question 2 cross-classified by gender and qualification in mathematics.

There appear to be gender differences, like those observed in performance scores, especially school mathematics performance, in the survey (see Chapter 3): 90 per cent (nine of ten) of the men, but only 50 per cent (six of twelve) of women get this question correct. However, *since the number of subjects is small*, it is important to remain sceptical about the reproducibility (in other similar samples) of even apparently large percentage differences like this one. In this case, the gender difference is merely on the borderline of statistical significance ($p < .06$).[9] So this result might

Table 9.3(a) Performance on Interview Question 2 (Abstract 10%): Cross-Tabulation of Proportion of Subjects Correct by Gender, Qualification in Mathematics and Predominant Positioning

Qualification/ Positioning	Men	Women	Total	
High (O/A level)	4 of 5	3 of 4	7 of 9	(78%)
(SM)	(3 of 4)	(2 of 3)	5 of 7	(71%)
(PM)	(1 of 1)	(1 of 1)	2 of 2	
Low (CSE/none)	5 of 5	3 of 8	8 of 13	(62%)
(SM)	(3 of 3)	(2 of 6)	5 of 9	(56%)
(PM)	(2 of 2)	(1 of 2)	3 of 4	(75%)
Total	9 of 10 (90%)	6 of 12 (50%)	15 of 22	(68%)

Note: This table excludes 2 of the 25 subjects who were not asked to attempt Question 2, and one further (male) whose response could not be classified as correct or not

result from sampling variation, and we cannot be very confident that there would have been a gender difference in performance in the whole population of first year social science students from which this sample was drawn.

However, it is interesting to analyse the sample further, for 'suggestive' – rather than substantial, or statistically significant – differences, that may be supported by other findings. If we control for qualification in mathematics, the gender differences largely vanish for 'high-qualified' students (four of five men correct and three of four women correct), but they remain – indeed, are heightened – for the 'low-qualified' (five of five men and three of eight women correct). That is, there appears to be a much lower level of performance among low-qualified women students, compared with other groups; this replicates another finding from the survey.

We can also check in Table 9.3(a) whether the correctness of subjects' reasonings seems to relate to their positionings in practice. In Question 2, only six subjects called up work maths, 'consumer maths' and so on ('PM'), perhaps not surprisingly, as this problem was constructed as a 'SM' type. However they produced five correct answers (83 per cent), a higher rate than those with a SM positioning (ten of sixteen, or 63 per cent). Despite the very small numbers, this gives some slight support to the idea that thinking and problem-solving performance are specific to the positioning of the subject.

If we hold constant the subject's positioning, the numbers (in parentheses in the interior of the table) are so small (consequent on the multiple controls) that they justify only rather speculative conclusions. However, the relation of performance to gender and qualification differences discussed earlier, now appears to hold only for those with positionings in school (or college) mathematics, but not for those few who were coded as calling up 'money' (or PM) practices. In any case, to assess the effect of differences in qualifications, it is perhaps more important to focus on those whose predominant positioning is SM, than on the whole sample, since one might argue the influence of qualifications in school mathematics should operate mainly for those who had called up SM to consider the problem.

Table 9.3(b) shows the performance levels on Question 2 by social class and predominant positioning. Taking social class first, it can be seen that performance for students with working class parents (67 per cent correct) is about the same as those with middle class parents (69 per cent). Thus there are no social class differences in performance on this problem in this sample. Since there are no class differences to explain, I did not control for differences in school mathematics qualification.

Table 9.3(b) Performance on Interview Question 2 (Abstract 10% Calculation): Cross-Tabulation of Proportion of Subjects Correct by Social Class and Predominant Positioning

Social class/ Positioning	Middle class	Working class	Total
College maths (SM)	8 of 12	2 of 4	10 of 16 (63%)
Everyday maths (PM)	1 of 1	4 of 5	5 of 6 (83%)
Total	9 of 13 (69%)	6 of 9 (67%)	15 of 22 (68%)

Note: This table excludes 2 of the 25 subjects who were not asked to attempt Question 2, and one (WC parents) whose response could not be classified as correct or not

Bringing in the subjects' predominant positioning shows, as we saw above, those calling up everyday practices to be in a minority (six of twenty-two). But they performed at a slightly higher level than those with positioning in SM. Interestingly, slightly over half of the working class subsample (five of nine) were classed as calling up everyday practices, but only one of the thirteen middle class students. And working class students were as effective at thinking within the everyday PM discourses (four of five correct) as the middle class students were in SM (eight of twelve correct).

Calculation of 10 per cent Tip

The full set of questions posed for this problem, also about a 10 per cent calculation, but in an everyday tipping situation, is given in Table 9.2(a). I first consider gender differences; see Table 9.4.

Again, it is important to remember the small numbers involved, which mean that almost any difference found will be suggestive only. Here the level of performance, among both women and men, was rather high (nineteen out of twenty-three correct overall). The gender difference was small. However, when we consider predominant positioning, there are two findings, which, *if they could be replicated*, would be very interesting. First, there was an apparently lower level of performance among those judged to have a predominant positioning in school mathematics than among those positioned in PM/eating out (60 per cent – three out of five – compared with 89 per cent). Second, almost half of the women – but no men – called up school mathematics. Taken together, these very tentative findings again pose the question whether gender differences in mathematics performance could be explained, at least partly, by differences in positioning.

There is no difference in performance between t hose with middle class parents (eleven of thirteen correct), and those from working class families (eight of ten correct). However, there was again a slight difference in predominant positioning: four of thirteen (or 31 per cent) middle class students called up SM, and only one in ten of working class students.

Approximate and Exact Calculations of 9 per cent Pay Increase

Let us consider Problem 5. The full set of questions posed for this problem were:

(Contexting question C)   Have you ever received a slip like this? [Show the copy of a

Table 9.4 Performance on Interview Question 4 (10% Tip): Cross-Tabulation of Proportion of Subjects Correct by Gender and Predominant Positioning

Positioning	Men	Women	Total
College maths (SM)	—	3 of 5	3 of 5 (60%)
Everyday maths (PM)	10 of 11	6 of 7	16 of 18 (89%)
Total	10 of 11 (91%)	9 of 12 (75%)	19 of 23 (83%)

Note: This table excludes 2 of the 25 subjects who were not asked to attempt Question 4
   For indicators of SM and PM positioning, see the text

payslip.] Jennifer is expecting a rise of 9 per cent on her gross pay.

(Question 5A)   About how much will that be?
(Question 5B)   (If (A) is answered) Can you work it out exactly?
(Contexting question R)   Does this remind you of any earlier experiences?

(For the interview problems, including the facsimiles used, see Appendix 2.)

This problem is more complex than earlier ones. Producing a solution requires the subject to follow several steps:

1   extract the correct information about gross (rather than net) pay
2   decide on an appropriate way of calculating, or modelling, 'about 9 per cent', as, say, 'about 10 per cent', or '10 per cent minus a bit', or '10 per cent minus 1 per cent'
3   do the approximate calculation, including estimating as appropriate
4   do the exact calculation, including:

- setting the operations up the right way (9/100 x the amount)
- multiplying correctly
- putting the decimal point in the right place.

First I consider positioning. For part A of the question, most subjects called up practices related to paid employment ('PM'), since almost all subjects had previously had work where they were paid with similar payslips (though often less complex). For both parts of the question, the indicators for calling up PM were:

- doing a mental calculation
- giving an approximate answer, for example to the nearest pound.

For having a predominant positioning in SM, they were:

- using extensive written calculations, and/or
- giving an exact answer, or
- giving an 'unrealistic' answer.

Thus, number 18 (a young male), who gave an answer of '62p' (that is about 0.9 per cent, rather than 9 per cent) for this part of the problem, was classed, exceptionally, as calling up school mathematics for part A, as well as for part B.

For part B, eleven of the seventeen subjects attempting the question were judged to have called up school mathematics. Of the six classed as calling up 'practical maths', all used written methods. Here using *written* methods does not necessarily indicate that the subject has called up school maths. For example, number 13 (a former electrician, aged 25+) makes a creative approximation:

9 per cent of 1,335.45 = 9 x 13

He then writes it down in order to do the multiplication only, apparently as a memory-aid. This was taken to indicate a positioning in practical maths, since we would not expect *all* calculations within PM to be done mentally. Using pencil and paper – or indeed a calculator – to do a calculation must be distinguished from writing it down in order to set up the original fraction, as in: 9/100 x 1,335. Writing down at the original 'setting up' stage would tend to indicate a positioning in school mathematics, but writing down later would not.

Now, to *performance*. On part A twelve of seventeen (71 per cent) subjects produced a 'close' approximation (that is between £5 and £6.65). Four made very close estimates (namely, £6). For part B, however, only five of seventeen (29 per cent) produced a correct answer (£5.99, or, in one case, very close). Six made errors related to 'setting up' the calculation (for example choosing division, rather than multiplication), in calculating, or in placing the decimal point, two of whom noticed (see later). Six others *refused* to complete the problem!

There do not appear to be gender differences in performance in part B; see Table 9.5(a) (nor in part A, see Evans 1993: 459). However, there seems to be a slight gender difference in positioning – as well as a more substantial difference in performance related to predominant positioning for part B ($p < .03$).[9] Thus, for those calling up PM, four of six were coded as correct – while, for those calling up SM, only one in eleven was correct. (The former may have tended to find the problem 'familiar', in terms of their previous (usually work) experience; see Evans (1993: 499–500, n.16).) However, eight of the ten women called up SM, the less 'successful' positioning, compared with only three of the seven men.

There is one aspect of performance, not apparent in Table 9.5(a), in which there may be suggestive gender differences, namely in the *types of errors* made: five of the six who refused to persevere with part B through to the end were women (Evans 1993: 459). As we have seen, the men's perseverance did not result in more correct answers, but they did produce *more answers* – even if some were clearly wrong, such as number 13's calculation of '207' for 9 per cent of £1,335 (simplified as 9x13), or number 18's '62p' for 9 per cent of £66.56 (see earlier). It might appear that the men's perseverance was based on greater confidence – or perhaps greater 'bluff' – whereas the women were diffident or anxious. These emotions will be discussed in 'Theme 5', this chapter and in Chapter 10, using case studies.

Turning to social class, in Table 9.5(b) there appears to be a slightly higher level of performance among middle class (40 per cent correct) than among working class students (14 per cent correct). I decided not to control for qualification in mathematics

Table 9.5(a) Performance on Interview Question 5B (Exact 9% Calculation): Cross-Tabulation of Proportion of Subjects Correct by Gender and Predominant Positioning

Positioning	Men	Women	Total
College maths (SM)	0 of 3	1 of 8	1 of 11 (9%)
Everyday maths (PM)	2 of 4	2 of 2	4 of 6 (67%)
Total	2 of 7 (29%)	3 of 10 (30%)	5 of 17 (29%)

Note: This table excludes 8 of the 25 subjects who were not asked to attempt Question 5

Table 9.5(b) Performance on Interview Question 5B (Exact 9% Calculation): Cross-Tabulation of Proportion of Subjects Correct by Social Class and Predominant Positioning

Social class/ positioning	Middle class	Working class	Total
College maths (SM)	1 of 6	0 of 5	1 of 11 (9%)
Everyday maths (PM)	3 of 4	1 of 2	4 of 6 (67%)
Total	4 of 10 (40%)	1 of 7 (14%)	5 of 17 (29%)

Note: This table excludes 8 of the 25 subjects who were not asked to attempt Question 5

here, because there was only one working class student left with 'high' qualifications (O/A level) in mathematics in a depleted sample (only seventeen respondents had time to attempt Question 5). However, in examining the effect of positioning, we see again the substantial advantage in performance level for subjects calling up PM. Here, unlike in the gender analysis, about the same proportion (60 to 70 per cent) of middle class and working class subjects called up SM, rather than PM.

To sum up, despite the very small numbers and the simplification involved in coding 'predominant' positioning, there does appear to be a higher level of correct performance for Question 5, part B among those calling up activities related to payslips, and so on, than among those calling up SM. These sorts of differences in positioning may help to explain what sometimes appear as gender or class differences in performance. At the same time, there appears to be a much higher proportion of women than men who were coded 'incorrect' because they refused to persevere with the problem to produce an answer, although there is no gender difference in the level of correct answers.

The findings concerning gender and social class differences in these three 'percentage of' questions are summarised in the conclusions to the chapter.

Theme 4: Numerate Thinking as Specific to the Subject's Positioning

The results of the problem-solving interviews shed light on the strategies and methods of subjects' thinking, as well as the answers produced. Here I consider the strategies used in Question 6 ('best buy' decision), and methods used by subjects for critical evaluation of answers in Question 5.

Strategies Used: Best Buy Shopping Decision

'Best buy' problems were earlier researched by Capon and Kuhn (1979 1982), and by Lave (1988). Capon and Kuhn distinguished six 'levels of reasoning strategy' typically used by adults in addressing such problems; see Table 9.6(a).

I distinguish a 'strategy' from a 'method' or procedure. A *method* can be seen as broadly the same as an 'operation' in Scribner's (1985) hierarchy of operation/ action/activity (see Chapter 6). A *strategy* can be seen as a higher level set of

Table 9.6(a) Levels of Reasoning Strategy Coded in Best Buy Studies of Capon and Kuhn

1	Extraneous, task-extrinsic: e.g. 'I always buy the large sizes, since I don't like to shop often.'
2	Extraneous, task-intrinsic: e.g. 'I always buy the large sizes (or those marked 'reduced'), since they must be cheaper.'
3	Partial, non-inferential: e.g. 'Four ounces added to the smaller jar equals half more.'
4	Subtraction/difference: 'With the large one, you get 32 more grams for 36c, so it's a better buy.'
5	Price/quantity ratio comparison: e.g. 'Twice as much for less than twice the price – the big one is cheaper.'
6	Unit price ratio comparison: e.g. 'The small one is 17c per ounce, and the large one is 17.5 c, so the small one is a better buy.'

Source: Capon and Kuhn (1979, 1982)

guidelines (explicit or implicit) for the choice of methods or concepts to use (cf. Bell et al. 1983).

Capon and Kuhn, working within a Piagetian developmental framework, *ranked* subjects' strategies by their proximity to formal operational reasoning: only strategies (5) and (6) were considered 'conceptually correct', with (6) more generalisable. Lave and her colleagues, on the other hand, expected subjects to respond flexibly, that is to use a strategy appropriate for the problem 'type' (see Evans (1993, sec. 7.4.3). We might say that Capon and Kuhn are 'normative in general', whereas Lave is 'normative in particular', since she expects an 'appropriate' strategy for the problem at hand. Capon and Kuhn's insistence on seeing unit price reasoning (6) as the pinnacle of logical thinking – and other strategies as inferior (5), or as 'conceptually incorrect' (all others) – is especially limiting. They do not allow that choosing another strategy might come from a realisation that that strategy might be more *convenient* to calculate – or from an evaluation that it is not *worthwhile* to base a particular decision on prices and quantities only – or from a general position that ignores all but the grossest aspects of value for money. These sorts of positions are illustrated in this section.

The full set of questions related to problem 6 in my interview were:

(Contexting question CA)	Do you ever go shopping for food? Where would you normally go?
(Contexting question CB)	Do you ever buy ketchup (or jam, etc.)?
(Contexting question CC)	If you were buying ketchup or [other food mentioned] and several jars were available, how would you decide which one to buy?
	[Show a picture of two bottles of tomato sauce, with prices and sizes, metric and Imperial marked.] The larger bottle in this picture holds 30 oz and costs 69p. The thinner bottle holds 20 oz and costs 52p.
(Question 6A)	Which of these two bottles would you buy? Why?
(Question 6B)	Which is better value for money?
(Contexting question R)	Does this remind you of any earlier experiences?

Thus I used a relatively long sequence of 'contexting questions' (CA, CB, and CC). This differed from Capon and Kuhn's interviews at supermarket entrances, at least. Since Lave et al. conducted their study in several contexts (shopping observations, home testing) over some time, they were presumably able to draw on informal conversations to produce similar contextual information.

Here the analysis will centre on the range of 'strategies' used by the sample for question B, the practice(s) called up by the subject, and the relation of these to performance. Of course, I had only one best buy problem in my interview, compared with Capon and Kuhn's use of two problems, and Lave et al.'s twelve. In addition the number attempting Question 6 had shrunk even more (to fourteen) by this stage of the interview.

Yet the strategies used by my subjects on this one problem were almost as varied as those found by the other two teams; see Table 9.6(b) for the results on strategies used, compared with earlier studies. No one in my sample used a straightforward version of strategy (1), 'extraneous, task-extrinsic' reasoning. However, there were two cases coded as using strategy (2), 'extraneous, task-intrinsic' reasoning. One was Alan. After about twenty seconds considering the problem, he justifies his refusal to engage with the problem:

> I'd probably go for that one [points to larger bottle], but I haven't worked it out [laughs] . . . [four lines] . . . It's just the preconceived idea that you just presume that the bigger one is cheaper . . .
>
> (Interview transcript: 11)

Also coded as using strategy (2) was number 9, who originally appears to follow strategy (4), but he is confused and makes an error. So he appears to fall back on his general rule to buy anything in bulk 'simply because the value works out enormously better' (interview transcript: 15); see also below.

I did not consider (3), 'partial, non-inferential', to be a strategy, but rather a state arrived at, during the process of problem-solving, by several subjects, especially those using strategy (4), 'differencing'. Strategy (4) was the most frequently chosen – by 50 per cent (seven of fourteen) of my subjects – more than in the other studies. Lave (1988) has given a convincing rationale for the use of these strategies: they allow a decision based on *actually-existing alternatives*, and using a mixture of 'mathematical' criteria, plus criteria based on non-quantitative aspects of the context of shopping: such as cash to hand, carrying capacity, storage space available. Also, very few of my interviewees had calculators with them, so differencing may have seemed a more convenient strategy to use than (5) or (6), both of which require division.

For example, number 23, a young woman, who was being supported financially by her father in a Mediterranean country (and 'regulated' financially by her brother here), explained in response to contexting question CC:

> it's better to buy the . . . larger one, if [. . . 1 line . . .] the difference between the two, . . . the money concerned is not too big, you try to choose the big one
>
> (Interview transcript: 23)

To question B, she responds almost immediately: 'the larger one', and explains: 'It's only a difference of 17p, and you get 10 oz, and that should turn out more.' She apparently does the calculation mentally and quickly: it seems to be one with which she is familiar. However her answer is coded 'correct/incomplete', since her reasoning has not been completely explicated, but she does not appear to be guessing.

Though strategy (5), 'price/quantity ratio comparison', was judged to be used frequently in the other studies, none of my sample used it. However, there are 36 per cent (five of fourteen) 'unit price' (6) solutions, if we include number 4 (Keith), who compared the cost of 60 oz of ketchup, based on buying either three 20 oz bottles, or two 30 oz bottles, a rather creative 'lowest common denominator of quantities / unit price' strategy. This frequency of using strategy (6) is comparable with the Capon and Kuhn and the Lave simulations.

Thus my subjects used (6) unit price strategies about the same as would be expected, and difference strategies (4) more than would be expected, in comparison with the other studies, and (5) price / quantity ratios not at all.

In considering the practices called up by Question 6, I judged that most subjects called up shopping, as their 'predominant positioning', while considering the problem. This was presumably because of the relatively long sequence of contexting questions (see earlier), the graphic nature of the representation of the two ketchup bottles, and the familiarity of the practice to most students. For example, number 23, the young woman from the Mediterranean called up shopping, though it was somewhat painful because of her constrained finances.

On the other hand, though Keith refers to shopping, he makes it clear that he does not do any best buy calculations in the supermarket: 'It's just a matter of looking for the cheapest and assuming that . . . you're making an economy with a larger packet' (interview transcript: 11–12). Therefore I coded his predominant positioning as in school mathematics, when he produced his somewhat spectacular strategy (6) solution (see earlier).

Two subjects showed they had called up multiple practices. One is number 9, whose general rule, in discussing his upper middle class family's purchasing

Table 9.6(b) Strategies Used for Solving Best Buy Problems: A Comparison of Capon and Kuhn's and Lave's Simulation Studies, Lave's Supermarket Observation, and Interviews in this Study

Strategies	Capon and Kuhn simulation 2 problems	Lave simulation 8 problems	Lave observation variable number of problems	This study simulation 1 problem
Unit price	30 %	39 %	5 %	36 % (5/14)
P/Q ratio	25	47	35	—
Difference	7	9	22	50 % (7/14)
Other	38	5	38	14 % (2/14)

Note: This table excludes 11 of the 25 subjects not asked to attempt Question 6

Sources: Capon and Kuhn (1979, 1982); Lave (1988)

practices (PM), is to buy everything in bulk. However, beginning the calculation, he gets confused:

S: My god, *in these conditions*, Jeff, one gets extra nervous . . . I can't . . .
JE: You're feeling that, are you?
S: A little tense, because I'm doing it under, almost, in front of a maths lecturer.
                                        (Interview transcript: 14; my emphasis)

So 'these conditions' are 'doing it in front of a maths lecturer', that is college maths (or maths testing). Thus I coded number 9 as the second subject calling up college maths as his predominant positioning.

Another man, number 13 the electrician, seemed to call up three practices, in which the meaning of comparing two jars of ketchup as to the 'better buy' differed. When shopping for himself, it would be 'a bit nit-picking to . . . work out the actual difference per ounce'; however, when taking a group from the youth club on an outing, you have 'got to account, when you're using someone's money . . . be tight on calculations' (interview transcript: 17). On the other hand, if he had to answer something as a SM calculation, 'I need to sit down and . . .' (ibid. 16). In the interview, he chooses the larger bottle as the best buy. When I ask why, he produces very clear unit-price calculations (done in his head), but then claims that 'it's quite close . . . probably about the same' (ibid.: 16). This is perhaps because his approximation is too rough to allow him to distinguish, or perhaps he feels it is inconsequential, or 'nit-picking'. His predominant positioning is thus coded as shopping (his own), and his answer was coded 'correct' relative to that positioning.

In considering performance, since the numbers responding were by now very small (n=14), instead of the sorts of cross-tabulations used for Questions 2, 4 and 5, I use mostly illustrations here. Social class is focused on, because it is relevant to practices dependent on the availability and spending of money.

All five subjects using strategy (6) scored correct except for number 21, who made a slip in calculating one of the unit prices. All but number 13 are from middle class families, and he had been in a well-paid engineering job for some time. This recalls Walkerdine's point about certain kinds of calculation being enjoyed as a game within middle class families, but not within poorer families where resource constraints are often too painful to be ignored (see Chapters 6 and 7). Keith's impressive, almost playful, comparison of the cost of 60 oz of ketchup using two large bottles or three small, is a good example of a middle class approach.

Two women from working class families chose the wrong bottle. Number 1 (ex-nurse) chose 'probably' the smaller. She went on to make a crucial approximation error:

Dunno . . . That's [the larger] got 10 more ounces, but it's a lot dearer, and 70p is a lot of money, whereas 50p – nearly 50p – isn't so much . . . . [JE: Umm, umm] 70p is nearer a pound, 52p is nearer half a pound – in money [JE: Right, right] – so I wouldn't . . . yeah . . . I'd much rather the pound than the extra weight of ketchup.

                                                        (Interview transcript: 11)

This was classed as strategy (4), differencing. An explanation for the error might be that she desperately wants to justify buying the smaller – and cheaper – bottle.

Number 3 (Jean) also chooses the smaller bottle:

> it's cheaper than that one, and [. . .] the 17p price difference [is] more relevant [. . .] because of the difference in the ounces, the 10 oz, I think it works out cheaper in the long run.

> (Interview transcript: 11)

She seems to be trying to compare the price and quantity differences, but she seems overwhelmed by the price difference, which might well be more 'relevant' for someone worried about money. This is a version of strategy (4), with an overriding emphasis on price. This displays the practical constraint of limited resources, or *absolute economy*, that is ignored by the 'relative economy' assumptions of the 'mathematical' best buy task.

Here both working class women choose the 'wrong' bottle of tomato sauce, apparently because they seem reluctant to tie up an additional amount of their money in a bigger bottle. This can be contrasted with number 7 (Alan), who buys 'what I want' and refuses to get involved in best buy calculations when shopping with his parents' money, since it is not worth his time. These responses may plausibly be related to social class positioning. For more discussion of the effects on positioning of constraint and abundance, and especially of Jean's and Alan's cases, see Chapter 10.

To sum up, my group of fourteen subjects, for one best buy question, used a wide range of strategies. They used unit price strategy (6) (in Piagetian terms, the most generalisable) as often as subjects in the Capon and Kuhn and the Lave simulations, and strategy (4), differencing, rather more.

In addition, I have shown through a number of illustrations that both the strategies used by subjects, and the answers given, can be shown to be specific to their positioning in practices. In particular, social class-specific positions are important in several ways. The suggestive relationship between social class background and successful best buy reasoning supports the idea that the ability to reason in certain ways presupposes a position of relative freedom from constraint and anxiety, from which to consider a range of alternatives. For those from more affluent backgrounds, this problem also raised the issue of whether a particular difference in price would be evaluated as being *worthwhile* to calculate.

Methods of Critical Evaluation

Problem 5, on calculating a 9 per cent pay rise, involved a series of stages. Some subjects who worked through to an answer for Question 5B illustrate critical evaluation of one's own thinking. Number 22 (the part-time accounts clerk) set up the calculation as

$100/9 \times 1335.45$

That is the inverse of the operation needed. However she critically evaluated her answer, from within what appeared to be a school maths perspective: 'It looks too much' (interview transcript: 14). Also, her later answer was given without units and with four decimal places. Elsewhere in the interview, she indicated that her approach was often 'that very – sort of – trial and error method I used to find an answer that looked reasonable'. This error, and her basis for 'noticing' it, were similar to the error and the evaluation made for Question 4, by Ellen, except that Ellen did not appear to be using trial and error (see Chapter 10). If number 22 really did not understand the algorithm, as she claimed, then critical evaluation of answers was crucial to her approach!

One or two other subjects noticed errors, also. For example, number 1 (the industrial nurse) attempted to divide 9 into £49.80 (net pay) – rather than multiplying – then noticed that that would give more than £4.98 (not less, as she expected), but she could not find what to do next. Similarly, Sam (number 17), was unhappy when he calculated a 9 per cent increase on £65.56 as .594: 'a pence increase . . . [. . .] it doesn't look right' (Interview transcript: 17–18), but he did not know what to do with that insight.

On the other hand, the other four subjects who made errors at various stages did not appear to notice: for example, number 13 and number 18 (see p. 163).

Theme 5: Emotion Pervades Mathematical Thinking – 'Mathematics is Hot'

Many people are surprised by any claim that mathematics relates to the emotions. However, the pervasiveness of emotion can be assessed by the fact that *all* interviewees in this sample expressed some emotion – and, many cases, multiple emotions – in discussing their current and previous experiences of mathematical activity. I looked first for expressions of anxiety or fear, then for expressions of confidence, pleasure, or anger.

I classified each of my twenty-five subjects on whether or not they *expressed anxiety* at any point in the interview, by using terms like 'anxious', 'scared', 'unsettled'. Twenty-two of the twenty-five subjects are coded as expressing anxiety: nine of twelve men, and all thirteen women (see Table 9.7, p. 175, discussed in the next section). One clear example was interviewee number 1 (a working class woman, aged 21–24) – in response to my request 'to take a look at a few questions':

S: I'm scared . . .
JE: You're scared? Okay, why's that?
S: I'm just apprehensive in case I get them all wrong [laughs].

(Interview transcript: 2)

Further examples come from the two students who did not complete an interview in the usual way. Number 8, a 25+ working class man with no mathematics qualifications, came rushing into the interview at the agreed time, full of the 'panic' he had felt the previous evening when, offered a job in a bar, he had felt overwhelmed by fear about how he would cope with needing to add up the cost of each order of drinks. Since he clearly needed to discuss that, I dispensed with my interview script, and conducted what might be called a 'free association' interview, without posing any numerate problems.

Another student did not complete a normal interview – number 25, a woman aged 21–24, of 'mixed' social class (see note 8). Although having agreed to an interview, she missed three appointments before I met her by chance one day and we arranged to meet right after lunch. On arriving, she was clearly uncomfortable, smoked several cigarettes over a short period (in my 'no smoking' room), and refused to have the interview recorded, three times! Thus she *exhibited anxiety* in all these ways, and also expressed anxiety about 'numbers'. She managed to attempt only one question, and the interview broke down after fifteen or twenty minutes. The difficulties of this interview were, fortunately, unique.

Other emotions were expressed by the three men not expressing anxiety – and by most of the others, too. For example, *confidence* was expressed, often in terms of feeling more so as a result of the 'Methods and Models' course in the first year. For example, number 11, a young working class man, averred:

> After this course, after these three terms, I feel far more confident about numbers. I can go to a formula and actually tackle the thing .... [...] I would not have tackled it before.
>
> (Interview transcript: 20)

Another subject, Ellen, expressed overwhelming confidence after almost every question in the interview, for example for Question 1 (reading a pie-chart): 'very familiar . . . know exactly what it means . . . don't have to think about it . . . (interview transcript: 3). However, in her case, I felt this first impression needed to be interpreted carefully; see her case study in Chapter 10.

Others expressed *pleasure* in, or *liking* or enjoyment of, mathematics. Harriet expresses pleasure a number of times (see Chapter 10). For example:

> having a formula and . . . working it out, I love that sort of thing, . . . the idea, the sheer pleasure of doing that is just really very nice . . . I used to enjoy binary numbers . . . I haven't thought about this 'til today [said with pleasure] . . . I used to enjoy geometry.
>
> (Interview transcript: 17–18)

Donald (see also Chapter 10) expresses a range of new-found positive feelings for mathematics:

> I found connections of something there to go from one thing to another, and I found it [college maths, during the second term] exciting, you know, I couldn't get bored with it at all . . . . [two lines] . . . I liked it.
>
> (Interview transcript: 15)

Keith (number 4) expresses liking of doing maths at school, and of using it in his work. For example, he describes working through relatively advanced text-books with one other pupil at school:

> That thing of solving a problem . . . apply what you know, try and . . . crack it, I quite like that, but you don't get too much of that in your life.
>
> (Interview transcript: 13–14)

He also enjoyed

> working out the detailed costings of how much everything [. . .] I used to quite enjoy presenting that, working all that out, presenting it nicely, and then going up and arguing the case. Yes, I liked doing that, and having time to do it.
>
> (Interview transcript: 18)

Discourses related to paid employment may have a powerful affective charge, as illustrated by interviewee number 2's clear articulation of the 'cash nexus' type of quantitative relationship that an unskilled manual worker has with his/her work:

> S: . . . we'd have a period when they'd got a new wages girl, and she was *not too hot*, it wasn't so much the working out, it was, like, losing [laughs ruefully] your overtime pay on the wage slip . . . . [JE: two lines] . . . I think most people [. . .] used to go over their wages paid – you had to!
> . . . [six lines] . . .
> . . . like, the only reason you work in a factory you don't like it, working in a factory – you're only there for the money, aren't you? If they aren't getting paid for the hours, most people get very irate, and if you're coming off night shift, and you have to go up and sort it out with them . . . .
>
> (Interview transcript: 7, his emphasis)

The subject's emphasis on certain words exhibits his feelings of *frustration* and *anger*, confirmed by the later expression 'irate'.

One emotion said to be widely experienced at school is *boredom*. It is mentioned in several case studies; for example, Sam, number 17, provides an illustration of feelings of boredom, dislike and anger about maths. When I show him Question 2 (abstract 10 per cent), he says 'I used to hate these'. He is reminded that a tutor engaged by his parents 'gave me thousands of 'em to do'. He expresses a great deal of anger ('hatred') towards the tutor, his parents and maths. It may be that some of the negative feelings are displaced onto mathematics from the hateful 'telltale' tutor. Boredom is also mentioned by Peter and Keith.

Thus the claim that mathematics relates to the emotions is confirmed by reference to each of the interviews, and illustrated here. There remains one subject, Alan (number 7), who seems to resist this straightforward interpretation. When I ask what Question 1 reminds him of, he replies abruptly:

> It reminds me of earlier maths – uh, concerning my feelings about maths, I'm – it's *very* neutral – I don't have any strong feelings about maths. I have difficulties in

> maths, but um, and, I can't concentrate for too long on maths . . . I don't have *worries* about it, I just stop and have another go later on [laughs quietly].
>
> (Interview transcript: 3, his emphases)

This 'statement' about mathematics and (no) emotion – repeated later in the interview – will need to be interpreted carefully; see Chapter 10.

Theme 6: Gender Differences in Expressing Anxiety

My hypotheses were set down in Chapter 8, and propose that, during the interview:

1   A higher proportion of women than men would tend to express anxiety.
2   Men would tend instead to exhibit anxiety.

The general idea, expressed as Theme (6), that emotion not expressed may instead be 'exhibited', was discussed in Chapter 8, and will be illustrated using case studies in Chapter 10.

I classified each of the twenty-five subjects on whether or not they expressed anxiety at any point in the interview, and whether they appeared to exhibit it. The signifiers for expressing anxiety included 'anxious', 'scared', 'worried', 'unsettled'; the indicators for exhibited anxiety included:

- 'Freudian' (surprising) slips
- behaving 'abnormally', for example laughing a lot
- repeated denials (of feeling anxious).

For discussion of the ideas behind the codings, see Chapter 8. Examples of these three indicators for exhibited anxiety are provided, respectively: by Ellen, by Fiona and by Alan (see also Chapter 10). The results are summarised in Table 9.7. Twenty-two of the twenty-five subjects are coded as positively expressing anxiety, and three are coded as 'uncertain' in their expression of anxiety. Thus, the first part of the hypothesis receives some confirmation: the percentage expressing anxiety in the interview among women is somewhat higher than for men: 100 per cent, against 75 to 100 per cent, allowing for three uncertain codings among the twelve men. In any case, in this setting, a high proportion of men, as well as of women, is coded as expressing anxiety, perhaps higher than might have been expected.

I consider the three men who were coded as 'uncertain' in terms of expressing anxiety, namely, numbers 7 (Alan), 11, and 17 (Sam). The closest Sam (a young, black, middle-class man) came to expressing anxiety was in response to my question 'How do you feel about the way [. . .] you're able to use numbers these days generally?':

> Not as *comfortable* as I was when I was younger . . . get some practice again . . . [two lines] My brain's not very sharp with numbers. Usually I wouldn't want to depend on a calculator, I would want to use my head.
>
> (Interview transcript: 23, my emphasis)

Table 9.7 Expressing and Exhibiting Anxiety in the Interview: Cross-Tabulation of Numbers with Gender (n = 25)

Coding	Males	Females	Total
Positive expression of anxiety	9	13	22
Expression of anxiety uncertain, but exhibiting likely	3	—	3
Total	12	13	25

This was initially coded as not expressing anxiety, because he did not use any of the terms set down above as signifying felt anxiety. However, 'uncomfortable' is rather close to 'unsettled' or 'worried', so this negative coding must be considered uncertain. As for exhibiting anxiety, after agreeing (unenthusiastically) to try some questions, he told several stories with no clear point (to me), about work-mates at his summer job, which may indicate avoidance of, and hence anxiety about the maths problems. After calculating the price and quantity differences for Question 6 (best buy) he admitted 'I wouldn't like to work it out though' (interview transcript: 19); this reluctance may well be based on anxiety, and hence indicative of exhibiting it.

Number 11 does not seem to express anxiety about maths at all, though he does express 'hatred'/dislike, irritation, and boredom (in at least two places). His long hesitations (around 15 seconds) before answering any of the questions, and his use of 'damn' several times in talking with me, a teacher (though one he knew fairly well) may be indications of anxiety exhibited.

Finally, Alan's 'statement' about having 'no feelings' about maths has already been cited in the previous section, and will be discussed further in Chapter 10. However, there may be some expression of anxiety in the following:

S: So, yes, I was pleased with the interview . . . .
JE: You didn't . . . Did you find it nerve . . . making?
S: No initially, with the tape-recorder there, I thought – it did increase my nerves a little bit (JE: uh huh), just – yes, it did – otherwise, no

(Interview transcript: 14)

Alan's response to my (somewhat 'leading') question may indicate expressed anxiety. In any case, his *repeated* denials of any feelings about mathematics, as well as his repeated claims of having difficulty concentrating, are *prima facie* indications of defences against feelings of some kind, and hence possibly indicators of exhibiting anxiety.

These three men illustrate the somewhat different difficulties of coding subjects as expressing anxiety or not, and as exhibiting anxiety or not. The indicators for expressing anxiety seem relatively straightforward, but Sam and Alan provide cases that were hard to code. As for exhibited anxiety, even coding behaviours as positively exhibiting anxiety is uncertain and contentious, as all three cases show, in slightly different ways. However, coding the *absence* of exhibited anxiety is *in principle uncertain,* because another reading of the transcript might produce a positive instance. One must rely on linking the episode being analysed with other 'related' incidents elsewhere in the interview, or ideally in a series of interactions with the subject, which allow the testing of emerging hypotheses (Meehl 1954). Though this cannot be done in this chapter, it will be undertaken in analysing the case studies. Indeed, the idea of exhibiting anxiety shows the need to consider the role of defences, and of the unconscious in understanding anxiety and affect, as discussed in Chapter 7.

Thus we can tentatively conclude, not that women experience more anxiety, or are more anxious than men, but that these women – in this sample, in these settings – were more 'able' or *more 'willing' to express anxiety* than men. This may help to provide an alternative explanation for the higher levels of reported anxiety amongst women in many surveys.

However, caution is in order, on two counts. First, expressed anxiety from a 'live' problem-solving interview is different in kind to reported anxiety concerning situations the survey respondent is not currently 'in'. Second, the tentative conclusion above may seem to suggest that this differential willingness between the genders to express or report anxiety is itself an 'essence' or characteristic of men and women in general. However, my approach to studying anxiety (and emotion) sees it as specific to the practice(s) in which the subject has a positioning. The analysis will be extended so that expressions and exhibiting of anxiety, and gender differences therein, can be examined as specific to practices; see the case studies.

The context-specific qualities of anxiety are further important for the consideration of gender differences, since it is argued in Chapter 7 that certain practices may make particular subject-positions differentially available to men and to women. For example, Hollway (1984, 1989) argues that, in many relationships, there is a pattern whereby the man 'holds' the rationality and the woman holds the feelings. Some insights from this sort of analysis of the 'history' of relationships will be illustrated in the discussion of aspects of 'family dynamics' related to several case studies in Chapter 10.

Conclusions

This chapter aims to show that mathematical thinking and performance in the interview are specific to – that is, depend in crucial ways on – the context. This context I understand as the subjects' positioning within discursive practices.

I have made the argument in stages. First, in a given situation, it is possible to specify the practices at play, and related discourses, in which subjects may be positioned; I argue that in this setting it was either:

- college mathematics/school mathematics, with positions teacher and student, or
- research interviewing – with positions researcher and interviewee.

Second, what I call the subject's 'positioning' depends on the practice(s) called up by the particular subject, from among the practices at play (already specified), or from other practices (and positions) available to the subject; in this interview, many of these practices will include elements of what I call 'everyday' or 'practical' mathematics. The relevant everyday mathematics of course varies for different problems, relating to 'eating out' for Question 4 and shopping for Question 6.

Third, I argue that it is possible to describe the subject's positioning in particular episodes of the interview. Indicators that may be used include the explicit form of the task, unscripted aspects of social interaction, and the subject's talk in the interview. These indicators are available from the transcript of the interview, my fieldnotes (e.g. about the setting) and from reflexive accounts. Finally, I expect the subject's positioning to be fluid over the course of the interview.

I now consider the six themes addressed in this chapter, using cross-subject analyses of the sample of interviewees. A *caveat* is in order about the small sample size (n = 25). This means that almost all findings are *suggestive only*, especially those based on the cross-tabulation of two (or more) variables. In a number of cases, a statistical significance test was used, as a guide to whether the relationship found in my sample would likely to be reproduced in the whole population of (first year social science) students from which the sample was drawn. The few results that attained statistical significance are noted.

Theme 1 explores the idea that context can be ascribed in terms of positioning in practices, as outlined above. For Question 1 (reading a pie-chart), Question 2 (abstract 10 per cent calculation), Question 3 (reading a graph), and Question 5B (exact 9 per cent calculation for payslip), most subjects were judged to have their positioning in college / school mathematics (SM). For Question 4 (10 per cent tip), Question 5A (approximate 9 per cent calculation for payslip) and Question 6 (best buy), most had their positioning in numerate practices from work and everyday life, outside college or school (PM). However, even with an apparently simple question like Question 1, judging the subjects' positioning is not always straightforward. Sometimes, a subject appears to have a positioning which is *multiple*, that is which might be said to be at the intersection of more than one discourse. There were a number of striking associations of the material of these 'mathematical' problems made by subjects with their parents, brothers and sisters, and work experience, as well as with school/college maths. Some subjects appeared to have an emotional 'investment' in calling up discourses from other school subjects or from 'everyday knowledge', thereby avoiding the terms of the 'mathematical' question and the information presented. In at least one case, Fiona's, it appeared that her avoidance might have a basis in anxiety or some other emotion (see her case study).

In considering subsequent themes in this chapter, the coding of the 'predominant positioning' of particular subjects for specific questions has been indispensable. It seems to have been accomplished satisfactorily in most cases. Sometimes, however, it is possibly an overly simplifying assumption, which will accordingly be relaxed in next chapter.

Theme 2 aimed to examine the evidence for the inseparability of task and context. The main conclusions were that the practice called up is not determined purely by the mathematical qualities of the problem, but rather depends on the language and representation of the problem, and also on the social relations of the context. Thus the task and the context, broadly understood as in Chapter 6, cannot be neatly separated, as argued by proficiency views of mathematics learning.

Theme 3 considers any apparent gender differences, or social class differences, in performance for the three 'percentage of' questions, and whether they might be illuminated by differences in positioning. The findings on gender differences are discussed separately for Question 2, the only problem designed as 'abstract', and for Question 4 and Question 5B. Only Question 2 shows a gender difference in performance approaching statistical significance (5 per cent level). This difference is partly explained by differences in qualification in mathematics, but there still appears to be a lower level of performance among low-qualified women students for Question 2, as there was for total school mathematics performance scores in the survey.

For Questions 4 and 5B, there are no substantial gender differences in performance. However, turning to predominant positioning, men tend to be more likely to call up everyday maths (PM) than women. And those who call up everyday maths perform better than those whose positioning is in SM. For Question 4, the small gender difference observed in performance can be explained by the fact that all those calling up school maths were women! That is, the gender difference in performance is less striking that the gender difference in positioning. The apparent consistency of the performance advantage for PM over SM positionings over the three questions is a suggestive feature of this data set, despite the small numbers, and merits further research.

As for social class, the pattern is less clear. There appeared to be slight performance differences for Question 5B only. In seeking to explain the differences, we found no social class differences in positioning, although, as already noted, there were noticeable performance advantages for those calling up everyday maths (PM), rather than school maths. However, the latter were in the majority for Question 5B. Basically the same pattern held for Question 2 and Question 4: no social class differences in performance, a slightly greater tendency for middle class students to call up school maths, but slight performance advantages for those calling up PM. The main difference is that most students called up everyday maths for Question 4, but SM for Questions 2 and 5B. This gives some support to the idea that social structural factors such as gender or class, are not determinant, in any simple sense, and need to be understood as *related to positioning in practices*.

Some of these results accord with those of Cooper and Dunne (1998), working in a framework informed by the work of Basil Bernstein (1996).They, too, report a greater tendency for middle class pupils to call up SM (in their terms, 'esoteric') discourses, in response not only to problems readily categorised as school maths, but also to 'realistic' problems like Questions 4 and 5. This is because the middle class children are more 'competent' at recognising and deploying rules that are appropriate to the demands made by mathematics problems in school. Their work also suggests that working class pupils are less able to apply mathematical 'realisation rules' to realistic problems (that is they are less able to 'transfer'). Given Cooper and Dunne's findings, what is unexpected in my results is that working class students appear to perform as well as the middle class, when they have a positioning *within PM*. This puzzle needs further study.

For Question 6, I show through illustrations that both the answers given and the strategies used appear to relate to social class position. Most (four out of five) subjects who used the 'most general' thinking strategy were from middle class families, and most (four of five) using this strategy got the question correct in a cogent way. In contrast, two (out of seven) beginning with strategy (4), differencing, and getting the question wrong, appeared to do so because of anxieties about financial constraint. These findings, even with the small numbers, give some support to Walkerdine's (1988) idea that the freedom to reason in abstract ways may be enhanced by the freedom to consider a range of alternatives in a way that is free of constraint and anxiety. Whether a particular difference in price would be evaluated, by a particular subject, as being worth bothering about, also seems likely to be related to social class position.

Theme 4's assertion of the specificity of numerate thinking to the subject's positioning, is supported by the discussion of the strategies used in Question 6 (calculation of the best buy). As a group, my subjects showed almost as wide a range of strategies as were used by subjects in the two studies of Capon and Kuhn (1979, 1982) and Lave (1988) (using two and twelve problems, respectively). My subjects used unit price strategy (6) (in Piagetian terms, the most generalisable) as often as subjects in the other best buy simulations, strategy (4) differencing rather more, and strategy (5) price/quantity ratio not at all.

Another issue related to Theme 4 was the capacity for *critical evaluation* of one's thinking. Since the solution to Question 5 required successful completion of a series of stages, an error or slip could be made at several points. Several subjects were sufficiently critically reflective to notice an error – and in one case to correct it – but some others did not. This critical evaluation was done both within school maths and within discourses underlying the reading of payslips. Further findings about the bases for making such critical evaluation would be valuable.

Besides those relevant to cognition in practice, in this chapter I have also produced findings on affect. Concerning Theme 5, 'mathematical thinking is hot', every single student expressed some emotion related to the doing of mathematics, or the use of numbers. Not only was anxiety expressed by many, as expected, but also confidence, pleasure, and sometimes dislike or anger.

For Theme 6, women were found to express anxiety more frequently than men, though the difference was surprisingly small. However, those few men who did not appear to express anxiety during the interview could be interpreted as *exhibiting* it, that is, as displaying 'surprising' slips, or 'abnormal' behaviour, such as often can be taken to indicate the operations of defences against anxiety.

There are still outstanding issues from the discussions of this chapter. The analysis of positioning needs to be broadened to consider 'multiple positioning', rather than simply predominant positioning. We need to consider also:

- how the expression, and exhibiting, of anxiety are specific to positioning (Theme 7)

- the idea of exhibiting diverse forms of emotion (Theme 8), showing the need to explore psychoanalytic ideas more fully
- instances of the specificity of the relationship between thinking and emotion to positioning (Theme 9)
- any episodes where a student shows an ability to transfer their learning (or ways of thinking) in school or college mathematics to other contexts (Theme 10).

These issues are taken up in the discussion of the case studies in Chapter 10.

10 The Learners' Stories

If race and class, poverty and wealth, mental and manual labour, produce differently regulated practices, then it is important to examine a multiplicity of subjectivities produced in such conditions.

(Walkerdine 1988: 215)

The analysis in Chapter 9, using cross-subject analyses, based on accounts of all twenty-five interviews, produced several findings:

- Proficiency and functional conceptions are not adequate to understand the context of subjects' thinking in specific situations
- Instead, it is necessary to determine which practice(s) a subject has called up, and hence his/her positioning, in the situation.
- Several aspects of thinking in problem-solving situations depend on the subject's positioning in practices.
- Mathematical activity and thinking are charged with emotion.
- Differences in performance or in expressing anxiety, due apparently to structural differences (gender, social class), in order to be satisfactorily theorised, require taking account of differences in positioning in practices.

The approach of this chapter is different from that of the previous one. Here case studies are used to discuss the remaining Themes for the qualitative part of the study, as well as to enrich the analysis of those already discussed. Here, life history and problem-solving material for each student is analysed more intensively and more holistically than could be done in Chapter 9, so as to develop the remaining Themes.

Overview of Themes to be Analysed

The Theme of numerate thinking being specific to positioning is considered further – to allow for multiple or 'interdiscursive' positioning, rather than simply predominant positioning. Further illustrations will be given relevant to the Themes of gender and social class differences in performance and positioning, and the pervasiveness of emotion in mathematical thinking.

For Theme 6 (gender differences in expressing anxiety), each interviewee was coded as expressing anxiety (or not) in the previous chapter. Here, for Theme 7 (anxiety specific to positioning), we need to consider the context, the positioning, within which anxiety is expressed (or exhibited), for, if the context of a particular action is seen not as 'mathematics' (or not as *only* mathematics), then any anxiety expressed is not (necessarily, or only) 'mathematics anxiety'. Therefore, in order to appreciate the meaning of the situation, including its emotional charge, it is important to specify the subject's positioning in practices. In this way, I analyse several cases of what might appear to be 'mathematics anxiety'.

Theme 8 (emotion may not be expressed, but 'exhibited' instead) explains how anxiety may on occasion be expressed as 'no feeling at all', or else expressed as its opposite, that is over-confidence – because of the operation of psychological defences. Hence it is necessary to attend to symptoms, or indicators of the functioning of defences, produced by the subject in particular episodes. This Theme also provides the basis for reading particular interviews using other psychoanalytic insights, concerning, for example, the displacement of emotion onto mathematics, transference reactions to teachers (or researchers), and fantasy.

For Theme 9 (the relationship between affect and cognition as specific to positioning), the statistical analysis reported in Chapter 4 aimed to produce a general relationship between anxiety and performance, using what I call Model A. In contrast I argue that it is also necessary to consider the relationship between emotion and thinking, for particular subjects. Typically, I shall give several readings of the relationship between cognition and affect for each case, in specific interview episodes, drawing on the models discussed in Chapter 7: Model B ('cognitive-constructivist' or 'process' approaches), Model C (using psychoanalytic insights), and Model D (using psychoanalytic and poststructuralist insights).

In these analyses, it will be useful to examine any surprising 'slips', as to whether they might be explained by considering affective processes, or unconscious ones. Also, the notion of familiarity with a practice can be seen as straddling the cognitive and the affective.

For Theme 10 (the possibilities of transfer), we need to look for:

1   subjects' sensitivity to similarities and differences in signification between different discursive practices that they call up
2   deliberate attention by subjects to the relating of signifiers that would support meaningful interrelations between practices and the related discourses, and hence 'translation' of learning and thinking (see Chapter 6)
3   illustration of how mathematics may be connected with other apparently unrelated discourses, by the unexpected flow of affective charge along particular chains of meaning (Chapter 7).

Case Studies

Here I present case studies based on seven interviews, selected as explained in Chapter 8. All quotations in the case study accounts, unless otherwise indicated, are from the interview transcripts (see Evans 1993). As a reminder, case numbers 1 to

9, including Jean, Fiona and Alan, were interviewed in year two of the study, and numbers 10–25 in year three, including Ellen, Harriet, Donald and Peter.

Jean's Story

Interviewee number 3, here called 'Jean', was aged 18 at entry and working class[1] with CSE passes[2] in both Mathematics and Arithmetic. She had a part time job in a pub at the time of the interview.

My reflexive account, written at the time of the interview was as follows:

> I knew this student only by sight. In the first two terms she (like interviewee number 1) was a member of an all-women seminar group, with whose members I had friendly relations, though I did not tutor them. Besides having me as a lecturer for first year maths, as an intending social worker, she would have expected to have lectures (and perhaps seminars) from me in the second year. She was recruited to an interview by the random sampling exercise (see Chapter 8). This was only my third interview, and the first with a student that I did not know fairly well.
>
> (Evans 1993)

On the questionnaire completed the previous October, her score (16 of 22 correct) was in the lowest third of this subsample of interviewees, with two percentage questions wrong (see later). Elsewhere in the questionnaire, she rated herself as 'not very capable' on percentages, and 'not at all capable' on decimals. She expected 'a great deal of difficulty with maths' in her polytechnic studies, and indicated her desire to learn 'percentages and decimals' in the first year. Her anxiety responses on the questionnaire for numerical anxiety (NA) and for maths test/course anxiety (TCA) were both relatively high.

A major issue concerning her mathematical thinking and performance is: Why does she always get percentages 'the wrong way round'? She seems to call up school (or college) maths for Question 2 (10 per cent of 6.65) and for Question 4, the everyday practice of eating out, as well as school maths, though I considered the latter was her 'predominant' positioning. In one of her attempts at Question 2, she tries

$$10 / 6.65 \times 100$$

but then realises that is incorrect, and gives the answer '0.65', which she says is 'just a guess'. How did she come up with that? 'I just moved the dot'.

For Question 4, after she has illustrated her tipping practices by saying that for a meal costing £2.75 she would leave £3, she attempts to respond to the question about a '10 per cent tip on £3.75'. First, she tries

$$10 / 3.75 \times 100 \text{ (as for Question 2]}$$

and then, she tries

$$3.75 / 10 \times 100$$

However she doesn't really know: 'it goes something like that'. The discussion of Question 4 ranged over tipping in general, not having sufficient money, having to tip 15 per cent in the USA (for which she was about to depart), having to re-learn how to calculate percentages, especially 15 per cent.

It is helpful to re-examine her questionnaire. Recall that subjects were responding under time pressure – ten minutes for twenty-four questions – in their first two weeks at college. Her answer to Question 18 (10 per cent tip on £3.72), '72p' is difficult to explain unless she is using some rule such as 'Take the *last two* digits'! She correctly sees that Question 24, an especially difficult percentage question (see Appendix 1), requires her to take 44 over 78. However when she sets up the formula, she gets 'stuck' at

$$78/100 \times 44$$

Thus evidence from questionnaire and interview points to a *conceptual problem* with percentages, which might be helped by clarifying the distinction between 'percentage of' problems (for example Question 18) and transformations of a fraction to a percentage (Question 24) – which she seems to muddle. However, is emotion likely to be involved also?

In the interview as a whole, the main affective themes are anxiety, diffidence, and worry, constantly expressed or exhibited. She begins diffidently: 'I sound horrible on tape'; this may be exhibiting anxiety about the interview. Then concerning her CSE Grade 3 in Mathematics: 'I wasn't very good at all'. She also expresses a great deal of anxiety about percentages; for example 'I always get the formula wrong', 'I'm going to have to learn percentages again' (for the USA); 'I always mix it up', and so on.

However she seems to express even more anxiety about money! For example, 'I've never, ever got enough money' (for tips); about the level of tipping required in the USA, and about being able to afford 15 per cent. She is also anxious about being able to afford the trip at all (see later). When shopping, 'I do always follow the prices . . . for fear of being ripped off', and with wage slips, 'Yeah, I follow them through as well'. What might have appeared at first to be 'mathematics anxiety' seems now to be part of the fabric of her constant worry about money and financial constraint.

This account could be analysed in a straightforward way. We could interpret Jean's responses to Question 2 and especially Question 4, as errors. We could read these errors as based on conceptual misunderstanding (rather than resulting from a mistake or slip, or from misapplication of an algorithm). Apart from purely conceptual problems, the subject is thinking in a context, influenced by a complex of factors (cf. Ginsburg and Asmussen 1988), which may include:

- affective factors, separate from cognitive ones
- beliefs, for example about mathematics and about herself as a solver of problems
- social class and gender.

Here the affective factors include several types of anxiety:

- mathematics anxiety, especially about percentages
- some anxiety about the interview
- anxiety about the relevant practice, namely tipping
- apparently chronic anxiety about money.

This *first reading* is the sort of analysis produced by Model B ('process') approaches; see Chapter 7.

However, we can reformulate certain factors to 're-read' the research problem, for example, by re-examining some of Jean's beliefs about mathematics. Having done both CSE Mathematics (Grade 3) and CSE Arithmetic (Grade 2), she distinguishes the Arithmetic course as 'useful', 'should be compulsory', from Mathematics as 'not useful', 'should be an option', and 'I don't like it'. She also distinguishes the topics in first-year Mathematics at the Polytechnic in the same way; for example, percentages, graphs and statistics are relevant, but gradients and algebra are not. This distinction is also expressed in slightly different forms by other subjects, as we shall see.

This distinction, expressed in Jean's beliefs and affect, has some affinities with the notion of *classification* (Bernstein 1996, see also Chapter 6) though there are important differences. Classification, as part of the structuralist reformulation of beliefs and affect around mathematics, refers to complex social and institutional processes constitutive of knowledge representations – inside and outside schools – and consequently the socially structured nature of what appears to be a student's understanding of mathematics. With structuralist approaches, it is the 'deep structure' of a social system, that is, its class character, expressed in linguistic codes, which shapes the specific identities of school subjects and the individual's experience of them. Thus, it is possible to analyse Jean's account in a second straightforward way: rather than a number of factors, we might focus on a *determining* factor, namely social class. Thus, for example, her chronic anxiety about money is likely related to a class-based position in a family beset by money problems and to her 'orientation to meaning'.[3]

However, there is something more in her talk which might attract the attention of the researcher. Listen: 'it goes something like that'; 'I sound horrible on tape'; 'I wasn't very good at all'; 'I always get the formula wrong'; 'I always mix it up'. We might hear 'anxiety' or 'uncertainty', but all we have are *indices* of something other than words. We might quickly disregard them, calling them 'self-defeating self-talk' (Tobias 1978). Or we might pay attention to these signifiers marking her talk. They point to distinctions such as: right/wrong; true/false; (aesthetically) good/bad; all 'mixed up'. Meanwhile, her talk is asserting the usefulness of mathematics as practical. Following up this thread of signifiers and the related distinctions attracts our attention, as researchers, to the discourses that hold these distinctions in place.

As we continue listening to her narrative about her imminent trip, and her anxiety about it:

I haven't had a holiday for three years, so this will get us to America: it's the only way I can ever make it.

(Interview transcript)

This, and other passages quoted earlier, speak of unsatisfied desire, deprivation, anxiety. Following it would initiate a shift from a language of conceptual divisions to a psychoanalytic language of 'desire', that is, towards the use of Model C, a psychoanalytic approach, and Model D, which includes insights from poststructuralism (see Chapter 7). Three examples are given.

First, some of her talk suggests a possible defence against mathematics anxiety indicated by what might be called 'insouciance' (not caring). For example, during her thinking about Question 7 (cost of cake ingredients):

> I couldn't remember how many ounces is in a pound: I might have put sixteen for the flour and twelve for the sugar, or vice versa; then again, I might not have.
>
> (Interview transcript)

This appears as a defence against possible failure to solve the problem.

Further, Jean makes two errors in questions later in the interview. For Question 6 (best buy), she chooses the smaller bottle of ketchup (see Chapter 9). This answer was considered 'incorrect' in the interview problem-solving context. However, for someone worried about money in shopping, in practice, it might well be 'correct' to buy the smaller bottle. (Interviewee number 1, another working class woman, made a similar 'error' in this problem.) For a later problem on the cost of sports kit (see Evans 1993), her response is coded as too small: though it is correct for *one of each* piece of clothing, it is not for the problem posed – about buying 'a change of' socks, shirt and shorts – that is, two of each. Again, we can see that this error might be 'motivated' at an unconscious level by money worries. These examples also show that inferences about her positioning point to difficulties in judging responses to certain mathematical problems as simply correct or incorrect.

Finally, in my interviewing of Jean, I was more 'regulating' than with other subjects, offering her paper for calculation before both Question 1 and Question 2, and making a slip myself when I refer to the interview problems as 'this test', just before presenting Question 3. These actions might be explained by my relative inexperience with these interviews when I did this one (see the reflexive account) and perhaps my anxiety about them. My actions might also explain, at least partly, Jean's calling up school maths early in the interview, even for Question 4 (see earlier).

To summarise, one can analyse Jean's case in terms of a complex of factors, seeing social class as the determining factor. However, if we attend to the play of her language, as a thread of indicators, we are led from a language of conceptual divisions and oppositions to a psychoanalytic language of desire.

Another example how the research process can take account of such a move comes from Ellen's case study.

Ellen's Story

Ellen was aged 19 at entry, middle class (by parent's occupation), with an A-level in Mathematics. A student of town planning, she worked part-time, currently as an electronics assembler, and previously in a shop.

Ellen's performance on the questionnaire the previous October had been strong, with all twenty-two questions correct, except for Question 18 (10 per cent tip on £3.72), considered a practical maths (PM) item, where her response '37.2p' was scored incorrect. Her mathematics anxiety responses, for both numerical anxiety and maths test/course anxiety were well below average.

Ellen expresses overwhelming confidence after almost every question, for example for Question 1 (reading a pie-chart): 'very familiar, know exactly what it means . . . don't have to think about it', and throughout the interview.

But not for Question 4! When I ask what a 15 per cent service charge on a meal would be, she says 'Well, I'd have to use pencil and paper'. Then:

S: [7 seconds] [something inaudible, coughs] . . . [6 seconds] Well, 23 ½ pou – no, that's wrong . . . [12 seconds] . . . what I've done wrong, oh —
JE: Is it wrong?
S: Yeah, umm [laughs nervously] . . . I don't know what I'm doing . . .
      [She realises she has *divided* 15 per cent into £3.53, instead of multiplying]
      (JE [2 lines])
S: [15 seconds] . . . 52.95, 53 pence.

(Interview transcript)

Thus she recovers from her slip through critical evaluation of her first answer (see also Chapter 9). She explains that she rejected the answer produced by dividing because 'I just saw that it was obviously not right . . . it was far too small'. A *first reading*, a Model B analysis, would interpret her slip as just an aberration. We might conjecture also that she is feeling some anxiety, probably 'maths anxiety'.

This episode deserves closer consideration, beginning with her answers to the contexting questions. When I ask if she ever goes to a restaurant with a menu like that shown, she seems to reply very quietly and hesitantly. After she chooses the seafood platter (£3.53), I ask how much she would tip for a restaurant meal: she replies, again somewhat hesitantly, 'well 15 per cent, I suppose'.

After calculating her answer, she responds to contexting question (R):

S: No . . . I don't usually pay . . . I mean, I usually look at prices and things, . . . add them up in my head . . .
JE: Even if you're not paying?
S: I don't want to be an expense.

(Interview transcript)

When she is first asked to 'choose from the menu', she seems to call up the practice of eating out at restaurants. However, she reverts to using pencil and paper – indicative of school mathematics (SM) positioning – to calculate a 15 per cent tip. Her 'critical evaluation' of her first answer as 'obviously [. . .] far too small', could be supported either by SM or eating out practices. Though I considered SM her 'predominant' positioning for Question 4 in Chapter 9, here we can ask if more than one practice was called up.

Her initial slip and her response to contexting question (R), are suggestive of anxiety. Indeed, this anxiety might be understood as specific to positioning in one *or more* of the following practices, as follows:

1   anxiety about doing the problem itself, requiring a slightly more complex calculation (15 per cent) than she has so far had to do
2   anxiety to do with the interview itself, experienced as an evaluative situation
3   anxiety about doing the right thing in a restaurant.

The anxiety evinced might appear at first to be 'mathematics anxiety', since it appears while doing something that an observer could choose to label as 'mathematical'. However, calling the anxiety 'mathematical' would be accurate only if we assume that her positioning was solely in a school maths discourse, as for (1). It would not be accurate if her positioning is *interdiscursive*: that is, if more than one of the discourses indicated is called up.

As already indicated, support for (1) comes from her use of written calculations. Her hesitation and so on, immediately on being presented with the restaurant menu, that is *before* any calculation is asked for, provides support for (3), and also for (2), although, if she were feeling anxiety about the interview, we might expect indications of that before Question 4. Further support for (3) comes from 'I don't want to be an expense'. However both (1) and (2) are questioned by her performance on the next problem: there she has to calculate a 9 per cent pay increase, another 'non-10 per cent' calculation – but she gets it right first time. On the basis of these interpretations, I argue that her positioning is interdiscursive – that is, in 'eating out', as well as in school maths, and possibly as interviewee.

We can note the specificity of her positioning in gender, age and social class terms. Her tipping rule of '15 per cent' was unusual for most people in the UK – especially students – at that time: it is mentioned by only two other subjects, one of whom, Jean (see earlier) was about to leave for the USA, where 15 per cent tipping was customary. Suggesting 15 per cent tips might signify 'wealth' and 'generosity', but, as a 19-year-old student, she had to work part-time, and to limit her spending at the supermarket. The wealth and generosity seem likely to be someone else's.

Indeed, it emerges that she 'doesn't usually pay' in restaurants, and she doesn't want to be 'an expense'. The position of not paying when you eat out is one which is relatively more available to women, and/or to people younger than their hosts. Also the restaurants where she ate were likely for middle class (or financially unconstrained) customers.

The term 'expense' signifies in different ways, as an amount which could be arithmetically calculated within a mathematical or related discourse, or as being a burden within a relationship with other(s) on whom the subject is dependent (a parent or partner). This signifier thus functions at the intersection of these two discourses. Its negative connotations in the latter context suggest anxiety – associated with eating out, with the relevant social relationship(s), and with the operations involved: choosing a dish, calculating the total cost of her meal, and so on. In addition, the play across these different senses, related to different positionings, itself may generate anxiety.[4]

Thus, my *second reading*: Ellen's positioning is in a 'mix' of discourses: related to eating out, school mathematics, and perhaps being interviewed. The eating-out discourse is linked with discourses around relationships by a semiotic chain, including the signifier 'expense'. Her performance, thinking, and critical reflection need to be understood in this context. Similarly, the anxiety she seems to exhibit is in context. The anxiety would not be simply 'mathematical', if her positioning includes discourses in addition to school maths or college maths. The anxiety apparently triggered by this question is, at least partly, anxiety about the context of eating out, about the relationship(s) in this context, and about being a burden in that relationship. This reading is based on Model D, but without psychoanalytic insights so far.

The need for psychoanalytic insights is suggested by the conflict between the picture of overwhelming confidence about mathematics and numbers, backed up by a very competent interview performance, and the anxiety suggested by the slip and other responses to Question 4. Might her expression of confidence be interpreted as a *defence against anxiety*?

When I ask, towards the end of the interview, how she feels 'about the way you're able to use numbers these days':

S: [6 seconds] I feel, sort of, confident, I suppose, 'cause I feel I should be confident . . .

JE: Why . . . ?

S: Well, um, [4 seconds] given the sort of qualifications that I've got, and that the course is aimed at people who have less [. . .] the numbers aren't, uh, don't really give me the problem, the working, you know, the calculations. But it's more sort of [3 seconds] [quietly] talking about them [laughs nervously] . . . I don't know if anything gives me problems but on the most recent worksheet, there's the bit which I've – groan when I think about it, about talking about how you might *use* numbers [3 lines] . . . I'd just be much more interested in doing a calculation, getting an answer you know, having that done . . .

 [6 lines] . . . that being *it* . . .

(Interview transcript, her emphasis)

Her repeated expressions of confidence may cover up some amount of anxiety – exhibited for example by the instances of nervous laughter – and these expressions can be seen as a defence against anxiety (Laplanche and Pontalis 1973: 103–11).[5]

Next, Hunt's (1989) assumption that unconscious images and thoughts sometimes appear in jokes, 'slips', and so on supports a conjecture that anxiety was triggered by the presentation of the menu, prior to being asked to calculate a 15 per cent tip. Thus it was anxiety (to a great extent, at least) about the context of eating out, perhaps about the relationship(s) in that context. We might next conjecture that the fact that she made a slip is related to these anxieties. The *content* of the slip might be related to these anxieties, too: the latter, involving division rather than multiplying, led to a result that was *smaller* than it should have been. When we remember that she later admits to wanting not to 'be an expense', we might say that her slip was 'motivated' by the anxiety. Thus we can note the role that unconscious anxiety may possibly have played in the Ellen's responses.[6]

Third, her anxiety about being a 'expense' illustrates the links between the linguistic ideas of metonymy and metaphor, and the phenomena of displacement and condensation respectively, as emphasised in Lacan's discussion of the importance of language/discourse in psychoanalysis. The idea of being an 'expense' may be linked – metaphorically – in this woman's history, with that of being a burden in a relationship, one that is infused with desire. Because of anxiety, guilt, pain, and so on associated with this, the signifier is likely to be 'suppressed' (Walkerdine 1988). When a problem is presented involving choosing from a menu – with prices attached! – and when she is asked to calculate the amount of a tip, this calculation will be linked – metonymically, through the idea of summing – with the signifier 'expense'. This key signifier is thus located at the intersection of two, at least, discourses, and there is a play of meaning across its different senses. We could say that multiple meanings are *condensed* on the signifier 'expense', as they were on 'peeling oranges' in Hollway's (1989) illustration (see Chapter 7). Also, the linkage between the two discourses allows the strong feelings based on desire in the discourse(s) around her relationship and 'eating out', to be associated with this particular problem, – which at first seems so simply mathematical! At the same time, we can say that this subject *displaces* her anxiety about being an expense by moving along the chain of signifiers – from 'burden' to 'expense' to the calculation – and focuses it on the calculation!

Finally, why did I depart from my normal script for Question 4 to give her a more difficult question (15 per cent) than other subjects (10 per cent)? Here is where the reflexive account for Ellen's interview is useful:

> I was not aware of having met this student before the interview; our contact would have been confined to the first year Maths lectures that I gave. In the interview, I was struck by the fact that she had A-level Mathematics (attained by only 7 per cent of that year's entry) and was convinced at first by her expressions of confidence. I was concerned about how the interview was going – especially that she might be bored with such easy questions. Then came her responses to Question 4. Here, given the 'opening' by her mentioning 15 per cent as a tip, for the earlier 'reasons' in play, I asked her what a 15 per cent tip would be, as a more 'challenging' problem than that [10 per cent] posed to other subjects.
>
> (Evans 1993)

With psychoanalytic assumptions in mind, we can see that this standard 'reflexive account' needs augmenting. On reflection, I recalled feeling some anxiety myself: I had not met Ellen before, unlike most of the other interviewees. The decision to give her a more difficult problem can be seen as 'motivated' *inter alia* by anxiety that, if she found the problems too boring, the interview would not be a 'success'. This can be seen as an example of *transference*, the subconscious reaction of the researcher to the interviewee, based on the imposition of images onto her – for example, boredom – for which there was no substantial evidence.

By bringing in Models C and D, we show the complex character of Ellen's case, by considering:

- context not as a natural setting 'out there' (classroom, restaurant), but as constituted by practices, which are also constitutive of the 'individual's' subjectivity

- feelings which are not prior to, or outside of, discursive practices which position her as woman, as student, as 'poor', as interviewee, etc.
- language, not as simply representing preconstituted states of affairs but as actively producing them, for example through the *inter-relation of discourses* via key signifiers, like 'expense'.

Thus, shifting attention from consciousness (intentions, beliefs, and so on) and rationality, to semiotics and the unconscious, has opened up possible interpretations. In this *third reading*, following poststructuralist ideas such as meaning as a play of signifiers, and 'interdiscursive' positioning – as well as psychoanalytic concepts such as condensation, displacement and transference – a much richer explanation is constructed.[7] This analysis must be related to specific discursive practices – indeed to specific 'key signifiers' – and to a positioning that is related to (though not determined by) social differences such as gender, social class, and age, and that is generally interdiscursive.

This analysis shows how to extend the usual qualitative methodology so as to provide a fuller discussion of affective issues, through the use of psychoanalytic insights (Hunt 1989).

Fiona's Story

Interviewee number 5, 'Fiona', was 26 at entry, and middle class (by her parents' occupations and by her own). Her qualifications in mathematics were 'a very poor CSE grade and a very poor [O-level] grade'. Previously, she had worked as an unqualified social worker, where she used numbers very little, and as a financial adviser, where she needed to use numbers 'a lot'.

On the questionnaire, her performance was relatively poor (7 out of 10 SM items correct, in the bottom quarter of the sample, on a not very difficult test). Given her O-level in Mathematics, this is rather surprising. However her maths test/course anxiety score was around the ninety-fifth percentile of the whole sample. We can use the 'inverted U' relationship between school mathematics performance and maths test/course anxiety, produced in Chapter 4, to alert us to possibly 'deviant' performances. Using this resource, we can interpret hers as a 'slight under-performance' in that her observed score on the SM scale was less than would have been expected, given her mathematics qualifications, age, maths test/course anxiety score, and so on.[8] So in the interview it is important to look for evidence of any particular influences or features of her situation, which might account for her under-performing on the SM scale at the start of the year.

In the life-history part of the interview, she mentions that she was very 'unlucky at school':

> I had a very, very good maths teacher. She was very, very aware of people's problems [. . .] she used to work through step by step . . . and then she left a few months before I actually sat both my, the CSE and the O-level, and I *went downhill very rapidly*. I don't know whether it was a question of confidence, or inability [two lines] . . . I just felt that once she'd left, it became – it sounds

funny – but it became very, very *mathematical* . . . [. . .] Nothing, after she left, nothing was explained. We were just given the formulas and told to get on with it.

(Interview transcript, my emphasis)

This theme of 'sudden decline' or discontinuity of previously good performance in school mathematics, is one which emerged in a number of interviews (see 'Conclusions' to this chapter).

When I ask Fiona to 'take a look at a few questions', she responds immediately: 'Oooh [quietly] [. . .] paranoia's struck [. . .] Are they simple?' This is an example of what I call 'mock-anxiety', in that these appear to be expressions of anxiety, but their manner raises a question about whether they are genuine (see later).

For Question 1, I have discussed (in Chapter 9) her avoiding of school maths (SM) approaches, and her positioning herself in 'general knowledge' discourses, leading her to 'refuse the terms' of the question, and to give an incorrect response. She goes on:

I always had difficulty with that, I didn't enjoy it at all. School wasn't a particularly happy time for me anyway, so you might well find that a lot of my answers are negative . . . [4 lines] . . .

I was never explained how to work through it step by step so it certainly makes me feel very anxious [. . .] I don't actually trust my own perception to actually give the correct answer, because I don't feel I [. . .] know how to work it out properly, so therefore I don't think I would give the right answer – if that makes sense.

(Interview transcript)

These passages suggest that school, and school mathematics, are sites of much confusion and 'negative' affect for her: lack of enjoyment, unhappiness, anxiety, lack of confidence, and self-mistrust.

For Question 3 (graph of the price of gold, see Appendix 2), she appears to get lost in the detail of the gold price changes, for reasons which emerge:

JE: . . . which part of the graph shows where the price was rising fastest?

S: Maybe it's me being ignorant . . . but there doesn't actually seem to be any time specification along the bottom [axis of the graph] – which I find quite confusing [. . .] my father's a stockbroker, so I do understand a little about opening and closing . . . [6 lines] . . .

I mean there actually appear to be two peaks here, but I should say maybe when gold is at 650, it seems to rise very rapidly in the afternoon until close, and afternoon business, you know, afternoon trading. . . [trails off]

(Interview transcript, her emphasis)

She then confirms that she considers the price to be rising faster in the afternoon (rather than the morning) – which is wrong. Again, it appears that she refuses the terms of the question, and draws on information from an 'outside' discourse to answer. She then goes on to read the lowest price of the day as £580 (rather than £590), also incorrect.

A *first reading* using a 'cognitive-constructivist' model (Model B) would explain that her expressed 'confusion' leads to unsettled thinking and problem-solving performance. This confusion comes from the discrepancy between the presentation of the problem and her expectations, and is constructed (at least partly) as emotion. This interpretation is supported by her expressed anxiety, and by her lack of confidence in not trusting 'my own perception to . . . give the correct answer'.

However, for this problem, though she also mentions college maths (SM) – 'graph work' – in response to contexting question (C), she actually seems to call up what might be called 'money maths' (PM), from the position of a stockbroker's daughter:

S: my father dealt with money all the time, um, because he was a stockbroker, and therefore it was the essence to him and his making a living, but it wasn't anything that we were allowed to sit down and discuss, or even talk about, or offer advice [. . .] we were always told we wouldn't understand [. . .] because time is money, money is time and he hasn't got time to explain to me the information that he thinks is going to be relevant to me at a later date because I'm a woman and I don't understand . . .

JE: Is it – a woman, or you're a child? . . .

S: I think it's very much both . . .

JE: What about your mother? Does she, is she allowed to ask questions?

S: Well, no, no, just the same. Family and business should never mix [. . .] my mother wasn't ever allowed to ask and it certainly affected her far more than it did us because as a stockbroker, your home and your material valuables are on the line all the time [. . .] on a couple of occasions the family home was under great threat [. . .] It wasn't something that family and children discuss . . . [two lines] . . . he was the man of the household and he could deal with it [. . .] . . . most of the time, it was like living under a time bomb (JE: mmm, mmm, I can appreciate that) especially if you don't quite know how the time bomb's made up or when it's going to explode.

(Interview transcript)

When I asked how she saw his work, to pick words, adjectives to describe his work, she replied:

capitalist, corrupt, business-like . . . um, mathematical, calculating, devious, unemotional.

(Interview transcript)

Here is a lot about her perception of her father and of family life and a lot of feeling in these passages (and others, not quoted here, but see Evans 1993). She seems *angry* at being positioned as a child who is deprived of information about her father and his work, because she 'wouldn't understand'. This lack of knowledge is linked to the anxiety she exhibits, perhaps most graphically in her comment that growing up with a stockbroker as father was 'like living under |sic| a time bomb . . . ' She uses a similarly striking metaphor in discussing Question 2 (abstract 10 per cent calculation):

I don't think that – if you haven't the knowledge fresh at hand, that you can then attain the next step because it's, it's like building a tower block without the foundation. Everything collapses underneath you, and I feel in some ways that's what happening now [in maths].

(Interview transcript)

And being positioned as a child – and also perhaps a woman – who 'wouldn't understand' is likely to have contributed to her *lack of confidence* in school, and in mathematics, mentioned in her list of epithets for her father's work.

Question 4 (10 per cent tip) seems to call up eating out:

S: [. . .] where I'm going to get something special [. . .] not something that I can cook at home. I don't believe in wasting money . . . I guess that's something I've been taught from an early age, not to waste money . . . . [2 lines] . . .

JE: [. . .] Would you mind choosing a dish from that menu?

S: [. . .] I'd go for the grilled trout . . .

JE: Right. Now supposing at this place, service is left to the customer—

S: —so when you take me out, that's what I'd like . . .

JE: I see, oh, I see, OK, I'll remember that [she is laughing] So, [. . .] what would you do about leaving something for service?

S: It's usually 10 per cent, isn't it? [. . .] so now you're going to try and ask me what 10 per cent of £3.81p is . . .

JE: Yes, could you?

S: I thought you would [laughter] – oh, bless you! Yeah, I could hate you. No is the answer to that [both laugh . . . subject laughs] Well, it would be about 38p, but I think that's awfully mean . . . [6 lines]
       . . . if ever I was taken out for a meal as a treat [. . .] it was always the male that was left to deal with the paying of the bill and the tipping, unless [. . .] I said I was 'going Dutch'.

(Interview transcript)

This episode reveals her familiarity with calculating tips, both as diner and as restaurant worker. Though she comes from a middle class family, as does Ellen, she is somewhat older, and she is positioned differently in the discourses around eating out: she sometimes pays for her own meal (and tips), and she has been taught to regulate her eating out, so that she does not 'waste money'.

She also shows herself as able to play with multiple positionings: she moves from being positioned as student, or as interviewee, to 'being flirtatious' – that is, to positioning herself as a woman and possible companion in eating out. She then takes up a position momentarily in the interview, by offering 'mock-resistance' – as a joke – to the 'maths' question: 'No is the answer to that'.[9] Finally, she calls up 'eating out' to calculate the tip, with much less effort than the mathematically similar Question 2 required.

There are several issues which would benefit from deeper analysis, using psycho-analytic insights. First, it was valuable to make a request – unscripted – that the subject associate a chain of words, adjectives to describe her father's work; see Figure 10.1. There is much emotion and much ambivalence exhibited here.

```
capitalist . . . corrupt . . . business-like . . .
                          . . . mathematical . . . calculating . . .
                                    . . . devious . . . unemotional
```

Figure 10.1 Chain of Signifiers in Fiona's Associations with her Father's Work

The overall chain of thinking begins of course with the graph showing the changes in the price of gold, including one substantial fall at the start of the day (see Question 3 in Appendix 2). She associates this with her father and his work several times; there may be a defensive displacement (of anger, and so on) from the father to his work. In this chain, we can expect the affective charge, based on *desire* for her father, to flow between signifiers. The final signifier is 'unemotional', which may signify her father's rejection of her questions about his work, and thus of her. The *key signifier* in this chain is 'calculating', located at the intersection of family discourses about the father/his work, and mathematical discourses. In the former, it exhibits or signifies disappointment and anger, which is suppressed; in the latter of course, it signifies a central activity of the practice, and this may explain her ambivalence about getting clear how to calculate, in school/college mathematics at least (see the reflexive account later).

In her family, it seems that there was a 'splitting' (Hollway 1984): the father practised the rationality, while the rest 'held' the emotions, especially the anxiety about what might happen if the rationality carefully arrogated to himself were not sufficient and the 'time bomb' exploded. We read the signified of the time bomb as the anxiety. Though it first seemed to be generated by the 'mathematical' graph, and therefore to be 'mathematics' anxiety, another, more useful, description is possible: namely, that certain elements of the problem signify practices, which in turn signify strong (perhaps suppressed) feelings – here, anxiety, anger – in a way which is particular to this subject, and specific to her positioning.

In the perspective described, being a child (as well as female) excused her from doing calculations, from being rational: her father did *that*. This extends into adulthood: if she was ever taken out for a meal, it was generally 'the male' that was 'left' to deal with the paying of the bill (see earlier). However, she is perfectly able to calculate a 10 per cent tip in the interview. Thus, while her father – or generally 'the male' – has to perform, she tends to be protected from that.

These practices and relationships are constructed on the basis of socially available discourses; for example, 'he was the man of the household' so 'he could deal with it' In the chain of signifiers about her father's work, there are echoes of a corrupt capitalism, and of popular discourses about ('unemotional') mathematics.

We must investigate possible transference with this subject. I begin with the reflexive account:

> I knew this student fairly well, as she was in my seminar group in the first two terms, and also attended the (optional) tutorials regularly, as did interviewees numbers 2 and 7 (Alan), and several others. She was a bright but sometimes difficult student, on occasion arriving late, fidgeting, dropping jokey asides in

class. She hosted a party for the seminar group (at the beginning of November in her first year), which I attended. She was one of the students selected by the random sampling exercise in late May, and she agreed to do the interview.

(Evans 1993)

Her being a 'difficult' student may relate to the *resistance* that can be better understood in the light of the associations of mathematics in the chain of signifiers with her father's work and her feelings about him. It would not be unexpected for her to transfer that resistance to myself or my colleagues, other older (mostly) men who worked with mathematics. We can hear reverberations of her disappointment and anger in the following 'complaint' about teachers generally at the Polytechnic:

> it was too fast for me [. . .] as I said before, it's essential that, to be able to use any knowledge, that you understand *every element* – and I find even now, here, people don't like being questioned. I'm very often being told off for interrupting, questioning, interjecting, and to me that's the process of learning. I think it's essential. I can't, I'm not, you know, a blank slate, I do have thoughts, I do have a brain, and I expect to be able to use it, in anything, whether it's Maths or Philosophy, or Psychology or what.

> (Interview transcript)

What I originally called 'mock anxiety', another aspect of her jokiness, can now be seen as a *defence against anxiety,* pervading the practices discussed here and what she says about them.

This interview allows me to explore the themes of the influence of family discourses, including age-specific and gender-specific positionings, in generating emotion – here, especially anger and anxiety – that may be displaced onto mathematics, and which may be studied by examining a 'chain of signifiers'. In the interview, this subject was able to play with multiple positionings. She also provides several illustrations of transference and 'resistance'.

Harriet's Story

Interviewee number 16, Harriet, was 28 at entry, working class by parents' occupation and middle class by her own (residential social worker). She had passed CSE Mathematics (grade 2), but had not been entered for O-level. She was an intending Social Work student. The reflexive account for her interview was as follows:

> I knew this student from Social Policy classes, during the first two terms. I had had several further contacts with her. Once, as coordinator of first-year Maths, when she claimed 'unfairness' in the different ways that groups were helped with their worksheets, I was resistant to meeting her about this (and later apologised). This latter event (from the previous term) did not seem to be on her mind at the interview, held on a hot summer afternoon.

> (Evans 1993)

On the questionnaire, she scored overall only 15 correct of 22, with three wrong (apparently 'slips'), and four not attempted. Her average responses for both mathematics anxiety dimensions were below average. Her 'unexpectedly low' performance score – given her mathematics qualification, gender, age, TCA score, and so on and using my version of Model A – means that she might be called an 'under-achiever', at least for that occasion, and those scales (see note 8).

So in analysing the interview, I was looking to her performance, and also for reasons why she might 'under-achieve'.[10] However, her performance in the interview was excellent. For all five questions tackled, she did the calculations in her head – indicative of her fluency and accuracy in mental arithmetic, and got them all right – until Question 5B (see later). Interestingly, she seems to call up school mathematics for very few of the interview problems.

For Question 2, she first calls up 'in a shop, working out, you know, the actual cost, how much you'd get off'. She then immediately displays her method of working out 15 per cent by adding 10 per cent to half as much again. Her recurrent discounting fantasies may support this creative method, though she does not give herself much credit for it. This fluent method of mental calculation can also be compared with the need of Ellen – much more successful in school mathematics – to calculate a 15 per cent tip on paper.

She then recalls helping the residential home children with their maths work – thus being positioned as 'teacher' and relative expert in school maths. Her answer '66p, 66 ½p, if it's pence' (correct), interestingly, transforms an abstract question to a practical form (as did several others, see Chapter 9).

Question 5 calls up 'payslips' for her, and she does the approximate calculation of 9 per cent of £66.56 quickly in her head: '10 per cent will be 6.65 ... just under 6', and very accurately! When she is asked to work it out exactly, she is reluctant to try to do so in the interview, before she can recall how it was done at school:

JE: What would come up when you tried to remember?
      [2 lines]
S: [4 seconds] something to do with insurance ... something to do with the figure times possibly 9 over 100 – I'm not sure – I think I'd have to ask actually, or look it up in a textbook ....
JE: Do, there's some paper there if you want ...
S: [laughs] No!! ... [6 lines] ... I always feel much easier if I can go and look something up and check it, ... than just try and guess if I don't really know ... [...] ... – which is why I wouldn't do it now [laughs].
                                                                    (Interview transcript)

Such a 'refusal' response to Question 5B was made by six subjects, five of them women (see Chapter 9). She explains her reluctance to 'guess', by the importance of 'getting things right' for credits awarded within the competitive 'house system' at secondary school:

S: things I was quite confident about I'd just get on and do, but something like this which I was never confident with, I'd sort of flounder about. I think I'd find them quite difficult to do as homework as well, partly because there wouldn't be anybody to ask, at home . . . Yeah, homework, you're on your own: unless you've got somebody right nearby who can give you a bit of advice or a book that you can refer to, then you're stuck, if you don't know what you're doing . . . and I used to feel that quite a lot when I was younger – and if I didn't get it right, my dad used to shout at me [inaudible] . . . [eleven lines, most quoted in the next extract] . . . so it was easier not to ask him for help, because then I wouldn't get told off . . . [3 lines] . . . and he didn't really agree to me taking it round to other people . . . [8 lines] . . . it was a bit awkward 'cause I lived in quite a small village and out of about fifteen of us, only four of us went to grammar school, one was a boy and so he went to the boys school which was very different . . . [7 lines] . . . and we didn't have a phone then, so [. . .] there wasn't many people actually around that I could get in touch with . . . . I only realised that now.

(Interview transcript)

Thus the school put the pressure on her to get things right, as (we shall see) did the family. In addition, she was isolated: with 'homework, you're on your own'.

It is important to consider her position in the family. She was the oldest child in a working class family and the only one to go to grammar school:

S: That *set me above them* anyway, and I was *supposed to know* . . . My mum was never very good at maths, and my dad – he had quite a good head for figures – but . . . he expected me to know, and if I didn't, he used to tell me off, and as far as he was concerned, I hadn't been listening in class – and I should know.

(Interview transcript)

Thus, she was positioned as 'knowing' or at least 'supposed to know' by her family, but also isolated, because of her attainments. At school, she had not very good termly reports – and mathematics was always the lowest – but surprisingly good exam results. However, her father took more notice of teachers' reports, and she used to 'dread' taking them home. When she got Grade 2 CSE: 'I remember being quite surprised . . . I was that *good*' (her emphasis). Towards the beginning of the interview, she recalls an especially poignant memory:

S: At the last open evening, the teacher [told] my mum and dad that I could go on and do an O-level, A-level in Pure Mathematics, she never said a word about that to me . . . I enjoyed maths, but I *never knew* I could do that well . . .

JE: So how did you feel . . . ?

S: . . . Cross!

(Interview transcript, her emphasis)

Apparently, her career in school mathematics was plagued by knowing a fair amount – but *not knowing that she did know*. She is caught between the teacher who tells the parents but not her, and her parents, who have been told that she knows but don't tell her either.

Here we could explain her 'under-achievement' in school (or in her question-naire) performance – using Model B, and referring to life-history material, and her feelings about it: she 'lacks confidence', and no wonder! Her talk seems to support this reading, on one of several occasions when she expresses anxiety about mathematics: she felt 'uncomfortable' about percentages, because 'I'm never quite sure whether or not it's quite right [laughs]'. She also exhibits a lack of confidence about mathematics, in her eagerness to know 'what's right ?', at the end of the interview. When I confirm that she got all of those she attempted correct, she muses:

S: It's really odd how I need to be reassured that I'd done it right . . . I'm sure that stems from years ago, . . . not having the confidence, I suppose . . .
[2 lines] . . . because the teacher just didn't say, that, you know, yes, well done, you can do this.

(Interview transcript)

Thus, she feels that she did not have sufficient reassurance or encouragement from school. In addition, her parents' failure to relay the teacher's positive evaluation might be read as 'lack of encouragement' from home. However we can see her case as more complex, and related to her positioning relative to knowledge and knowing, in the two contexts.

In the home discourses, according to her father principally, she was expected *actually to know*, as a schooled person, fortunate and rare in that family. For him, you come to know, by paying attention in class, and doing your homework seriously, that is, on your own. You know that you know according to what the *teacher* says, the legitimate authority. In the school discourses, she was positioned as a candidate for knowing, by passing the eleven-plus exam and being selected for grammar school. Indeed you *have to know*, in order to survive in the competitive ethos. However, she remembers being awarded end-of-term marks and teacher's reports that always seemed 'quite low', especially in mathematics – compared with her exam marks; she 'could never work out why' – and does not recall being told, or 'reassured'. Further, the teacher, knowing the student could do O-level or A-level Mathematics, perhaps because of school protocol, tells only her parents, and not her.[11] The parents defer to the authority of the teacher to say who knows, and the daughter is kept 'out of the information loop'.

Thus, she seems to have been caught in a *double-bind* (Bateson 1973), at home and probably also at school, where she does not remember getting the reassurance she longed for. Further, there is inadequate communication (and possibly conflict) between the two domains. This caused confusion and ambivalence.

Yet her activities since school have afforded her new, more authoritative, posi-tions. In the residential home, she was responsible for the children as 'a mum and dad would be', and she had done remedial mathematics (and other subjects) with them. As a student at the polytechnic, she was also positioned as an expert in college mathematics: 'A lot of the maths that I've been doing has been stuff that I already knew, that I had to know, to help the children'. She helps other students. She even had an argument with another lecturer 'about whether or not you do it [logarithms] in such a way', and 'was quite pleased, 'cause I knew I was right, 'cause I love working with logs, I love messing about with tables, I really enjoy that'.

Besides anxiety and diffidence, another emotion often expressed in this interview is *pleasure*. Her description of the 'argument' with another lecturer shows this. It was also 'pleasantly surprising' that she 'could do quite a lot' of maths at college:

S: having a formula and . . . working it out, I love that sort of thing, . . . the idea, the sheer pleasure of doing that is just really very nice . . . I used to enjoy binary numbers . . . I haven't thought about this 'til today [said with pleasure] . . . . I used to enjoy geometry . . . .

(Interview transcript: 17–18)

This student, like Donald (see later) and as much as any other interviewee, reports experiencing pleasure from academic activities, including 'maths problems'. This apparently involves mostly the application of general, fairly technical, methods or formulae.[12] She is able to recall – and to reclaim – this enjoyment in the interview, though it has been overshadowed by her failure to reach her potential at school.

We could analyse some of these descriptions at a deeper level, using psychoanalytical insights.

Her 'diffidence', generally in maths, and particularly in attempting Question 5B, has been interpreted in terms of family pressure, such as her father's 'telling her off' when she couldn't do her homework, in terms of school pressure to perform, and in terms of 'lack of communication' between home and school (Brown 1990). However this lack of confidence might be also seen as 'intrapsychic conflict' (Hunt 1989) about knowing. Her father shows her that he wants her to know. However 'knowing' – which is how they read her going to grammar school – isolates her: 'it set me above them'. It may lead to 'fragmentation, having to be "somebody different" at home and school' (Walkerdine et al. 1989: 111–13). This knowing and the consequent educational success may also lead to

the trauma of leaving and isolation, the disdain with which one is supposed to view the place from which one has come and the terrible guilt that we and not they have got out, have made it, and will work in conditions which they can never know.

(Walkerdine and Lucey 1989: 12)

This would explain Harriet's ambivalence about 'knowing'. However, her father may also have experienced conflicts over his daughter's knowing. He appears to have had a strong 'investment' (Hollway 1984) in her 'knowing', but the latter also distanced her from the family, and from him. This might explain her parents' not communicating the teacher's view that she could do O-level Mathematics.

Might her *pleasure* in applying formulae be interpreted otherwise?

S: somebody gives me a way of doing something and then says you've got this bit, now go and do it, I mean I love doing that, the idea, just the sheer pleasure of doing that is really very nice . . .

(Interview transcript: 17)

Thus the formula seems to be empowering, to give her access to authority, security: but perhaps not to the full 'rationality' of mathematics. The pleasure seems limited to a relatively technical orientation, rather than being derived from 'mastery' (Walkerdine 1988). The formula may keep the anxiety, the discomfort, the panic at bay.[13]

As for her reported pleasure from confidently challenging the other maths tutor, might this come from *anger displaced* from the mathematics teacher who failed to let her know that she knew?

Her resistance to attempting Question 5B (9 per cent of £66.56) in front of me might be related to her experiences of trying to solve mathematics problems in front of her father. It might be an example of *transference*. However, none of the indicators that Hunt (1989: 57ff.) mentions for transference – dream material, fantasies, slips, jokes, or strong emotions expressed by the subject towards the researcher – has been observed in the interview material, or in the reflexive account.

Harriet's case allows exploration of a number of themes:

* explanation for 'under-achievement' in school in terms of anxiety or 'lack of confidence'; possible double-binds within, and conflict between, home and school; and/or ambivalence
* the importance of social class, and her family relationships, especially with her father
' descriptions of instances of pleasure, and the use of 'creative fantasy', in numerate activities, along with psychoanalytic interpretations of these.

This interview functioned not only as data production for my research, but also as *consciousness-raising* for the subject (Carr-Hill 1984). It gives a space for her to recall earlier achievements and satisfactions: 'I haven't thought about this 'til today', and to make new connections: 'I only realised that now.'

Alan's Story

Interviewee number 7, here called 'Alan', aged 20 years at entry, had a CSE Grade 2 in Mathematics, having 'failed O-level completely'. A student of psychology, I judged him to be middle class, given his accent, and his having attended public school. The reflexive account for his interview was as follows:

Alan was a member of the group I had taught during the first two terms – as were Fiona and number 2 – and attended both seminars and tutorials regularly. He was often very diffident about maths (but eventually performed well in the final assessment). I found him easy to get on with, and had met him socially, at the party held by Fiona in October. There my friend had a long talk with him about his background and schooling. When I found I was short of male interviewees in his year, I asked him if he would give me an interview.

(Evans 1993)

On the questionnaire, his performance on the items classed as school mathematics was 10 out of 10 correct. He omitted all practical maths items from Question 19 onwards, which suggests he was working slowly and deliberately. His anxiety responses, most strikingly, were mostly '4'. Thus perhaps more than any other interviewee, his responses are clustered around the 'neutral' point (on the seven-point scale). Alan thus seems a 'typical man'.

Because of his 'unexpectedly high' score on the SM subscale (given his gender, age, mathematics qualification, maths test anxiety score, and so on), using my version of Model A, he appears to be an 'over-achiever', at least for his performance on the questionnaire in October of his first year (see note 8).

In the interview, when I ask what Question 1 reminds him of, he is impatient to give me a 'statement' about his feelings about maths:

S   [abruptly] It reminds me of earlier maths – uh, concerning my feelings about maths, I'm – it's *very* neutral – I don't have any strong feelings about maths. I have difficulties in maths, but, um – and I can't concentrate for too long on maths . . . I don't have *worries* about it, I just stop and have another go later on [laughs quietly]

JE: . . . Why is it that you can't stick with it? . . .

S: [. . .] it's not usually related to something that I want to know. Say, for example, psychology – I cope with the maths because I'm interested, I want to find out something through it, *via* it, therefore I'll stick with it a lot, lot longer than just maths on its own . . . [2 lines] . . . If I need it, I'll just use it, but otherwise I just don't want to study it as a subject.

(Interview transcript, his emphasis)

He appears clear that he doesn't have any strong feelings – and certainly no 'worries' – about maths. Another 'statement' about mathematics follows Question 3 (graph of gold price):

JE: Right, does that remind you of anything, anything you do these days?

S: Not particularly, no – I wouldn't use that, wouldn't look at it, wouldn't be interested in it, [3 seconds] unless I was interested in gold [laughs].

(Interview transcript)

Of the six questions he attempted, all but one were correct. For Question 5A (9 per cent wage rise on £66.56, approximately), his response – 'roughly about £6' – was very close to the exact answer (£5.99). His response to Part B ('exact calculation of 9 per cent increase), after a fifteen second pause was: 'No (I couldn't tell exactly) . . . I'd have to look it up again', in what he calls his 'comprehensive' mathematics textbook.

When I ask which of the two bottles depicted for Question 6 he would buy, he points to the larger, but admits:

S: It's not the sort of thing I'd work out unless I was [. . .] living on my own, [. . .] if I was short of cash; but as I don't normally buy the food, I'd just probably buy that [the larger jar] presuming it was cheaper. I wouldn't bother working it out and spending the time . . . the time's more valuable to me than the money at the moment . . . [3 lines]

. . . sometimes my parents go away, so I have to buy my own shopping; it's not usually my own money, it's usually theirs [laughs] . . . [4 lines] . . . I don't spend too much time working out, I just get what I want . . . . [10 lines] . . .
Sometimes, it's the sort of problem that does come up in mathematics, in O-level or CSE [. . .] In that case, I would spend time trying to work it out, in the correct manner [. . .] But at the moment, I'd probably be pretty lazy [laughs].

(Interview transcript)

Alan also seems to have a clear idea of the difference in approach required in the different activities of shopping and school maths. In practical problems, he is more calculating about time – and effort – than about money. This is perhaps heightened by its being *his* time whereas the money is provided by his parents, through family routines, themselves based on family discourses about spending money. Here these discourses are based on a material foundation of plenty, a lack of constraint: 'I just get what I want'!

School maths problems, in contrast, need you to 'spend time trying to work it out, in the correct manner'. This might help to explain why he was one of the estimated 5 per cent of students who did not get beyond Question 20 in the questionnaire performance scale: he was 'spending time' on questions he perceived to be school maths. (For the eighteen questions actually attempted on the questionnaire, he made only one mistake, a slip.)

Thus, a 'Model B' explanation for his 'over performance' on the questionnaire might be that he works painstakingly and carefully, because of anxiety or other emotion. However, I was unable to find any clearly expressed feelings about mathematics in the transcript (see Chapter 9, Theme 5). For example, he reports feeling basically neutral about the interview :

S: So, I was pleased with the interview . . .
JE: You didn't, did you find it nerve . . . -making?
S: No, initially with the tape recorder there, I think it did increase my nerves a little bit, just – yes it did – otherwise, no.

(Interview transcript)

This is the only point where he gets at all close to expressing any anxiety in the interview – and it is pretty restrained – especially since he was the only subject asked such a leading question about his feelings in the interview.

We might be tempted to simplify our Model B explanation earlier, by following the subject's own account. His performances on the questionnaire and in the interview are basically competent, if sometimes painstakingly slow. He seems 'rational' and lucid about the need for 'mathematical' calculation in various contexts: for school maths, he will 'spend time working it out, in the correct manner', but otherwise he will not spend his time. Indeed, Alan appears very shrewdly to assess just how much time and effort he needed to spend on the interview problems: for example, his approximate calculation of 'about £6' for

Question 5's wage rise appears to be very close to the exact result (£5.99)! Finally, how could one doubt his claims to be 'neutral' in his feelings about maths?

However, there are several aspects suggesting exhibited anxiety, unacknowledged by him:

- the way he insists that he is 'very neutral' about mathematics: strikingly forcefully in the first 'statement' earlier, and repetitively in the second
- the frequent, sometimes nervous, laughter
- the slow manner in which he answers several items, for example almost forty seconds to answer both parts of Question 3, parallelling his apparent slowness on the Performance Scale of the questionnaire
- repeated claims that he 'can't concentrate' on maths.

Is he perhaps 'protesting too much'? Using psychoanalytic insights, we can begin with what might be seen as his *denial* of strong feelings towards maths. This clashes strikingly with what may be ways of exhibiting feelings. He seems to be defensive, perhaps trying to control strong feelings.

He certainly seems to be concerned with control in some senses. He has a 'comprehensive' O-level Mathematics textbook available when confronted with a question which he doesn't know how to do. It may serve as a defence against anxiety.[14] At home his needs are catered for: his parents routinely do the shopping, and leave him money when they go away. At college, when I ask what would be the best course arrangements for him, he would like 'an hour per week of individual attention' on demand – that is under his control – or else 'my mind will just wander off' (interview transcript: 11). Here is 'can't concentrate', again!

Let us examine his inability to concentrate. In the first 'statement' quoted earlier, he allows that he has 'difficulties' in mathematics and 'I can't concentrate for too long on maths'. In the next part of the statement, he continues: 'If I need it, I'll use it, but otherwise, I just don't want to study it as a subject.' That is, 'can't concentrate' has become 'won't concentrate'. Could what seems at first a reasonable and coherent argument turn out as a an example of *rationalisation*, based on avoidance?

This 'rationale' may be supported by class and family discourses on time and money: 'The time's more valuable to me than the money at the moment . . . .' His family's relative affluence allows Alan to say this, because he has some access to his parent's money.[15] These insights, based on Models C and D, are suggestive, but, as with all these single interviews, only further material would allow us to pursue them more fully.

This interview allows me to explore the account of an apparently 'typical man', who describes himself as a 'neutral', without feelings about mathematics. However, as a a reaction to puzzling aspects of his own account, we find that psychoanalytical insights may help us to understand more deeply and more critically such self-reports as possible instances of denial, avoidance and rationalisation.

Donald's Story

'Donald' (interviewee number 10) was in his forties, with an O-level equivalent in Mathematics. His parents were working class; his own occupation was middle class, as he had worked on the money markets in London. Having originally applied for social work, he had switched to the town planning track. His reflexive account was as follows:

> This student was a member of my Methods and Models Maths seminar group, which worked exceedingly well, in the first two terms. He seemed satisfied with the group, and was interested in the philosophical aspects of mathematics;[16] in fact, he had almost switched to a philosophy degree. I was pleased when his name came up in the sample (and said so in the interview).
>
> (Evans 1993)

On the questionnaire, his performance on SM items was 9 of 10 correct, and 9 of 12 for PM items. The two PM answers marked wrong – 'prices would have gone down' for Question 14 on inflation and '37.2p' for Question 18 (10 per cent tip on £3.72) – were surprising for someone with his work experience, though neither error was unusual in the whole sample. His average scores on the mathematics anxiety scales were exceedingly low (in the 3rd percentile only, for maths test/course anxiety). Further, since the residual for my version of Model A was very small, he was considered neither an 'over-achiever' nor an 'under-achiever' for the SM results from the questionnaire.

In the interview, for Question 1 (reading pie-chart on water use), in response to contexting question (R) about 'any sorts of earlier experiences with numbers that it reminds you of', he replies:

3. No, nothing comes straight to my mind at all . . .
JE: Okay. Do you remember those from school at all, or from work?
S: I find it very difficult to remember school at all – not only just school, but anything in my childhood really, so . . . I know I didn't really understand maths . . . [3 lines]
    . . . I couldn't see the point of it at all to my real life, you know . . . .
JE: What were you mainly interested in those days?
S: Literature . . . [JE: 2 lines] . . . Oh, Shakespeare, Bronte sisters, and I read a lot of books, I read escapism books . . .
JE: Any particular favourites there?
S: I remember the first book I read was a book called *Red Cloud*, about mid-Western America, chasing buffaloes . . . .
JE: That's pretty good memory.
S: Yes, I wanted to be with them, I'm sure.

(Interview transcript, his emphasis)

Here is a contrast between school, of which he can remember nothing, and his 'escapist' reading, which comes back with striking clarity, after many years.

He attempted Questions 1 to 4, all done in his head, 'correctly' – except for Question 3 (reading the graph of gold price changes). When the question and graph (Appendix 2) are presented:

JE: Does that remind you of anything that you do these days, or you've done recently?

S: Er, some of the work we done in Phase One [the first semester of the college course], but if you ask me straight out of my head, what it reminds me of – I worked once with a credit company and we had charts on the wall, trying to galvanise each of us to do better than the other . . . . and these soddin' things were always there and we seemed to be slaves to the charts . . .
[2 lines] . . . I found it impossible to ignore them, even though you know that they're just getting you at it . . . [2 lines] . . .
That's what that reminds me of – a bad feeling in a way – I felt that a human being was being judged by that bit of paper.

(Interview transcript)

Here we notice that Donald calls up both 'college maths', and business practices, his earlier managing of a sales team, so I would class his positioning as interdiscursive (though his positioning was classed as predominantly financial practices in Chapter 9). Note that calling up the business practices brings ('bad') feelings with it.

Next in the same episode, he responds to my question about college maths, seeming to link 'financial maths' with it:

JE: . . . you mentioned [. . .] Phase One – does it remind you of Phase One?
S: Yeah, well, we done some of the questions like this, and er, the run over the rise and that kind of thing . . . [5 seconds] . . . trends, I suppose if you were judging a trend . . .
[2 lines] . . .
I find good, I like the fact I can do a chart now . . . (JE : uh huh), but [. . .] I couldn't sit down and do it straight away . . . [2 lines] . . . With maths I have to go back to the basic things all the time.

(Interview transcript)

Here he uses the language of college maths, describing the gradient as a 'run' over 'rise' (whereas it is the inverse. We cannot say if this is simply a slip, or evidence of a misconception.) He then shifts into work discourses, as a evidenced by his use of the terms 'trend' (rather than 'gradient') and 'chart' (rather than 'graph'), which were not used in the college teaching.

Next I ask specific questions about the graph:

JE: Right, okay, may I ask you which part of the graph shows where the price was rising fastest?
S: If I was to make an instant decision, I'd say that one [before midday], but obviously want to make it on a count of the line, wouldn't I?
JE: You'd? . . .
S: I'd count a line – as a it goes up . . . [25 seconds] . . . eleven over six [for the increase before midday, see graph] and ten over six [for the increase before the close], so that one's right – in the first one . . .
JE: . . . [2 lines] And um, what was the lowest price that day?
S: This one here, 580 . . . went higher at the close, for some reason.

(Interview transcript)

When asked to compare the gradient of two lines, he makes a perfectly accurate 'instant decision', presumably drawing on his work experience. However, he also feels impelled to 'count a line', which I take to mean calculating the gradient by counting squares on the graph, as in college maths. There he gets the correct answer, confirming his earlier decision based on work practices – and confirming that he can use the formula for gradient correctly – though his calculations are approximate. His reading of the lowest price is not quite right: it should be $590, not $580. At the end of the episode, he is back in the 'money market' practice, as is shown by his speculating about why the graph 'went higher at the close'.

A simple interpretation of his misreading of the lowest price on Question 3B is that it was a 'slip'. Admittedly, the photocopy of the graph was not perfect, and also the graph was slightly 'tricky' (in that each vertical division represented a $5 difference, not $10). Altogether nine of twenty-three subjects misread the graph here. However, *Donald's* slip is rather surprising: we might expect some transfer of learning from his work practices and familiarity with 'charts', to college maths and familiarity with graphs.

I consider the positioning within which he addressed this problem as interdiscursive, according to a full analysis (see earlier) – but the predominant positioning would be 'financial maths', as it was for three other students: number 9 (an ex-stockbroker), number 12 (an ex-manager), and Fiona (a stockbroker's daughter). All four made an error in reading the lowest point on the graph. This puzzling result *might* be explained by suggesting that, in certain financial practices, the readings of graphs are regulated differently than in SM, since they are made for different purposes, for rough comparisons, rather than for precise individual readings. On this *first reading*, addressing the problem within business discourses, rather than within academic mathematics, reduces the need for precision.

However, on reflection, this contextualism may be too simple. Would Donald, at least, not have needed, in his sales positioning, to be very precise, to compare his performance with others?

Might his slip be related to the range of feelings he has expressed in this episode, and elsewhere in the interview? Looking for indicators of affect, we find that Donald ranges between 'good' and 'bad' feelings. He again expresses mixed feelings about work later in the interview:

S: Once you're in there you do perform – you wouldn't do a bad deal in a million years 'cause it's yourself's on the line . . .
     [5 lines]
JE: . . . that sounds like pressure, doesn't it . . .
S: Oh, dreadful!
JE: . . . did you feel the pressure or the anxiety?
S: Oh, very much so, yeah . . . Sometimes I got a pain in your chest [2 lines] – you had, the form gets stuck to your hand, the tension, the sweat . . . But once you do a good deal, somehow, it could kill you somehow, you just feel good or something, as if it's your own money . . .

                                                              (Interview transcript)

However, he expresses confidence about his numeracy at work: 'I'd no confidence with figures when I started . . . sheer use made me good at them'. But now 'I can read figures [. . .] I just had a gift for that'.

He also expresses a range of feelings about college maths, including new-found positive feelings for mathematics:

S: I found *connections of something there to go from one thing to another* [that is between financial maths and college maths], and I found it [maths, during the second term] exciting, you know, I couldn't get bored with it at all . . . [two lines] . . . I liked it.

(Interview transcript, emphasis added)

However these new feelings are still tentative. He is 'not [. . .] afraid of figures, but the *formulas* and things still frighten me really . . .' (emphasis added). He also describes an experience of feeling a 'block', when he first attempts the current maths worksheet, followed by 'panic':

JE: Panic, uh huh. So when you look at a question, what happens? . . . [1 line]
S: Some kind of inferiority inside of me says I can't do it . . . My brains tell me I can do it, but something says I can't.

(Interview transcript)

We can check the questionnaire, for other indications of anxiety, or lack of confidence related to graphs, or formulae. On the 'Experience' Scale, he rates himself as 'very capable' in all areas of basic maths, except for algebra and graphs. In response to Question 18 ('anything special you would like to learn about maths . . . this year?'), he writes: 'practise with graphs'. Thus a *second reading* of his slips on Question 3 is that they may relate to what appears to be a chronic lack of confidence, and also occasional acute fear and panic, about graphs and algebra: a Model B explanation based on his talk in the interview, and confirmed by the questionnaire.

The interview suggests links between this reading in terms of affect and emotion, and the previous reading, in terms of different contexts. In several places, Donald appears to suggest a distinction between what might seem to be two different 'types' of mathematics: school maths and a particular kind of practical maths.[17] These occur in different contexts and are marked by a range of different feelings. For example, 'banking and figures were a job, or something, but maths were there to trip me up or something'; and 'I feel – not afraid of *figures* – but the *formulas* and things still frighten me really' (interview transcript, my emphasis).

He finds the formula frightening, because 'it's divorced from reality in my mind'. At school, he would 'get an answer to pass an exam, but I'd no idea what it was all about. I [. . .] couldn't see the point at all – to my *real* life, you know' (interview transcript, his emphasis). This distinction seems to have a strong emotional basis for Donald. It appears similar to that made by several other subjects; see 'Conclusions', this chapter.

Let us reconsider several issues, using insights from psychoanalysis. It is striking that he finds it 'difficult to remember school, . . . anything in my childhood really'.

The memories appear to be *repressed*. This may be related to some earlier experiences with mathematics in particular: he has 'an inferiority' about mathematics; and 'I was frightened of maths really' as if 'maths were there to trip me up'. It may also have something to do with his anxiety about a lack of control generally in school, compared with his good feelings about literature, especially 'escapism' books. It is difficult to be sure what the basis of this amnesia might be: this would have been an area to pursue systematically, if this interview had been part of a series, rather than a one-off.

Fantasies provide a site where the subject can be in control, and Donald produces several. Reading 'escapist' literature led for example to a desire to be with the nineteenth century buffalo-hunters described in the stories (see p. 205) – which is far away from school in twentieth century Europe. At the end of the interview, when we return momentarily to Question 2, he recalls his younger days:

S: . . . in the shop [. . .] my mind just would make prices up . . . [2 lines] . . . If say [inaudible] somewhere, say, reduced by 15 per cent, I could do it in my head without thinking, almost.

(Interview transcript)

Thus he clearly had insistent fantasies involving making up prices, and calculating discounts in shops. Could these have been on goods which the young Donald desired, but his family couldn't afford?

Later, he works in selling money, where it sometimes feels 'as if it's your own money' (in the context of feeling good after doing a good deal). This practice, and the figures, and so on may relate to deep fantasies: it is exciting to play with money. They may also be a defence against anxiety about not being in control, about not having something. So far in his life, school mathematics has failed to relieve this set of anxieties for him – in contrast with those boys attracted to the 'mastery of reason' (Walkerdine 1988) – though it is interesting to note that college mathematics has given him a taste, and he now feels it would be 'exciting' to do a mathematics degree at the Open University. Thus, he has different affective 'investments' in school maths and 'formulas', from those in his money-market practices and 'figures'.

Returning to his slip on Question 3B, where he read the lowest point on the graph as $580, less than the correct value of $590. We can recall that he says that the graph reminds him of 'a bad feeling', and of how he 'found it impossible to ignore' the graphs showing his performance on the wall. Is it possible then that his misreading of the graph might be motivated by his desire to *avoid* such 'charts'? Though admittedly only very suggestive, this *third reading*, based on Model C, may provide the affective/psychic basis for the chronic lack of confidence about graphs, the basis of the second reading.

Finally, we can note that Donald's fear of mathematics has changed over time: he finds 'connections' in college maths – 'Methods and Models', the Polytechnic's first-year course – and connections between college mathematics and his earlier work maths. He finds this 'exciting', and a source of *pleasure* and *enjoyment*, similarly to Harriet. Thus he is beginning to have a 'second chance' with mathematics, as are some other interviewees. In his case, the basis of the second chance is a college course that seems to be appropriate for him.

Donald's sensitivity to 'connections' shows up in the episode with the graph. This illustrates several points made in my discussion of context, and especially of *transfer* or 'translation' across discourses (Chapter 6):

• Donald appears able and willing to use both college maths and financial maths; further he seems able to choose which practice to use to address the problem posed, that is, to decide whether to apply his (more precise) college maths methods of calculating gradients to indicate when the price was rising faster.

• He is also aware of the *different goals* of the two practices, relating to different objectives in using the graph. In business, the objectives are implicitly competitive, to compare persons or groups, and growth-orientated, to make comparisons over time; in college mathematics, the aim is to analyse the qualities of the curve, including the rate of change. He is aware of *different values* and standards of regulation, in particular of precision, required in the two discourses.

• He is also open about the *different feelings* evoked by the two practices. For example, his awareness of the goals of business practices is sometimes painful, whereas he gets pleasure from college maths.

• Donald is apparently able to focus on *discursive similarities and differences*: he seems able to read the diagram as a a 'chart' (business maths) or as a a 'graph' (college maths), and to recognise the connections between a 'trend' and a 'gradient' (respectively). Though not certain, it appears that Donald is able to bridge the two practices, that is to transfer his college mathematics methods to help solve a problem involving charts.[18]

• In this analysis, attention is drawn to the diagram, and to the role visual representation might play in either discourse. Here the diagram seems to provide a crucial representation, facilitating the setting up of chains of signification across discourses, as it did with other students, for example Fiona.

This *fourth reading* takes us towards a Model D account of the episode.

This interview allows me to explore a number of themes, including:

• the relevance of the context of numerical thinking, here contrasting business/financial practices and school/college mathematics

• illustrations of mathematical and numerate activities suffused with a whole range of acute emotions – including excitement, enjoyment and pleasure – as well as chronic anxiety and lack of confidence

• the effect of social class background, and level of material affluence

• illustrations of the meaning and possible influence on mathematical thinking of fantasy, defence, avoidance and repression

• an account of transfer of learning across discursive practices.

Peter's Story

Interviewee number 19, 'Peter', was 20 years old at entry, and from a middle class background. He had passed O-level Mathematics and was specialising in Economics. His reflexive account was as follows:

I had not taught Peter, nor even met him, as far as I could recall. When I wrote inviting him to interview, he accepted, but then did not respond to my offer of a time. However, he did show up at the time offered – meanwhile booked for number 25. When she did not show up, we began.

(Evans 1993)

On the questionnaire, he scored 9 of 10 correct on school mathematics items. His mathematics anxiety responses on both scales were close to the median for the whole sample. Using my version of Model A, his score on the SM scale (relative to his gender, age, mathematics qualification, TCA score, and so on) suggested neither 'over-achievement' nor 'under-achievement' (see note 8).

Beginning the interview, he tells me he had to keep taking O-level Mathematics until he passed it on the third try: 'So I was [. . .] under a little bit of pressure . . . [at grammar school]'. He then explains how he came to be taking Mathematics O-level:

S: See, my father's an engineer and all of my brothers, bar one, are teachers – and I was not pushed, but I was *gently persuaded*, in the area of taking sciences at O-level, then another two sciences at least at A-level, going on and doing some kind of teacher training degree like, probably in something like physics [. . .], hoping for an easy job at the end – not an easy job, but an easily attained job – so, you know, I was *always* pushed towards taking Maths at O-level and A-level, and also Physics at O-level and A-level, and unfortunately I just didn't master either of them.

(Interview transcript, his emphasis)

Being the youngest of five sons in such a 'mathematical family' can be a mixed blessing, especially when it comes to *homework*, the activity which quintessentially involves inter-discursive positioning across family and school practices:

S: My father's [. . .] very good at maths – which is a shame really because he's always tried to teach me, and I just couldn't ever have much success [. . .] to get me through my O-levels, both he and my oldest brother almost constantly tutored me in my homework and everything . . . And my dad is the sort of person who will, if you ask him a question, instead of giving you straight answers, he says – well, hold on a minute, I'll go and find a book – and there's another book, and then another book, and hopefully a five minute explanation turns into a half hour looking through [books] – and you're getting a very complicated explanation. And I found from that I don't think I was ever really interested in mathematics.

(Interview transcript)

Being tutored by his oldest brother (himself a qualified mathematics teacher), was even worse:

S: if I didn't get anything right, then . . . it was even more – you know, lecturing and er, you know, sort of, not exactly saying that I was stupid, but getting onto the old intelligence bit – so I suppose I became a bit scared of maths in general as a subject as well as physics, and as I say, it was a relief to take something else as non-numerical, or easy to grasp, as Law or Economics or History [subjects he did at A-level] (JE: Right, sure) – there are numbers involved, but they're just not *forced* on you in the same way, or not quite as ['quickly'?] . . . . And I never really found mathematics in economics hard to handle.

(Interview transcript, his emphasis)

Thus, Peter makes a distinction between mathematics/physics and 'non-numerical' subjects. First, the mathematics used in economics is not 'hard to handle' since he can understand the symbols; this differs from algebra where, in his view, there is 'no excuse to make up symbols to replace something [which has] an actual reality'. Second, after being 'pushed' to take subjects like Mathematics and Physics at O-level, it was a 'relief' to take 'non-numerical' subjects like Economics, Law and History at A-level: they are interesting and useful, unlike Mathematics.

He deploys different working habits in different areas. Though he mentions appreciatively that you can look up a date or a statute in a (history or law) book, he seems resistant to doing the same thing in first-year Maths, when he misses a lecture and doesn't understand something (indicated at three different points in the interview). Later he makes a statement about what he has to do when he doesn't understand something in mathematics: 'think about it [. . .] – but I don't reach for a book!' Of course, this may have a lot to do with the way his father used books, when this son asked a question (see earlier).

Currently he has a number of 'difficulties' with mathematics in particular. He loses *concentration* often, and often finds things don't 'stick in his mind'. Nevertheless, he hopes to '*master* the approach' in his second year mathematics option for economists (my emphasis).

His methods in mathematics often involve working from first principles. For Question 2 (10 per cent of 6.65), Peter offered two methods – reconstructed as (1) and (2):

1   10 per cent of 665 (having moved decimal point two places to the right) = 66.5, so (moving decimal point two places to the left) 10 per cent of 6.65 = .665
2   10 per cent of 1 is .1; 10 per cent of 6 is .6; 10 per cent of 6.65 is .665.

He began with (1), then seemed to become confused, and moved to (2), then seemed to decide to move back to (1). He explains:

S: And I try to do it from either end, to make it as easy as I can, without just doing the problem straight off . . . . (JE: right), 'cause I can't usually do it straight off . . . [3 lines]
      . . . percentages [. . .] at secondary school [. . .] was one of the things I was able to do, and you can work it out on a piece of paper, if you want [. . .] . . . *I prefer to work things out in my head* [. . .] particularly with maths.

(Interview transcript, emphasis added)

Here 'doing it from either end' seems to mean that (1) involves 'magnifying' the original number so that there are no decimal places: *viz.* 6.65 to 665, taking the answer from 10 per cent of the latter, and 'reducing' it correspondingly. Method (2) is a type of 'decomposition' (Nunes et al. 1993), which is especially suited for mental calculation, and which he insists on doing for reasons given here.

All the questions are done in his head – not what is expected from someone with a predominant positioning in SM, for most of the interview (except for Question 4). He continues, giving some insight into why:

S: maybe people can't see what you've, what's going on in your head – and they can see what you're writing on a piece of paper.

(Interview transcript)

For him, *being observed* signifies being regulated, pressured, 'pushed' by his father (and brothers), without being helped very much. Not surprisingly, there are no traces of calculation evident on his questionnaire, unlike say Alan or Donald. Thus, using mental methods does not necessarily preclude a school mathematics positioning.

He expresses much anxiety about mathematics in general. For example, after his brother's 'help' with homework, 'I suppose I became a bit scared of maths in general' (quoted earlier). And at school, 'I seem to remember in my first year of secondary school having long multiplication questions, and not being able to do them, being scared out of my head at being marked.'

He also exhibits lack of confidence about mathematics, perhaps most notably in his *resistance* when I ask him to try some problems:

JE: . . . if I give you a few questions to try. Would you be happy about that?
S: Well, reasonably . . .
JE: Reasonably . . .
S: Yes, I've been to a couple of job interviews where there's been some pathetically easy sums on a piece of paper, but because they've been on a sheet, set out in front of me *with someone looking over me,* I haven't been able to do them (JE: yes). I think that's something that's come from having been taught by my father in that way.

(Interview transcript, emphasis added)

His resistance seems related to anxiety shown by the story about his job interview. He returns to this job interview towards the end, when it emerges that he had not read the instructions carefully for the arithmetic problems in the selection test. Note that this failure is mirrored by his not reading Question 1 carefully in this interview (see p. 155).

A range of feelings are expressed in the following story about school:

S: I remember [. . .] in maths lessons being caught day-dreaming [. . .] And then I would have to [. . .] admit that I wasn't paying any attention . . . [4 lines] . . . I think that was something that was unique to maths lessons, [. . .] being asked questions and not knowing the answer, and . . . being very, very – first of all, you know, embarrassed and ashamed – and then a little bit angry at being asked a question in the first place. What am I doing here, y'know, sitting in front of these useless numbers? – they'll never be any use to me – and why would I want to know how long the side of a triangle is?

(Interview transcript)

This chain of feelings on the part of a student – embarrassment, humiliation, anger, leading to resistance – is not uncommon in my experience as a teacher.

Peter gives two unsolicited 'performances' during the interview in what seem to be practical maths. The first followed Question 2, for which his labourious method (2) (see earlier) involved decomposition:

S: I've always thought – well, how would I cope working behind a bar, when you have to add up a round and I'm always left doing it in my head, and I do [. . .] whole numbers first . . . 65p plus 50p plus 20p, then I would do sort of . . . 60p + 50p + 20p + 5p . . . much easier if it's a more awkward number like 68 or 64 or something.

(Interview transcript)

Interestingly, this strategy also involves decomposition (though he doesn't remark on this). The second 'display' comes up when I present Question 4 (10 per cent tip):

S: Well, just adding up a full meal would be [. . .] coffee [27p] and the chicken [£3.75] . . . would be [20p?] £4.02.

(Interview transcript)

Both of these performances – that working behind the bar and that in the restaurant – are based on *imagining* a situation which he is rarely or never in, and then performing competently in it. The restaurant example, if the partially inaudible '20p' is correct, illustrates more convincingly the power of decomposition for mental addition than the bar example.[19]

Towards the end, when I ask how he feels 'about the way you're able to use numbers in general these days':

S: I think I'm alright, as far as I can see, I'm okay using numbers in my head. I find I'm far more comfortable [. . .] working out [. . .] my gross handicap [in golf] in my head, rather than having to put it down on a piece of paper. Or if I'm playing darts . . . [3 lines]
. . . But it gets [. . .] worse the more I have to write things on paper . . . . And I find the first time I do a sum, I hate it; and the second time I do it, it's not quite so bad . . . And as soon as I've actually figured out the simplicity involved in it, and finding out how actually, how *simple* it really is – once I've done that, then I'm all right, I'm coasting . . .

(Interview transcript, emphasis added)

Peter expresses confidence in these examples – which are actual examples, more convincing than the imagined ones given earlier. The second part of the passage seems to illustrate a sort of *accomplished*, if limited, 'mastery' in his 'coasting', despite having earlier given examples of lacking mastery in school mathematics, or in the selection test.

We can consider this case, using Model B. Peter experienced pressure to do mathematics and physics by his father and brothers, and O-level Mathematics by the school. These pressures have produced a chain of negative feelings for him to deal with, for example in the story about 'daydreaming at school'. He seems to deal with these feelings by developing a resistance to mathematics. He expresses (and exhibits) much anxiety and lack of confidence, but also some confidence, though the latter is somewhat fragile. He gets 'relief' (from anxiety, from pressure), through a few topics within maths, for example pie-charts, and by choosing relatively 'non-numerical' subjects, such as Economics, his degree subject. He experiences a seeming 'mastery', or at least relief, from practising a 'sum', and 'figuring out (its) simplicity'.

For a deeper level of analysis, using Models C and D, we need to explore his anxieties, resistances, fantasies, what he means by 'mastery', and how all these relate to his positionings, within his family especially.

In this family, being able to do mathematics signifies *intelligence*, and *rationality*, for the father and five sons at least. For example, his father puts a lot of effort into knowing about mathematics. If Peter made mistakes, his older brother, trying to teach him, 'would get onto the old intelligence bit'.

Peter is certainly the interviewee who mentioned 'mastery' most often.[20] It eluded him in mathematics and physics at school, but he hopes to 'master' the approach in the Mathematics for Economists option just begun. He describes two situations where he is in control of whatever comes up, but he is not actually involved in these imaginary (for him) practices. There is much *fantasising* here.

At the end, he refers to repeating a sum to the point where he has 'figured out the simplicity' involved in doing it, and is 'coasting'. But what sort of 'mastery' is this? He rejects the way symbols are used in algebra. His vision of what he lacks seems to be the grim sort of mastery possessed by his father, struggled for over long years, and needing constant updating from books. For Peter, the 'coasting' he desires seems more like 'relief': he is unlikely to feel pleasure, as do some of the boys in *The Mastery of Reason* (Walkerdine 1988), or even as Harriet does in reclaiming, tentatively, her ability in school maths.

Furthermore, though he fantasises about mastery, and success, his apparently resistant actions militate *against* success. The interview provides examples of overt resistance to mathematics: not reading his father's books, not using textbooks for first year maths, not attending lectures on time. However, there are also examples of what may be unconscious resistance: 'losing concentration', not reading the instructions on the arithmetic selection test, not reading the instructions for Question 1 (pie chart) in the interview.

This shows considerable ambivalence. It also leads to his appearing incompetent, useless in mathematics – whereas he has developed some useful insights (for example via decomposition – see earlier).

In the story about the selection test, he exhibits much anxiety. He explains this as due to the sums being 'set out in front of me with someone looking over me', and makes a connection between the job interview, and his sessions with his father doing mathematics homework. He also seems to connect the sessions with his father with doing problems for me in the interview. When I ask if he would be happy about 'a few questions to try', he immediately tells me the first part of the job interview story. Thus the anxiety seems to be displaced, to flow from the homework sessions, to the job interview, to the interview here. His resistance in those situations seems an attempt to avoid, or to manage, anxiety.

We might explain Peter's unconscious reaction to me as interviewer as *transference*, since he may be displacing feelings for his father, his brothers, onto me, as a mathematics teacher (Hunt 1989). He also does the same with the tester at the job interview. Transference might also help to explain the perhaps 'provocative' frequency with which he mentions missing first year Maths lectures, some of which were given by myself. The idea of transference helps to explain some episodes in the interview, and in Peter's life, that might be puzzling otherwise.

Peter presents his father as anxious for him to know, especially about mathematics; in this sense, Peter is like Harriet. However, unlike Harriet's father, who does not assume that he himself can know, Peter's father not only expects the son to know, but expects to know – and does know – himself: what you do is get out a book. However Peter's father still seems anxious himself, as is shown by his always wanting to look things up, and needing to give long explanations. Peter explains that his father had an unconventional, drawn-out engineering education, and also that 'he just likes to explain things very methodically and making sure he gets everything right before he tells you something, instead of jumping in at the deep end' (Interview transcript). This father deals with his anxiety by continually needing to know more, and the anxiety may be passed on to the sons.

Indeed, despite his resistance to his father's methods, Peter seems, sometimes, at one with him, in his 'need to know everything about a maths problem, everything about . . . an idea in maths, for me to understand it'. Does the son *identify* with the father in this respect?

In the only mention of his mother, she is presented as having an 'awful habit': she adds up her shop receipts only when she gets home and, if she's overcharged, it is too late to do anything but go into an impotent rage. Perhaps significantly, this story followed immediately after his fantasy of summing the total cost of a meal, including coffee (27p). Do his imaginary stories of mastery allow him to identify in some way with his mother? In some ways, he is at one with her in his tendency to fail to do the right calculation at the right time. And perhaps in his isolation from the knowledgable, mathematical men in the family?

Why does she do this; what do these small amounts of money mean for her? For him? There seems to be a strong moralism, and perhaps strong anxiety, around money in this family. We can compare this with Alan's lack of worry about money, which we can infer was shared by his parents. It is also different from the fantasies of Donald and Harriet, about discounts on things which they desired but could not buy. In Peter's family, there seems to be an anxiety about losing what you already have, rather than anxiety about never having things.This suggests a 'lower' middle class position, compared with that occupied by Alan's family.[21]

This case allows me to explore a number of themes:

- the effects of a positioning in family (and school) discourses as incompetent in mathematics, leading to pressure through 'homework', and to ambivalence and resistance
- the involvement in fantasy, in a somewhat different way from Donald and Harriet
- possible indications of identification (with both parents) and transference.

Conclusions

My analyses of the seven case studies in this chapter address the ten themes of the 'qualitative', interview-based part of this study – and build on the cross-subject analyses in Chapter 9. I employ and develop my central concepts of *context as positioning in practices*, and *emotion as charges attached to signifiers* in semiotic chains, forming an integral part of mathematical thinking.

Before summarising the findings, I should stress the limitations of the case study data, in terms of the *number* of cases, and also in terms of the *amount of research material* for each case – one questionnaire and one interview of a half to three-quarters of an hour per subject.[22] However, in all cases, these are supplemented by notes of other interaction; see the general reflexive account in Chapter 8, and the individual reflexive accounts earlier. The aim has been to argue as convincingly as possible for the interpretations of each case offered, to suggest more general motifs through constant comparisons, and to reflect critically on more traditional accounts of 'mathematical thinking' and 'mathematical affect'.

The analyses support Themes 1 and 2 (developed in Chapter 9), that context should be understood as integral to activity, as positioning in practice.

Concerning Theme 4, my analysis supports the idea that the subject's thinking and performance on 'mathematical' or numerate problems depends on his/her positioning in practice(s). Jean's problems with percentages relate to 'conceptual difficulties' within the school mathematics (SM) discourse, whereas she solves the problem of giving a tip on a restaurant meal without such difficulties. Fiona, purporting to call on everyday discourses, 'refuses the terms' of two problems, and hence performs incorrectly. Peter calls up SM for most problems, but generally calculates in his head – rather than on paper as would be expected for SM – because of the way he has been positioned in SM and related practices, especially in doing homework with his father. Ellen draws on her familiarity with SM, or perhaps eating out practices, in order to be 'critical' of a calculation where she has made a slip. Donald is aware of the greater precision 'normal' (that is, required) within SM, compared with standard financial practices, for comparing rates of increase, and is able to choose to deploy one or the other practice.

Similarly, for Theme 7, this analysis supports the idea that the emotions experienced, especially anxiety, are also specific to the practices called up. Thus the responses of Fiona and Ellen to Questions 3 and 4 respectively may well signify anxiety, but I question whether this anxiety, which might at first seem 'mathematical', is not instead (or at least additionally) related to the subject's positioning in another practice. For Fiona, this is as a child who 'wouldn't understand' her father's work which entailed financial risks threatening the whole family. For Ellen, the positioning seems to be as someone who 'doesn't usually pay' while eating out with someone with whom she is afraid of being 'an expense'. Harriet and Peter appear to lack confidence in maths; this seems to relate to their experiences of being watched, and thus regulated, by their fathers in doing mathematics homework. In fact, for many students – and for all seven cases (except perhaps Ellen) in their different ways, school mathematics and (sometimes) college mathematics are related to 'negative' affect: dislike, anger, boredom, diffidence, and especially anxiety; this reinforces the findings on Theme 5, that emotion pervades mathematical thinking (Chapter 9). For Donald, in addition, 'bad feeling' is associated with some aspects of work practices, for example the competitiveness.

Theme 9 points to the relationship between thinking and emotion as being specific to the subject's positioning, too. To begin with, more confused, less cogent performance may be observed when school (or college) mathematics is called up. This is not only because of misconceptions, memory failure and so on – though sometimes these may be crucial, as in the case of Jean, but it may also be related to the negative emotional charges that are in many cases specific to school mathematics practices. Harriet, Fiona and Alan, in their refusal to attempt Question 5B (9 per cent of 66.56, exact calculation), and Peter throughout illustrate this.

In contrast, for Donald, the bad feelings associated with work do not seem to have interfered generally with the numerate aspects of his performance there (despite his puzzling slip in reading the gold price graph). Ellen's case shows that emotion (here, anxiety) associated with a relationship, itself linked with everyday practices like eating out, can interfere with thinking, including numerate thinking that looks like 'mathematics'. Overall, I argue that the relationship between anxiety and performance, rather than being taken as general across subjects, can only be fully grasped through analyses of particular cases, with reference to positioning in practices, as is done here. (See also the discussion of psychoanalytic insights later.)

Themes 3 and 6 point to gender, class and indeed other differences in numerate thinking and expressing emotion. My analysis shows the importance of such social differences, and also their relation to the subject's positioning in practices. For example, the positioning of Ellen, a young middle class woman, as someone 'who doesn't usually pay' has effects both on her thinking around calculating a tip, and on her feelings about it. Fiona, another middle class woman, is older, and can consider the problem from a position within several practices: being 'invited out' (as with Ellen), 'going Dutch', and/or as a former restaurant worker.

These interviews also illustrate a number of ways in which social class cultures position subjects. For example, Peter, as the youngest son in a middle class family, was subjected to pressure within family discourses to take mathematics (and science) GCE exams, and to know about these subjects in a certain way. Harriet, the first in her (working class) family to go to grammar school, was denied knowledge of her capabilities, by both family and school, that might have allowed her to take, and succeed in, O-level Mathematics. Though there are similarities between the way that Harriet's and Peter's fathers regulated their homework in maths, there are also crucial differences: Harriet's father was anxious about knowing, and about his own

– and her – ability to know; Peter's father seemed (from his son's account, at least) to be less anxious about how you get to know, and wanted his son to follow his example, for example by consulting books on the subject.

In addition, these interviews show all the subjects to be very specifically positioned in terms of relative affluence or relative poverty in a range of practices – from Jean's worry and pain about never having enough money, to Alan's lack of concern about best buys in shopping. Also striking are Harriet's and Donald's fantasies, using numeracy, about reductions in prices (see later).[23]

At the same time, the comparison between Alan's parents' and Peter's mother's practices around money shows the scope for variation, even within the middle classes. What I have shown in this chapter and the last, is both the *effectivity* of 'structural' differences – gender and social class – but also their *limitations*. The case studies show the need to take account of the particular subject's positioning within specific discourses.

Part of the way affect has effects is through making the discourse of mathematics – or some parts of it – familiar. The idea of 'familiarity' with a discourse is a crucial straddling concept between cognitive and affective. As illustrated principally by Donald – and mostly for financial practices, with which he still feels much more familiar than with those of mathematics – familiarity embraces a number of aspects:

- It requires *knowing the language* of the discourse: see the discussion of the 'graph' episode where Donald solves the problem within both business maths and school maths discourses.
- It involves the memory, and may be developed by repeated practice: see Peter's statement about 'coasting' once he has 'figured out the simplicity' of a sum.
- It requires a 'feeling' for it (Carraher et al. 1986, Willis 1984), which may sometimes seem 'intuitive':

  S: I had a gift . . . [2 lines] . . . I could run down the columns and [. . .] know what I'd be expecting and if something wasn't right, I'd say there's something wrong there . . . Not always, but I had a feeling for it . . . I just let my feelings go . . .
  You didn't have to think about it [. . .] you took a chance [. . .] and you were right six times out of ten . . .
  When your intuition is good, you can smell it . . . .

  (Donald's transcript)

- It is partly about having an idea of possible 'solution shapes' (Lave 1988): see Donald's 'expecting' a figure in a certain range (see previous quote); and other subjects' 'critical evaluation' of implausible answers to problems (Chapter 9).
- It may be related to *confidence*, as with Donald's 'things I know I'm good at and can do – like columns of figures . . . ' – but it may also be linked with bad, or mixed, feelings, as with the 'charts' at work.
- It may be developed within *relationships* with positive affective charge (Scribner and Cole 1973, D'Andrade 1981) – though not necessarily, as Donald recalls: 'In business, most people are very jealous of their expertise, and sometimes they don't want to pass it on to you . . . .'
- It involves seeing the activity or action as meaningful or appropriate for

yourself, as not other or 'alien' (Murphy 1989), as something you 'can do' (HMI 1989, Coben and Thumpston 1995): Donald distinguishes mathematics during his time at school from 'my real life' as a boy, and from his working life later.

My characterisation of 'familiarity', because of its *strongly affective* aspect represents a development from the basically cognitive sense used in explanation by Scribner and Lave. In my approach, familiarity contributes to considering the cognitive and the affective as part of a whole. For example, the idea may help us to understand the phenomenon of 'sudden decline' in the school mathematics career of certain students, as a rupture of an established familiarity.[24]

The last feature of 'familiarity' suggests a distinction between 'different 'types' of mathematics, made by about half of the twenty-five interviewees. Besides having substantial cognitive consequences, this process is also profoundly affective, and particular to the person concerned. For example, Donald uses a deeply-felt distinction between 'figures' on the one hand, and 'formulas and rules' on the other. For him, this relates to several distinctions:

- between work practices and school/college maths
- between being meaningful, 'having a point', and being 'divorced from reality'
- between confidence, 'feeling good', and being frightened, in 'a bit of a panic'.

In contrast, Jean articulates her distinction between two exam courses, CSE Mathematics and CSE Arithmetic (and also between topics in first-year Maths) on the basis of:

- 'usefulness'
- whether it should be optional or compulsory.

These divisions recall Bernstein's (1996) concept of 'classification', used to describe the strength of boundaries between school subject contents in educational knowledge (see Chapter 6). These subject contents, and their boundaries, are *socially recognised*, and *have effects on* the 'identity' of teacher and student, who are positioned to accept the authority and values of these school subjects. Bernstein's principle can be contrasted with the boundaries involved in making the sort of 'distinctions' I am describing, which appear to relate to the particular student, and – to some extent at least – to *flow from* his/her subjectivity.

The hierarchical organisation of mathematics seems especially marked, and hence the impenetrability of the subject, or at least parts of it, may provide the basis of seeing those parts as alien. The emotional charge of that alienation, however, is provided by the formative experiences and the investments that are part of the history/biography of the particular subject. Hence, I would argue that the boundaries between academic subjects are both cognitive and profoundly affective.

Insights from psychoanalysis have been fundamental to the account of affect and emotion produced here; this is because so much affect is linked to ideas repressed into the unconscious. In considering Themes (6) and (8), I have used the distinction between expressing anxiety and exhibiting anxiety – in order to take into account the idea that anxiety may be unconscious, because of defences against it. I have been able thereby to sensitise myself to cases where anxiety has not been expressed, but where there are indications of defensiveness. Hence my analysis can move beyond the constraints of depending *only on subjects' reports* of their feelings – which limits much conventional qualitative research.

I have undertaken to describe all types of emotion, expressed or exhibited, since anxiety may apparently be transformed by the operation of defences such as denial and 'reversal into its opposite' (Laplanche and Pontalis 1973: 399-400). Significantly, it has been possible to describe what may be defences against anxiety in all seven cases here:

- Ellen's repeated expressions of confidence
- Jean's 'insouciance'
- Fiona's jokiness
- Harriet's pleasure in formulae
- Alan's denial of having *any* feelings about mathematics
- Donald's forgetting of school
- Peter's denial of his failings in mathematics, through fantasy.

Theme 8 provides the opportunity to use a range of psychoanalytic insights that go beyond the notions of repression and defence. As well as the emotion being transformed, the object or focus of anxiety, anger, and so on may also be transformed by the operation of unconscious processes. I have illustrated the use of displacement/metonymy and condensation/metaphor in the interpretation of chains of signifiers, particularly in Ellen's case around the key signifier 'expense'. Fiona and Peter also express anger about their regulation by their fathers in various discourses, which anger I would argue is likely to be displaced onto mathematics or mathematics teachers.

We have also seen the possible effect of subjects' emotional investments on answers given to problems, especially as 'motivation' for 'slips'. For example, Ellen, after choosing the least expensive dish on the menu, made a slip in calculating the tip that would make her less of an 'expense'. Jean made a slip in interpreting a problem on calculating the cost of sports kit that led to a lower cost, and may have lessened her evident worries about money.

Fantasies provide a site where the subject can be in control. Donald and Harriet had fantasies in which they used numeracy to make up prices, and to calculate discounts in shops. If these were about goods which the young child desired, but the family couldn't afford, then they may have had effects – not only in providing wish-fulfilment (see Chapter 7) – but also possibly in motivating Donald to 'work with money', and in supporting Harriet's creation of shortcuts to aid mental calculation, for example of 15 per cent as (10 per cent + ½ of 10 per cent). Some of Peter's fantasies, of a similar type, seem less successful, in supporting the development of his numerate skills.

There are striking differences in the prominence given to description of family relationships by interviewees (cf. Legault 1987), and these seem not to correlate with age, gender or class. Ellen, Jean and Donald make no mention of parents, and

Alan very little. Fiona, Harriet and Peter, however, mention their fathers a great deal. The two women mention their mothers slightly, whereas Peter mentions both his mother and his oldest brother more fully. For these three, in different ways, the father represented knowing mathematics, doing mathematical work, and/or the need to know mathematics.

These three subjects also show the most *resistance* to doing mathematics problems in the interview. This resistance needs to be understood as an effect of each one's (different) history of positionings in learning mathematics.

Psychoanalytic insights allow us to appreciate the emotional charges which motivate much of social life in sometimes surprising ways. For example, I have argued that the concept of splitting (a defence, in the Kleinian approach) may help to explain a division of labour in Fiona's family when she was growing up: her father practised the calculating, the rationality, while the rest of the family, including Fiona and her mother, 'held' the emotions: anxiety and anger, in Fiona's case. Thus this particular psychic defence may underpin, or infuse, at the level of family or culture, the dimensions of difference – here gender and age – which form the basis for certain arrangements which may be seen as sexist or 'adultist'. (See also Chapter 7.)

A psychoanalytical approach alerts us to the possibility, in research settings, of *transference*, the subconscious reactions of either subject or researcher to the other (Hunt 1989). In these interviews, indications of possible transference reactions towards me, as interviewer, as mathematics teacher, were clearest in the cases of Fiona, Harriet and Peter. This is not surprising, since all three subjects had fathers who were associated with their learning of mathematics and I was an older man associated with the learning of mathematics. Being aware of possible transference alerts us to powerful forces shaping what may otherwise be less comprehensible behaviour on the part of subjects, in particular in these cases, to their resistance.

Hunt points out that transference can work in the other direction as well. For example, in constructing the usual 'reflexive' account for Ellen's interview, I concluded that, as a defence against my own anxieties, I may have been 'motivated' to give her a more difficult problem than the other subjects, for Question 4: namely, calculating a tip of 15 per cent rather than 10 per cent. The examination of my decision to deviate from procedure with Ellen shows how to investigate the *relational dynamics* of the interview in a way that takes account of the importance of unconscious defences against anxiety, and of transference.[25]

Thus my analysis drawing on psychoanalytical insights shows the limitations of the clinical interviews used widely in mathematics education, and of traditional qualitative methodology, more generally. But I also show how to extend these approaches to provide a fuller discussion of affective issues – through the use of concepts such as the unconscious, defences, and transference – as well as a fuller account of mathematical thinking generally.

The interview also gave space to subjects, especially Donald and Harriet, to describe the beginnings of a 'second chance' with mathematics. In Harriet's case, the interview itself seems to have played a powerful 'consciousness-raising' role.

Finally, Theme 10 considers the possibilities of *transfer*, understood, after these analyses, as dependent on crucial similarities and differences in signification, and on emotion, as well. Sometimes the subject attempts this 'boundary-crossing' *deliberately.* Other times it occurs apparently *involuntarily*, due to the flow of possibly unconscious emotion, or the flow of meaning across a signifier, or along a semiotic chain.

In the first case, I argue in Chapters 6 and 7 that, for 'transfer' (of learning or knowledge) to occur, a 'translation', a making of meaning, across discourses would have to be accomplished through careful attention to the relating of signifiers (and signifieds), and other linguistic devices used in each discourse, so as to find *key signifiers* that provide points of 'articulation' or inter-relation. This interrelating of practices is not 'risk-free', since it may be misleading when a signifier functions in two discourses in ways that are different (see Walkerdine's example of 'more'). Alternatively, the relations between the discourses might be emotionally distracting or distressing.

There are examples of both 'deliberate' and 'involuntary' transfer in this corpus of interviews. Donald is apparently able to focus deliberately on *discursive similarities and differences:* he seems able to read the diagram as a 'chart' (business maths) or as a 'graph' (college maths), and to recognise the connections between a 'trend' and a 'gradient' (respectively). Though not certain, it appears that Donald might be able to bridge the two practices, that is, to 'transfer' his college maths methods to help solve a problem involving charts.

Ellen, less voluntarily, appears to be drawn, by the presentation of the 'restaurant menu', and the request for a calculated tip, back to anxiety about a relationship in which the signifier 'expense' speaks of her pain. This displacement is exceedingly distracting and distressing, and hence likely to interfere with transfer. However, Ellen recovers to critically evaluate her answer, and to revise it, possibly because her interdiscursive positioning includes school maths (as well as 'eating out'), and given her skill and confidence in that practice.

Fiona, less comfortable in mathematics, and also distressed by the anger and anxiety displaced from her relationship with her 'calculating' father onto mathematics, makes errors in both parts of the graph-reading problem that calls up, for her, his stockbroking practice and their difficult relationship. Here it seems that the key signifier 'calculating', rather than facilitating the bridging of practices, builds up the boundaries between her father's work, from which she is side-lined, and the mathematics which he uses.

Jean's distress seems to come from never 'having enough money'. This likely means that she is blocked from interrelating school mathematics and everyday practices (especially those involving money), and distracted from deploying school mathematics, because of her anxiety and partial isolation at school (see note 3), as well as conceptual problems. For Harriet, too, the boundaries charged by pain seem too much to surmount: the pain of being denied (by parents, and by teacher) the reassurance she desired that her knowledge in mathematics was sufficient to succeed.

However, fantasies are a way to cross a crucial boundary, that between reality and the unattainable. Harriet remembers fantasies of calculating discounts for desirable goods, as does Donald. And she engages, too, in fantasies of successful calculations that may supplant the painful memories of opportunity foregone – as does Peter.

Alan, for Question 6 (best buy) calls up shopping practices – in which for him great value is not placed on operations aimed at saving a few pence, or indeed on saving money at all – and is not 'motivated' to call up school maths. Here, we may conclude that a "mathematical" signifier is not recognised as such, whereas it may be recognised but its mathematical meaning be undermined by competing values related to other discourses (see Chapter 6 and note 18, this chapter).

Sometimes the associations between the problems posed and the subject's responses, for example to the contexting questions, may *seem* bizarre; for example Ellen's association of dividing recipes for four into recipes for two, with Question 2 (10 per cent of 6.65). It is not always possible to pull out a comprehensible chain of meanings for any particular reported association, because of the limitations of the interview material, and of course the basis of the associations may not be fully conscious. However, psychoanalytical insights may help: the discussion of the meanings attached to 'being an expense' in Ellen's case, and to 'peeling oranges' by Hollway (see Chapter 7), show how meaningful associations may be recovered.

This chapter's analysis has suggested three novel findings with respect to the phenomenon of transfer. It seems to require *interdiscursive positioning*, that is, the subject is able to call up and consider more than one practice. This multiple positioning may be *deliberately* produced by the subject, or *involuntarily* experienced. And transfer of a kind may be *facilitated by fantasy*. Illustrations of all three points are given here.

In this chapter I have developed my alternative views of mathematical thinking, and especially of emotion, and have aimed to show the value of Models C and D, based on insights from psychoanalysis and poststructuralism. In the final chapter, I bring together the main conclusions of the study.

11 Conclusions and Contributions

So that is what we can use the multiplication table for.
(Chilean worker, at a class on reading blueprints,
during the Popular Unity period,
quoted in Zaslavsky 1975: 232)

This final chapter gives an overview of the most important findings and areas of relevance of the study.

Mathematical Thinking and Emotions in Practice: Contributions to Theory

Mathematical Thinking and Performance in School and Everyday Contexts

At the beginning of the intellectual journey involved in this study, I attempted to specify the context of a mathematics problem by its wording and format. I produced findings within this framework (Chapters 2–4) and evaluated them (Chapter 5). Limitations and anomalies in my own findings, as well as concerns emerging in the literature about a narrow definition of context (see Chapter 6), led me to seek to formulate a broader notion of context.

Tine Wedege's (1999) distinction between two fundamental meanings of 'context' – the *task-context* and the *situation-context* – is helpful here. The task-context means the linguistic features, the wording of the task, and also the assumptions that the adult must make so that the problem can be solved mathematically. The situation-context concerns the social, historical, psychological and other circumstances in which the problem is considered or learning takes place. In the first phase of the study, I tried to vary the task-context between 'school maths' (SM) and 'practical maths' (PM) tasks, while the situation-context (survey) remained constant. In the second phase, I aimed to vary some dimensions (see later) of the situation-context. Analyses of context in both of these senses are useful.

In my reconsidered approach to the idea, any context is based in, and given meaning by, one or more practices, and related discourses. *Discourses* are characterised as systems of ideas expressed as signs. This allows us to draw on ideas from linguistics and semiotics to analyse discourses, and hence contexts (see Chapter 6).

An important aspect of practices is the variety of *subject-positions* made available to be taken up by different people. This concept subsumes ideas of social relations, power, social difference, institutional authority. In a particular setting, it is possible to analyse the practices that are at play, and that position the participants. This is what a situated cognition or structuralist analysis would give us.

However, the positioning of a subject is based on more than the one (or more) practices that have been analysed as being 'at play' in the setting, generally, for all participants. I use a special term to indicate that the particular subject's *positioning* depends not only on the practices at play generally in the situation but also on the practice(s) called up by the subject, because of cultural meanings, and his/her own emotional investments.

In the interview analyses I have shown how to describe the context by describing the discursive practices in which the subject has his/her positioning (see illustrations in Chapters 9 and 10). Results from the cross-subject analyses of the interviews showed that performance on items where the subject had a 'predominant' positioning (see Chapter 9, pp. 152–6) in practical maths was generally better than that where school mathematics was called up; this can be seen for an abstract 10 per cent calculation, a '10 per cent tip', and a 9 per cent wage increase. Though the results from these cross-subject analyses are only suggestive in most cases because of the small sample size, the idea that differences in performance might be explained as at least partly due to the context, converges with findings from many other studies, such as Lave (1988) and Nunes et al. (1993). However, as I have shown, the notion of context differs in these different studies (see Chapter 6).

The understanding of context that I am using in the second part of my study is multi-dimensional. Because the context of subjects' thinking and actions is seen as constituted by discursive practices, it is infused with the features that characterise the practice(s):

- the language, organised as discourse
- the affective charges attached to terms, symbols, diagrams and so on
- the goals and standards
- the social relations.

The context is also supported and constrained by:

- the material and other resources available.

This conception of context has similarities to that researched by Nunes et al. (1993) in the ethnographic phase of many of their studies, and to Lave's (1988) idea of structuring resources for settings, and I have drawn on their work.

However, the experimental phase of several of Nunes et al.'s studies allows variation only in background to the problem-solving, or task-context. This leads them to narrow their focus to emphasising 'symbolic systems', and to play down aspects of the context, such as social relations, and affect or emotions, for example confidence (Nunes et al. 1993, Chapter 3). These are aspects of the context of problem-solving that I have found important. Further, much of Nunes et al.'s (1993) discussion of symbolic systems is based on the distinction between 'oral' (including mental) and written procedures when solving mathematical problems. This distinguishes clearly between school and out-of-school mathematical thinking. However, if used as a simple dualism, it seems unable to categorise the procedures evinced in some of my interview episodes. For example, Peter uses school algorithms sometimes, but, at the same time, usually calculates in his head ('orally') for reasons revealed by his case study; others I judge to have called up practical maths do part of their calculations in written form (see Chapter 9).

My approach, like situated cognition, recognises different practices as in principle distinct, as discontinuous: for example, school mathematics and calculation in everyday practices like street selling, or shopping. However, I aim to avoid the trap of the *strongly situated* approach, of seeing a proliferation of different types of 'situated mathematics' – say, one for each work practice and leisure activity – with each one claimed to be disjoint and non-overlapping.

This points to the need to analyse the differences – and the similarities – between practices, rather than simply proclaiming them as distinct. This in turn requires a systematic way to describe practices. My approach draws on Saussurean structural linguistics, to produce semiotic ideas on meaning-making, particularly ways of analysing signifier–signified relations, and the use of devices such as metaphor and metonymy. I further draw on the poststructuralist idea of the inevitable tendency of signifiers to 'slip' into making links with other discourses, thereby producing multiple meanings for certain *key signifiers* (see Chapter 10).

There is a tendency in some sociocultural work to assume subjects are positioned normally in one basic activity (often a work practice) at any one time. In contrast, Jean Lave (1988) proposes the idea of the 'proportional articulation of structuring resources' from multiple practices, but does not use its full potential. In my case studies, I relax my earlier assumption of the subject's having a *predominant positioning* in one single practice or another, to consider *positioning in multiple practices* (see Chapters 9 and 10).

My work builds very much on that of Valerie Walkerdine, as I have indicated, particularly her ideas of discursive practice, relations and chains of signification, and the importance of subjectivity and emotion. In developing my own approach, I have aimed to emphasise the idea of *positioning*, rather than *being positioned* (in subject-positions), thereby signalling my desire to avoid a strong discourse determinism (see Chapters 6 and 7).

The spark of unpredictability, and hence the basis for a partial determination (only) of, say, a subject's response to a problem, comes from the unexpected ways in which signifiers may slip across into other discourses – because of cultural (or local) meanings – and from the related flow of emotional charges along these chains of signifiers, because of the person's emotional investments. This unpredictability is the basis for the potential to interrelate practices – and hence for any possibility of learning transfer. It also challenges any strong form of situated cognition, structuralism, or other position emphasising strict disjunctions between practices.

My approach thus aims to understand how subjectivity is formed: this includes the ways that subjects examine and think about specific problems to be solved, and the emotions they feel.

The Relationship Between Thinking and Emotion: Mathematics is 'Hot'

I see affect and emotion, as inseparable from thinking, including mathematical thinking. Therefore, in specifying the subject's 'positioning', that is the practice(s) within which the subject is addressing the problem, I aim to move beyond the structuralist analysis of the one (or several) practices that are at play in the situation, and to allow for the subject's emotional investments that will influence him/her to call up one or more practices that seem 'related' to the situation faced. Such a relation is accomplished via the flow of meaning along a semiotic chain, energised by emotion and desire; see Ellen's and Fiona's case studies in Chapter 10.[1]

My empirical study of affect and emotion focused mostly on anxiety. My initial attempt to measure anxiety in different contexts as different *types* or dimensions – namely, 'maths test/course anxiety' (TCA) and 'numerical anxiety' (NA) – was broadly supported by the factor analysis of the students' responses to the items (though I found several anomalies; see Chapter 4). On reflection, however, I questioned the limitations involved in using self-report items about feelings in situations that were only briefly described in general terms. Thus, in the second phase of the study, a subgroup of students was presented with problems to solve in an interview, where their emotional reactions could be expressed directly, observed and discussed, in the context of my judgements of their positionings in practice.

Now, if a subject has a multiple positioning, his/her anxiety may relate to more than one context or practice – and hence the 'type' of anxiety becomes ambiguous. Thus Ellen, when asked to calculate a 15 per cent tip, exhibited anxiety which I judged to relate to being 'an expense' within a relationship (involving some amount of eating out), rather than being simply mathematics anxiety.

Once it is accepted that anxiety may be unconscious, it becomes necessary to look instead for *indicators of defences* against anxiety: anxiety may then be exhibited as confidence, or as a 'slip', or as nothing at all. At least some of these indicators can be specified (Hunt 1989). Thus, despite my initial emphasis on it, the study of anxiety cannot be separated from the study of emotion more broadly, for at least two reasons:

1   Anxiety and other emotions may be difficult to separate if they occur together in a string of emotions; see for example Fiona's story about the anxieties, uncertainties and disappointments in growing up at home.
2   Because of defences and so on, anxiety may sometimes be presented as another emotion, say anger or confidence.

Indeed, every one of the twenty-five students interviewed expressed some emotion related to the doing of mathematics, or the use of numbers. There was a wide range of feelings, often intense: not only anxiety, but also confidence, diffidence, pleasure, dislike, anger, and boredom (see Chapters 9 and 10).

In approaches emphasising the emotions, including those using psychoanalytic insights, there can be difficulty in giving sufficient explanatory emphasis to *the social*. Nevertheless, if we link, to these psychoanalytic insights, ideas from post-structuralism, we can understand particular motivations in terms of social difference and deprivation, as well as desire and early family dynamics. This can help us to understand, say, Donald's pleasure in controlling large sums of money at work, as related to a need to overcome early deprivation of material goods, themselves objects of his desires and his fantasies of discounts in shops. Hence it is possible to acknowledge the likely complexity of the bases of his feelings, and to examine how the apparently different emotional, social and material bases may be linked through chains of meaning, infused with desire.

The relationship between cognition and affect was considered in the quantitative part of the study, using statistical modelling. The confirmation of an 'inverted U' relationship was exciting, since it suggested that the relationship between school mathematics performance and maths test/course anxiety, for example, might be interpreted as producing an optimal level of performance for moderate levels of anxiety, and lower levels of performance for higher *and lower* levels of anxiety. This would question the assumption in most research over the last twenty-five years that an increase in anxiety is debilitating for performance at all levels (see Chapter 4). However, such a conclusion would raise questions about the theorisation of anxiety, and its relationship with performance: for example, anxiety might need to be reconceptualised as attention or 'arousal' (see Teigen 1994).[2]

In considering the interview results, I came to see the relationship of *emotion* in general to thinking, not as always simply interfering (as in one of the versions of Model B, see Chapter 7), but, instead, as dependent on the subject's positioning. Nevertheless, the case studies do provide a number of examples of emotional interference, many of which show anxiety interfering with numerate thinking.

The research also illustrates a number of ways in which emotion can *support* numerate or mathematical cognition. For example, both Donald and Harriet have gained pleasure from playing with figures, which I argue may relate to earlier fantasies involving (calculations of) price reductions of goods in shops. Keith (interviewee number 4) claims to have derived pleasure from both school maths and college maths, and he illustrates this with a creative response to the 'best buy' problem (see Chapter 9). However, the subject's positive affect does not necessarily support the mathematical thinking or performance expected by a teacher or interviewer, as Harriet's difficulties with a percentage calculation, and refusal to engage with it, in the interview show.[3] Similarly, bad feelings do not appear necessarily to interfere with cognition; for example, Donald's bad feeling towards the competition-inducing wall charts at work does not prevent him from clearly explaining solutions in both financial maths and school mathematics.

The idea of *critical incidents* affecting students' emotions about mathematics, and consequently the development of their mathematical thinking, is based on a number of writers discussed in Chapter 7 (Tobias, Buxton, Nimier). In some of the interviews reported as case studies (e.g. Jean, Harriet), it was possible to designate one of the incidents related by the student as possibly 'critical' – on the basis of the interview; in the rest, either continuing (often, family) factors seemed to be more influential (e.g. Peter, Ellen) or a mixture of critical incidents and ongoing factors (e.g. Fiona) were important.

The earlier descriptions of the relationship between cognition and emotion, in terms of emotion 'supporting' cognition, or 'interfering with' it, are in line with views which mark the affective as the 'other' of, as separate and distinct from, the cognitive. However, I have attempted to inform my analysis with insights from psychoanalysis that see emotion in terms of *charges of feeling attached to* (or infusing) ideas, and as thus related to the cognitive. In this sense the affective is not entirely 'other' to cognition.

As we have seen, however, emotion can be *displaced* onto ideas different from those to which it was originally attached. This means that, though affect is not entirely 'other' to cognition, neither is it completely 'at one with' cognition. These ideas – the fluidity of emotional charges and the associated slipping of meaning along a chain of signifiers – have been used in the analysis of the case studies, notably Ellen's and Fiona's.

Gender and Other Social Differences

One of the objectives of the research was to investigate social differences in mathematics performance and in affective responses to mathematics, and to consider critically the findings produced in some earlier studies. The fact that some of these ideas may act as 'myths' was discussed earlier (see Chapter 7).

One such possible myth was the idea that 'men are better than women at mathematics'. This was widely believed, and continued to receive confirmation in studies published in industrial societies at the beginning of the 1990s (cf. Burton 1990).[4] However, my survey results, using the controls provided by the statistical models, by no means straightforwardly replicated these conventional findings; for example, the male advantage for school mathematics (SM) performance held only for mature students, and not for school-leavers (see Chapter 3). However, my finding a pocket of 'low-performers'[5] – 'mature' (21+) women without O-level in Mathematics – in both the quantitative and qualitative analyses (see Chapters 4 and 9), is noteworthy, and suggests further research (see later).

A related and possibly important tendency, evident for interview Qu 4 (a 10 per cent tip), was for women to call up school maths, rather than 'practical maths' associated with eating out, more often than men. Again, this receives only suggestive support because, as must be remembered, the number of interviews was small (n = 25). However, the idea that gender (or other) differences in performance might be explained through the practice called up in response to the problem is emphasised in other studies. For example, Cooper and Dunne's (1998, 2000) analyses, based on a much larger interview sample, point to the influence of *social class* on the calling up of 'esoteric' (school) or 'realistic' (practical) approaches to problems taken from national tests, and not dissimilar to mine (see later).

Despite the lack of confirmation for the influence of social class in the quantitative and the qualitative cross-subject analyses, some of the case study accounts strongly suggest its importance. Sometimes social class is implicated in a critical incident, for example in Harriet's not being told by her teacher or her parents that she was capable of doing O-level Mathematics, and sometimes it is (also) an ongoing factor, as in Harriet's diffidence resulting from her being the first one in her family to go to grammar school (see Chapter 10). In addition, the resistance to schooling or working class culture in general is illustrated by my interview with number 8, a mature working class male:

> I think we got very resentful towards [. . .] the system, towards school, in a way . . . It was – they can't be bothered with us, so why should we be bothered with them?
>
> (Number 8's transcript)

and

> When I actually got into the factory, and you work with some of the guys there, they get you through it, you can sit down and chat with them, they sort of say, forget what they taught you there [i.e. in school] – you can imagine, like some of these old boys [we both laugh] – just forget what they taught you there, it's complete rubbish, y'know, we'll teach you.
>
> (Interview transcript)

In seeking to explain performance, gender differences – or any other social difference,[6] or indeed set of differences – cannot be expected to tell the whole story. Using the idea of thinking as specific to positioning (and not simply *position*), we would of course expect *variation within* a sample of women, or a sample of middle class students, and that has been amply demonstrated, in both quantitative and qualitative findings. For example, Fiona's response to the tipping problem differs from that of Ellen, as Fiona can address it from several positions: waitress and 'going Dutch', as well as someone invited out.

Next, the idea that 'women are more anxious about mathematics than men', another possible myth, received confirmation from the analysis of the questionnaire (self-report) items (see Chapter 4). My first response that, rather than *being* more anxious, women are simply more likely to *express* anxiety, received some confirmation from cross-subject analysis of the interviews (see Chapter 9).

However, this risks a new kind of gender essentialism – that women are more able to express their feelings. Against this, the idea that anxiety is specific to positioning within practices means that anxiety is produced within the same practices as those in which thinking, including numerate thinking, is produced. That is, we need to consider gender in relation to positioning, in order to understand how its effects are produced. Ellen's anxiety in the context of the interview tipping problem provides a good example (see Chapter 10). It is not on her gender alone that we must focus, so as to understand her anxieties, but rather on the meanings generated by the discourses providing the basis for her positioning; these discourses concern not only school mathematics, but also eating out and involvement in relationships. Gender is implicated in these positionings, but it is not determinant in any ultimate, or exclusive, sense.

Rethinking the 'Transfer' of Learning

My discussion of transfer, the (attempted) application of school mathematics learning to non-school contexts aims to clarify the problems with several current approaches to it, and to show how my approach helps to reconceptualise and resolve the problem, at a theoretical level (for more detail, see Chapters 6 and 7). In particular, I criticise 'utilitarians' (including, on this issue, both proficiency and functional views), for their overly simple faith in transfer as relatively unproblematical, and on the basis of my findings that task always needs to be understood as dependent on the context, including the social relations of its presentation, for its meaning (Theme 2, see Chapter 9).

On the other hand, what I call the 'strong form' of situated cognition (based on Lave 1988) argues that transfer as a goal is basically hopeless, because of the high boundaries between practices (Chapter 6). Yet other sociocultural researchers have proposed solutions to the transfer problem. In particular, Saxe resists seeing transfer as an 'immediate generalization' of prior knowledge to a new context (as expected by utilitarians), that flows from the special power of mathematics, due to its generality and abstraction. He proposes instead that transfer be conceived as 'an extended process of repeated appropriation and specialisation' (Saxe 1991b): that is, as a series of reconstructions of ideas and methods from the context of learning, so that they are appropriate for the target setting.

However, it is important to supplement this important move, so as to be able to develop a full explanation for both the successes and failures of 'transfer', by taking account of two further aspects of the process. First, affect and emotion (in addition to ideas, strategies and so on). The potential importance of emotion in processes of transfer is illustrated by Fiona's errors in interview problems, that can be attributed to her being distressed and distracted by associations of the graph of the gold price with feelings for her father (see Chapter 10).

Second, despite acknowledging the importance of meaningfulness in mathematics learning and use, most other researchers do not take account in sufficient detail of the ways in which meanings are carried by semiotic chains, in particular, the capacity of a signifier to provide unexpected links between a mathematical term or problem, and a non-mathematical practice. This is illustrated by Ellen, for whom the task of calculating 15 per cent tip is linked unexpectedly with the total cost of a meal, and in turn to anxieties about a relationship within which eating out is a familiar activity.

Thus transfer is difficult to predict or control, not only because ideas and strategies are formulated differently in different contexts, nor solely because of the differences in goals and values, social relations and regulation, and standard language used, in different discursive practices. It is also due to:

*   the unpredictability of the *flows of meaning along chains of signifiers*
*   the sometimes unexpected *flow of emotional charges*.

For example, responses to the 'contexting questions' used in the interviews revealed a wealth of unexpected associations between the 'mathematical' problems presented and the subject's memories and accounts of experiences, which speak of the meanings these problems have for him/her.[7] That is why I argue that the ability of a signifier to form different signs, to take different meanings, within different discursive practices, constitutes a *severe limitation* on the possibilities of transfer (as conventionally understood), since the attempt to 'build a bridge' to a specified target practice might be diverted or blocked.

Yet this capacity also provides *the basis for any possibilities* of transfer/ translation. Any kind of interrelation or 'articulation' between practices will depend on signifier–signified relations within the two practices. This would require a process of what we might better call 'translation' across discourses.

Examining several different approaches discussed earlier points to a potential for convergence of views on transfer. Two views, from developmental psychologists, emphasise a similar need for careful attention to the relating of 'quantities' (Schliemann 1995), or 'elements' (signifiers and signifieds – Walkerdine 1988), in the two contexts across which transfer is to be promoted. Further, these views may possibly converge with those of some cognitive psychologists, who argue that transfer between tasks is a function of the degree to which the tasks share 'symbolic components' (Anderson et al. 1996, Singley and Anderson 1989, but see Cobb and Bowers 1999). I emphasise the need to broaden the discussion by attending not only to similarities of elements, but also to *differences which can be specified* (see Chapter 6).

Bringing in insights from psychoanalysis and poststructuralism is helpful with the problems raised by the 'unpredictable flows' of meaning and emotion. The analysis of the case studies suggests several novel ideas about transfer. First, it would seem to require *interdiscursive positioning*, that is, that the subject is able to call up and consider two or more discourses at one time. This multiple positioning may be *deliberately* promoted by knowledgeable subjects who desire to extend their mathematical gaze to new areas of application (cf. Noss and Hoyles 1996a). Or the multiple positioning may be *involuntarily* experienced, because of the 'unpredictable flows' of meaning and emotion. Finally, transfer of a kind may be *facilitated by fantasy*, which bridges the actual and the unattainable (see Chapter 10).

Although I am calling my 'sceptical-optimist' approach to transfer 'translation', it is important to realise that translation generally involves *transformation*, since the sign, the relation between signifier and signified, between word and meaning, also depends on the rest of the discourse. For example, 'mathematics education' does not translate simply into French.[8] Boaler (1997) also declines to describe the process as 'transfer', preferring to call it 'adaptation' (recalling Saxe) and 'reforming' (cf. 'transforming') of ideas and methods that the students in her more 'progressive' school had learned in their classes. Calling the process translation/transformation reminds us that the translation can be 'free', as well as 'strict', and the mathematical tools (such as procedures for calculating) may themselves be changed in the process.

Ideas for Pedagogy and Practice

My approach to the description of context and transfer aims to introduce some scepticism, in particular towards conventional views that appear to be in the ascendancy currently in the UK. For example, the currently fashionable buzz-word 'transferable skills', used to describe *capabilities claimed to be essentially suited for immediate application* in other activities, is founded on a misconception. As argued in Chapter 6, there is no such thing as intrinsically transferable terms, ideas or subject contents. What matters is how meaningful relations are set up between them, in pedagogic (or other) situations.

Adult-Friendly Contexts for Learning

It is not possible to present blueprints for 'teaching for transfer', but there are signs of a possible convergence of thinking in this area (see earlier), and related practical suggestions are emerging; see Anderson et al. (1996), Carraher and Schliemann (1998), Masingila et al. (1996), and Bessot and Ridgway (2000). Here I propose a provisional set of hopefully stimulating guidelines for teaching/learning for transfer, based on the analysis in Chapters 6 and 7 (see also Evans 2000).

1   Show the learners how to perform a detailed analysis of the shared or similar components – and the specifiably different aspects – of the initial and target tasks. For example, the aim may be to transfer the use of the derivative to optimise the value of a function, from modelling ballistic motion to modelling profit or surplus value. Similarities may include the 'inverted U' quadratic form of both functions (see also the model relating mathematics anxiety and performance in Chapter 4), and the finding of a maximum by setting the derivative equal to zero, in both cases. However, there are differences in terms of whether the independent variable is continuous (e.g. time), or discrete (e.g. units of output); this difference is important, since the discrete case may undermine the use of the calculus, which is assumed to be about infinitesimal changes, though if the relevant levels of output are high the infinitesimal model may 'fit' acceptably. There are also evocative differences in language in the discourses around the two problems.

2   Include the ability to transfer as a specific, and explicit goal, by establishing inter-relations between the two situations and the related discourses and practices, by *translating* between the terms/languages used, and by *generalising* the methods used across contexts.

3   In teaching the initial task, seek to incorporate a balance of generality and situational features. That is, anticipate what will come to be seen as *similarities* across situations, and *differences*, respectively.

4   Teach the initial task in more than one context: for example, use examples of both a child throwing a ball vertically into the air, and of a goalkeeper in football kicking the ball into the air and down the field.

5   Allow practice in recognising the *cues* that signal the relevance or 'applicability' of an available skill: in mathematics these cues are recurrent features of pattern, structure, or relationship. For example, both situations in point 4 have been analysed (and simplified by assumptions) to depict a quadratic relationship between two variables (that holds independently of place and time).

6   Allow repetition or practice on the target task, to help the student to appreciate

the possible range of generalisation, and the constraints on it, resulting from crucial differences in discourses/areas of application.

7   Encourage the students to seek to understand what might be emotional blocks to actively seeking out possible applications of their learning.

In considering the teaching and learning of adults, it is vital to consider transfer 'in the opposite direction', namely harnessing ideas or skills from non-school practices to use in school or college contexts; see Schliemann (1995, 1999)

Two proposals based on ideas from this study can be summarised here. First, Evans and Rappaport (1998) suggest the creation of a new context for teaching social research methods and statistics, called *community research*. This context would be positioned between, and aim to draw on, both the course members' outside activities, as students and as members of various communities, and a college-based course in social research methods and statistics. The aim is to encourage students to bring problems from their everyday activities, to be investigated in the course: for example, the adequacy of facilities at their local hospital or library, or views of community members on the quality of the environment. The meaningfulness of this created learning context might be expected to be shared to a reasonable extent across certain groups of students, and the students' relationship to the academic discourses of social research methods and statistics might be seen as being apprentice 'barefoot statisticians' (Evans 1992).

The second proposal (Evans and Thorstad 1995) for younger and/or more differentiated groups of students, is to seek to build up a relatively generally shared discourse around activities, in which the learners as 'citizens', present and future, are highly likely to participate:

- purchase and/or growing, and consumption of food, and other necessities
- involvement with and/or raising of children
- paying for (perhaps building) and maintaining a dwelling and surroundings
- engagement with discussions and debates about personal, family and public well-being and about describing, evaluating, deciding on future directions.

A thoughtful engagement with these activities, in particular the last, might be called *critical citizenship*. The 'skills' necessary for its practice might provide the basis for a course offered as, say, statistics or 'mathematics across the curriculum', or as 'civics', or 'responsible citizenship'.

Finally, mention must be made of approaches to the teaching of mathematics and numeracy to adults, which emphasise *inclusiveness*. These use ethnomathematical, feminist (e.g. Harris 1997) and working class and adult-friendly approaches (e.g. Schliemann 1998).

Reclaiming Numeracy

In the post-Cockcroft period, there was much interest in teaching mathematics in a functional way, and wide discussion of Cockcroft's suggestions for classroom practice (1982, paras 242–52). However, by the early 1990s these ideas of numeracy were undermined by the decline of the area of 'using and applying mathematics', in the UK National Curriculum.[9] In the late 1990s, the work of the Numeracy Task Force has led to the original concept of numeracy being drained of much of its vitality, so that it is now 'shallow' (Noss 1997). Further, the term 'numeracy' appears to have been detached from having its main position in discourses of functional numeracy, and has slipped into use also within the numerical skills proficiency conception of mathematics learning that appears to be the basis for the work of the Numeracy Task Force (1998).

My work suggests some bases for questioning this limited proficiency vision of numeracy. Earlier I have questioned several features of this view, for example the neat division of the mathematical task from its context, the consequent tendency to play down the importance of the distinctive features of the context, as described in this study, and the resultant simplistic assumptions that the transfer of learning from educational to everyday settings is relatively straightforward (see Chapters 2 and 6). I am also concerned as to how these changes, notably the 'return to abstraction' may threaten affective responses of students to mathematics.

The problem is how to rescue numeracy from this dead-end. The solution is clearly not to advocate a return uncritically to functional numeracy, of which I have shown the limitations (see Chapters 2 and 6). The position I am developing here suggests a provisional working definition for a reconstituted idea of numeracy:

> *Numeracy is the ability to process, interpret and communicate numerical, quantitative, spatial, statistical, even mathematical, information, in ways that are appropriate for a variety of contexts, and that will enable a typical member of the culture or subculture to participate effectively in activities that they value.*

This characterisation draws on those of Askew et al. (1997) and Bishop (1992). It is also in accord with the *social practice model* of numeracy (Baker and Street 1993), the third type of numeracy, besides proficiency and functional numeracy, described by Brown et al. (1998). The key points in this working definition are:

- Numeracy's raw material is *information*, which has *meaning within a practice*, rather than being restricted to abstract numbers as used in academic mathematics, or (sometimes) in school mathematics.
- Not only is calculation, or processing of information, important, but also *interpretation* of its meaning, and ability to communicate.
- Nor is the information to do only with number, since quantity, shape, statistical indicators, chance and so on are relevant to many practices.[10]
- The context of numerate thinking is important, within a range of school, everyday and work practices, and some way of describing the context, as discussed throughout this book, is crucial.
- There is an idea here of the range of activities that are *relevant to a typical citizen*; see the discussion of 'critical citizenship' earlier.
- These activities embody the goals, values and passions of the citizen, so that numeracy is both charged with commitments, and hopefully empowering.

Noss (1997) makes a number of crucial points. He argues that the notion of numeracy must be broadened, not narrowed (as over the last twenty years or so): the Crowther Report (1959) saw numeracy as the mirror-image of literacy, a broad introduction to mathematical and scientific reasoning 'an understanding of the scientific approach to the study of phenomena – observation, hypothesis, experiment, verification' (Ministry of Education 1959: 270). Noss warns that school mathematics has recently been cut off from its 'broader roots in science and technology'.[11] Concerning technology, an issue not addressed explicitly by my working definition, Noss emphasises the use of the computer as a means to build and to use 'numeracies in which learnability and knowledge are not antagonistic' (Noss 1997: 31).

This is a sketch of several key aspects of a reinvigorated numeracy. At the same time, it must be recognised that currently the concept of numeracy is heavily contested, given that it features in several competing discourses.

Questions of Methodology and Further Research

Fruitfully Combining Quantitative and Qualitative Approaches

In recent years, quantitative methods have been used less, and qualitative methods have been in greater favour, in mathematics education research, and educational research generally (for example Delamont 1997). Whatever the reasons for these trends, my aim here – rather than polarising the discussion by asking which method is 'best' – is to note the relative strengths of each, and to show how I attempted to combine the different approaches effectively for my research aims.[12]

In this research, both quantitative and qualitative methodologies were used. The quantitative phase used questionnaires (including a test), analysed with statistical modelling. The qualitative phase used semi-structured interviews analysed in two ways (see later). Because the sample was from an institution of higher education with an unusually high percentage of mature students – most of whom had previously done full-time work and/or child-care, I argue that the study was more representative of the population at large than most college or university-based samples would have been (see Chapter 2).

The second part of my study used a qualitative data-production phase, combined with two approaches to data analysis:

- a *cross-subject* approach (drawing on Miles and Huberman 1994)
- a 'within-subject' *case study* approach.

These two types of 'qualitative' approach will be distinguished here.

The *quantitative approach* is useful for producing a general overview, for example of levels of performance, or for making general comparisons across groups of subjects, such as ones about gender differences in expressing mathematics anxiety. The statistical modelling allowed hypothesised differences, say in performance between men and women, to be examined while controls were operated simultaneously for a range of other relevant factors, such as qualification in mathematics and age. In addition, the subgroups in which the differences were strongest could be specified (using interaction terms). The modelling also allowed for the exploration of an 'inverted U' relationship between performance and mathematics anxiety. Thus the inclusion of a quantitative component in this study has allowed me to question previous ideas that were limiting in theoretical or policy terms, and to contribute my own findings to debate in these areas.

The qualitative data production used a specially developed form of semi-structured interview which elicited life history material, as well as including problem-solving situations. The interviews afforded several things:

- the opportunity to subjects to describe experiences, including critical incidents with school mathematics and other numerate activities
- the space for subjects to express anxiety, or other feelings, about these experiences and/or about the problems presented
- the occasion for me to observe and describe numerate thinking and problem-solving processes, and the experiencing of emotions, in a way not possible with the questionnaires.

Two further features of my interviews were important:

- the innovative use of *contexting questions*, the answers to which provided indicators for the subject's positioning in response to problems presented (see Chapters 8 and 9)
- their use to tap experience with the numerate aspects of a *range of practices* (called up by subjects), rather than attempting a full ethnography of one or two practices at a time, offering the advantages of economy with research time, as well as widening the variety of practices studied.

The *qualitative cross-subject approach* provides 'rich' data, for example on the subject's positioning or expression of emotion, but it was nonetheless possible to categorise subjects, for example, as to 'predominant positioning', so as to produce comparability and hence the potential for some generality in findings. The sample was chosen in a representative way, but it was small relative to that for the survey.

The *qualitative case study approach* was useful to explore the coherence, and process of development, of a limited number of cases. I was able to consider the full 'multiple' positioning of subjects; to trace the multiple meanings of 'key signifiers' (as with Ellen's fear of being an 'expense'); and to produce different readings of the interview transcripts.

This research uses the relative strengths of the different strategies of data production. The quantitative aims to produce dependably *generalisable* findings based on comparability within populations of interest (via representative samples), backed up by (triangulation with) the qualitative cross-subject findings. The qualitative case studies aim to produce *subjectively meaningful* accounts grounded in specific contexts and particular subjects' perspectives.

Thus this research shows ways of bringing together quantitative and qualitative approaches to research in a mutually supportive way. The questionnaire threw up some surprises – for example, the frequency of the answer '37.2p' for a restaurant tip – which could be investigated in the interviews. It provided the sampling frame for choosing a stratified random sample for the qualitative interviews. In addition, the quantitative modelling allowed certain students' results to be pinpointed as 'deviant' from the expected, as 'under-performers' or 'over-achievers', and worthy of further investigation.[13] The interview could investigate possible reasons for anomalous questionnaire responses at a group level (see earlier), and for an individual, for example an 'unusual' pattern of responses to the mathematics anxiety questions. It also investigated areas that the questionnaire could not: because they required dialogue in order to produce indicators, such as those for positioning; because they were highly emotive, due to, say, the exhibiting of anger or panic; or because they required support for recollection or exploration, for example early memories of school or family.

Finally, the aim of the interviews was 'data production' rather than 'consciousness raising' (Carr-Hill 1984). However, the interview clearly helped several subjects (such as Harriet, Fiona and Keith) to remember experiences with mathematics and numeracy, including pleasurable ones. It was apparently 'therapeutic', too, in allowing Harriet to express herself about a number of past sources of frustration, disappointment and anger.

The interviews also allowed subjects to describe – and to celebrate – their 're-emergence' from mathematics anxiety and maths blocks, towards a 'second chance' in mathematics. In some cases, a subject developed numeracy through work practices: e.g. for Donald by working in the money markets; for interviewee number 9, stockbroking. Sometimes, they developed a facility in college mathematics: for Harriet, having to teach children school mathematics in the residential home helped later with college maths; for Donald, the first year Methods and Models course helped him discover an interest in mathematics for the first time.

The success of these interviews with several subjects points to possibilities for using similar interviews in mathematics anxiety intervention, or other learning support, programmes. For example, such interviews could be used to encourage the development of 'self-reflexivity' in students about their experiences in learning, and about their feelings and beliefs that may block their developing learning, or their attempts to apply it.

Further Research

A study like this raises many questions for further research. Here I can only mention briefly what seem to be the most promising issues.

First, the findings have opened up some further questions. For example, in the survey, I found a pocket of 'low performers' among the low-qualified, older females (and also in the interview cross-subject analyses). Would the same thing be found in other institutions admitting a high proportion of mature students to higher education? What reasons might there be for this in terms of the 'numerate careers' of adult men and women? Is it illuminated by the 'anomalous' finding from the national cohort studies that adult women in their early 20s produce self-reports claiming no more 'difficulty in using numbers' than men, but score less well on numeracy tests (see Chapter 2)?

The other main puzzle from the study, from the qualitative cross-subject analyses and hence less dependable because of the small numbers, was the superior performance, of those with positioning in practical mathematics over those in school mathematics, on several of my interview problems (for example the three involving percentages – see Chapter 9). First, this needs replication, as do all the findings from my cross-subject analyses. In any case, this was what would be expected from several earlier studies (such as Lave 1988, Nunes et al. 1993), but not from an approach based on structures of social class. For example, Cooper and Dunne (1998, 2000) found an advantage in performance terms for middle class children, from their being able more often than working class children, to call up *both* esoteric (school maths) and realistic (practical) approaches; see also Bernstein (1996) and Holland (1981). Cooper and Dunne's work, using a broader range of problems than my study, throws up fascinating insights, and suggests fruitful comparisons with my approach to assessing context; both studies attempt to assess the type of context called up by the subject, rather than using the researcher's *a priori* judgement about what type of problem it is.

Cooper and Dunne have also used their approach to study gender differences. In this study, the interview analysis suggested a tendency – only very suggestive, because of small numbers – for women to call up school mathematics (rather than everyday practices) more often than men, for a particular problem (tipping); it is worth exploring this possible tendency for a wider range of practical problems (see also Murphy 1989).

Following the valuable work using direct observation of work and other everyday practices (see Chapter 6), there are still particular practices in which we need to know more about the use of mathematics, especially 'more numerate' practices such as those of engineers, technicians and accountants (cf. the recent work of Noss and Hoyles, see e.g. Noss et al. 1998). This will provide an opportunity to develop the ideas of interrelating (or 'webbing') among practices, and (multiple) positioning in practices.

Similarly, we need to know about adults' numeracy needs in connection with *everyday* practices. For example, how many adults *actually use* the sorts of 'best buy' comparisons (studied by Capon and Kuhn, Lave et al., and myself) while shopping? Is it 'worthwhile' for them? Does that judgement change with changes in the available computation technology? This area is more difficult because of the tendency for people to consider as 'everyday', to *take for granted*, the things they *can do* – even if others would see those things as 'mathematical' (e.g. Harris 1991), combined with the tendency to *see as 'other'* what they *can't do* – for example, mathematics (Coben and Thumpston 1995).

The relationship between 'hot' emotions of anger, frustration, panic, enjoyment and so on described in the second part of the study, and 'cooler' attitudes of anxiety, confidence, avoidance, dislike and so on, studied in the first part, merits further research (cf. McLeod 1992). The ideas of confidence (Chapters 4 and 10) and familiarity (Chapter 10) figure in suggestive findings here, and need development through studies in other settings, and with different types of subject.

Finally, a number of writers on mathematics difficulties and mathematics anxiety have raised, explicitly or implicitly, the question: 'What's so special about mathematics?' Here I have shown that 'mathematical' terms may show up in unexpected ways, and that 'mathematical' activity and 'mathematics anxiety' can be read in quite different ways. That is, I have shown the complex ways in which the *apparently* simple and powerful signifiers of mathematics function also as elements in other discourses (and vice versa), thereby producing – and transforming – meaning. Investigation is needed of the ways in which mathematics – in its multiple intersections with, say, mathematics education, economics (Dow 1999), politics, computing, marketing, or graphics in television documentaries and so on – constitutes itself as a field, and how whatever specificity it has is produced, and delimited.

Appendix 1
Questionnaire Design and Fieldwork

Here I give further details about the design of the questionnaire, and its completion during the fieldwork, additional to those given in Chapters 2–4. A copy of the questionnaire to BASS students in cohort 3 is appended.

Questionnaire Design, Indicators and Coding

Basically the same version of the questionnaire was used for each group of students, with updating each year and slight differences in wording appropriate for the two courses, BA Social Science (BASS) and the DipHE (see Evans 1993, Appendices Q1–Q3). It was organised into three sections or 'scales'.

The Experience Scale included questions corresponding to social and affective variables in the conceptual map. The social variables were gender, age, social class (in cohort 3 only), and mathematics qualifications (exams passed). Age at entry (as at 1 September) was asked for (and coded) in years. For most age analyses, however, the students were divided into two groups: 18–20 (younger), and 21+ ('mature').

For social class, students were asked to report their father's and mother's occupation, when they began secondary school, and their own most recent paid occupation (if any) before joining the Polytechnic. On this basis, three indicators for social class were assigned. For all three occupations, the responses were coded into the Registrar General's six categories, using the CODOT classification (OPCS 1980); these were collapsed into 'non-manual' (RG's categories I, II and III-NM) or 'manual' (categories III-M, IV and V) for most analyses. For establishing a joint 'parental' occupational class, a symmetric categorisation was used: parental occupation was 'middle class' if both parents had non-manual occupations, 'working class' if both parents (or the only parent categorised) had manual occupations, and 'mixed' otherwise.

The affective variables, other than mathematics anxiety (see later), were:

- self-rating of confidence in mathematics
- use of numbers – in everyday life and at work
- difficulties with maths – in everyday life, at work, and on previous courses.

The Performance Scale included questions in both school maths (SM) and practical maths (PM) categories. This was based on the distinction made in Chapter 2 between the two, using the overt context (wording and format) of the problems attempted.

The Situational Attitude Scale (SAS) aimed to reproduce a sufficient number of items from both of Rounds and Hendel's dimensions of mathematics anxiety, maths test/course anxiety (TCA) and numerical anxiety (NA), in order to tap maths anxiety in school and college contexts and in practical contexts in a valid and reliable way – thirteen items for each subscale. It also included ten general anxiety items (constructed by a colleague in Psychology) and one 'state anxiety' item (constructed by myself). The major differences between the set of MARS items in the Polytechnic questionnaire and that used in the US research (see Chapter 4) were: first, we used twenty-six maths anxiety items instead of ninety-eight; and, second, we used a symmetric seven-point scale of responses, rather than an asymmetric five-point scale.

For more detail about the questions used as indicators for the variables included in the conceptual map and about coding, see Evans (1993: Appendices Q4, Q8 and Q9).

Fieldwork

The three versions of the questionnaire were given pilot runs on a total of twenty-five people, including students at the Polytechnic, colleagues, friends, and young people of college age living in my community (Crouch End, North London). I asked them to record the time taken for each section, as well as giving feedback on the questionnaire. They suggested revisions to the layout, to the ordering (such as not to have the age question first, as it might be disconcerting to some respondents), and to the wording. Many suggestions were taken on board (but not, for example, that made by several for a more specific description of the context for some anxiety items).

The questionnaire was introduced for BASS students at the end of the first Psychology lecture of the academic year – and for DipHE students, at the end of the first Quantitative Methods or Communication Studies lectures (depending on the relevant subsample) – by a member of the teaching team as an example of research relevant to that particular field. The questionnaire was not distributed at the BASS Mathematics lecture, since the Maths teaching staff were concerned that such a questionnaire, including a part likely to be perceived as a test, might distress some students, and might therefore interfere with the strategy, of reassuring incoming students about their ability to do mathematics successfully, that was crucial to the Methods and Models course approach.

Then the Psychology colleague, or myself, introduced the questionnaire with a standard introductory script (see Evans 1993: Appendices Q5 and Q6). In particular, the students were reassured that their performance on the questionnaire would not affect their position on the course because the questionnaires would be made anonymous. We then asked the students to complete the Experience Scale, but not to start the Performance Scale yet. After four or five minutes, we asked them to leave the Experience Scale (but to return to finish it, if necessary), and to go on to the Performance Scale. Here they were to try to do as many items – in order – as possible in the allowed time of ten minutes, but they were reassured that they were not expected to complete the whole scale. After ten minutes, we asked them to take as long as they wished to complete the Situational Attitude Scale, and to return to the Experience Scale, if necessary, but not to return to the Performance Scale. (This instruction was reinforced, on one occasion, when a few students were seen to try to return to completing the Performance Scale, and were asked to stop.) Generally, on all occasions, the overwhelming majority – if not all – of the questionnaires were completed and returned by the end of the lecture period, without apparent distress to the vast majority.

The numbers returning the questionnaire for each session were as shown in Table A1.1.

It is possible to consider the issue of *non-response*, and the related issue of the representativeness of the samples, in general terms. The samples of students completing the questionnaires might not have been completely representative of the populations of all students on the relevant courses, for several reasons:

1   Since the students' choices of overall course (or of modules, within the DipHE), were not yet finalised, the list of students registered for the course or module might not have been precisely representative of the group of students who would eventually be registered for it.
2   Some non-attendance of registered students at the particular lecture where the questionnaire was circulated was to be expected.
3   Some students at the lecture might possibly have refused to complete the questionnaire, or to return the completed questionnaire. (There were several cases of the latter, at the BASS Cohort 3 presentation of the questionnaire.)

The first problem was unlikely to be strongly related to the focus of the research and therefore unlikely to be very important, whatever the numbers of students involved. Similarly, the second problem – failure to contact students because of their non-attendance at the relevant lecture – was also presumably only randomly related to the focus of the survey, at least for the BASS group (where the questionnaire was completed in a Psychology lecture), and for the DipHE Communication Studies students. The third problem – non-cooperation of some attending students – *was*

Table A1.1 Presentations of the Questionnaire: Numbers of Respondents

Course/year	BA Social Science	DipHE QM100	DipHE CM100	Total
Cohort 1 (1983)	192	124	—	316
Cohort 2 (1984)	136	120	81	328*
Cohort 3 (1985)	160	55	82	291*
Total	488	299	163	935*

Note: * These totals are adjusted to avoid 'double counting' of the nine students in cohort 1 and the six students in cohort 3 who took both Quantitative Methods and Communication Studies

likely to be systematically related to the material of the survey, but as far as I could see, it formed a relatively small part of the non-response. (One student in cohort 3 protested about being asked for her parents' occupations. Another asked whether we always gave people such a 'rude shock' on the first day.)

Overall, we can be reasonably confident that systematic non-response (type (iii)) was very infrequent, and that the samples could be considered as acceptably representative of the populations of BA Social Science and DipHE science and social science students.

The Questionnaire

Anonymous Code Number . . . .

DO NOT WRITE YOUR NAME ON ANY OF THE ATTACHED
SHEETS

NAME . . . .

The following questions were designed to find out about your experiences with, skills in, and feelings about numbers or "maths". The results for the group as a whole will be used in helping this year's courses to work better, and for further studies by several members of staff aimed at making maths less painful for students in general. Results will be fed back to your group of students as soon as possible. Your name will be detached from the answer sheet when it is analysed and stored; results will be used only only on an anonymous basis.

Experience Scale

These questions ask you about your experience with mathematics so far.
(PLEASE TICK OR FILL IN YOUR ANSWER AS REQUIRED)

1. What Track(s) on B.A. Social Science (e.g. Psychology, Social Policy) are you most interested in? . . . .

2. What is your age (as at 1st Sept.)? . . . . . . years.

3. Male . . . . . . or Female . . . . . .

4. What, if any, has been your most recent paid work?

    (PLEASE SPECIFY FULLY AND IN PRECISE DETAIL)

    . . . . or . . . . . . None to date.

    If 'None to date' PLEASE GO DIRECTLY TO QUESTION 5.

    How long ago did you last do this job? . . . .

    Was it . . . . . . full time or . . . . . . part time?

5. What is/was your father's paid work when you began secondary school?

    . . . . or . . . . . . No paid work or . . . . . . Don't know.

6. What is/was your mother's paid work when you began secondary school?

    . . . . or . . . . . . No paid work or . . . . . . Don't know.

7. What was the main language you spoke at home while you were growing up?

    . . . . . . English or Other (PLEASE SPECIFY) . . . .

8. How much would you say you use   Not at all . . . .
    numbers generally in your everyday life? A small amount . . . .
    e.g. in checking your change, in   A moderate amount . . . . . .
    measuring a room to buy a carpet or paint? A great deal . . . .
                                              (PLEASE TICK)

9. How much difficulty have you   None . . . .
    experienced in using numbers   A small amount . . . .
    generally in your everyday life?   A moderate amount . . . . . .

                                              A great deal . . . .

10. How much have you used numbers   Question not applicable
    in work situations (including
     housework)?   . . . .
                                              Not at all . . . .
                                              A small amount . . . .
                                              A moderate amount . . . . . .
                                              A great deal . . . .

11. Would it have been useful to use   Question not applicable
    numbers in your work more
    than you did?   . . . .
                                              No, not at all . . . .
                                              Yes, a small amount . . . . .
                                              Yes, a moderate " . . . .
                                              Yes, a great deal . . . .

12. How much difficulty have you experienced in using numbers in work situations (including housework)?

None . . . .
A small amount . . . .
A moderate amount . . . . . .
A great deal . . . .

13. Qualification you have in Maths? A level . . . . . O level . . . . . CSE . . . . . None . . . . . Other . . . . . (PLEASE SPECIFY) . . . .

14. How much difficulty have you experienced in Maths courses generally before attending the Poly-technic? i.e. at school or F.E. college?

None . . . .
A small amount . . . .
A moderate amount . . . . . .
A great deal . . . .
(PLEASE TICK ONE)

15. How would you rate yourself <u>now</u> in each of the following areas?
(PLEASE CIRCLE THE MOST APPROPRIATE LETTER IN EACH CASE)
A = Very capable   B = Fairly capable   C= not very capable
D = Not at all capable

i   BASIC OPERATIONS (+, -, x, ÷)   A   B   C   D
    ON WHOLE NUMBERS
ii FRACTIONS   A   B   C   D
iii PERCENTAGES   A   B   C   D
iv DECIMALS   A   B   C   D
v   BASIC ALGEBRA   A   B   C   D
vi GRAPHS   A   B   C   D

16. How much do you expect to use Maths, and/or numbers generally, in your studies at the Polytechnic?

None . . . .
A small amount . . . .
A moderate amount . . . .
A great deal . . . .

17. How much difficulty do you expect to have with Maths, and/or with using numbers generally, in your studies at the Polytechnic?

Question not applicable
. . . .
None . . . .
A small amount . . . .
A moderate amount. . . . . .
A great deal . . . .

18. Is there anything special you would like to learn about Maths or about using numbers during this year of study? . . . .
. . . .
. . . .
. . . .

PLEASE DO NOT GO ON TO THE NEXT PART, UNTIL WE ASK YOU TO DO SO

Performance Scale

This scale is a measure of how people cope with numerical problems when they have to answer them under test conditions. It is NOT a test of underlying mathematical ability or potential. You should not attempt to complete all of the items on this scale. Please attempt the items in the order they are presented and do not skip an item unless you feel 'stuck'.

You may use any blank spaces on this sheet for calculations.

1. How much would it cost you altogether to buy a cup of coffee at 17p and a sandwich at 24p?

   ANSWER . . . .

2. How much does it cost to buy eight 14p stamps?

   ANSWER . . . .

3. Which is bigger (a) three hundred thousand or (b) a quarter of a million?

   ANSWER . . . .

4. If you buy five Xmas cards for 65p, how much is each card costing you?

   ANSWER . . . .

5. If you bought a raincoat in the 'summer sales' reduced from £44 to £29.50, how much would you save?

   ANSWER . . . .

6. **25% OFF**

   ALL MARKED PRICES

   If you saw this sign in a shop, would you expect to pay:
   - (a) a half, or
   - (b) three quarters, or
   - (c) a quarter, or
   - (d) a third

   of the original price?

   ANSWER . . . .

7. 27 + 33 = ...

8. 56 − 23.5 = ...

9. 91 ÷ 7 = ...

10. 13 x 9 = ...

11. If Z = 4, then Z + 7 = ...

12. If Z = 3, then 9Z = ...

13. If 4Y = 16, then Y = ...

14. Suppose the rate of inflation had dropped from 10% to 6%, which one of these results would you have expected:
    - (a) Prices would have gone down, <u>or</u>
    - (b) Prices would have stayed the same, <u>or</u>
    - (c) Prices would still be rising but not as fast as before, <u>or</u>
    - (d) Prices ought to have gone down but didn't.

    ANSWER . . . .

15. Which of the following numbers is the greatest (a) 0.76   (b) 0.768
    (c) 0.08. ANSWER . . . .

16. $13.8 - 0.73 + 5.9 =$ . . . .

17. The age (in years) of five people in a group are 18, 20, 22, 25, 30. What is the average age of the group?
    ANSWER . . . .

18. Suppose you go to a restaurant and the bill comes to a total of £3.72p. If you wanted to leave a 10% tip, how much would the tip be?
    ANSWER . . . .

19. This shows how the temperature changed during a hot day last summer.

    What was the hottest time of day? ANSWER . . . .

20. How hot was it at the hottest time of day? ANSWER . . . .

21. In an opinion poll for a by-election where there were two candidates 44% of those polled said they would vote for Jones, 34% said they would vote for Smith, the rest said they did not know. What percentage said they did not know?
    ANSWER . . . .

22. Whom would you expect to win?
    ANSWER . . . .

23. How certain do you think this prediction is?
    (a) totally certain   (b) fairly certain   (c) not very certain   (d) not at all certain.
    ANSWER . . . .
    WHY? . . . .
    . . . .

24. If the 'don't knows' are excluded, what are the percentages of the remaining voters (i.e. of those who have expressed a preference) who say they will vote for the two candidates?
    (a) Jones . . . . . .
    (b) Smith . . . . . .

Situational Attitude Scale

For each of the following items please indicate to what extent you would generally feel either relaxed or anxious in the situations they describe. Please rate the situation according to your immediate feelings, on the following scale:

1. I would be very relaxed.
2. I would be relaxed.
3. I would be fairly relaxed.
4. I would be neither relaxed nor anxious.
5. I would be a little anxious.
6. I would be moderately anxious.
7. I would be very anxious.

PLEASE WRITE THE APPROPRIATE NUMBER IN FOR EACH QUESTION

	Very relaxed	Relaxed	Fairly relaxed	Neither relaxed nor anxious	A little anxious	Moderately anxious	Very anxious
	1	2	3	4	5	6	7
1. Determining the amount of change you should get from a purchase involving several items.							
2. Asking a stranger which bus to catch in a strange town.							
3. Enrolling for a course which includes a compulsory mathematics component.							
4. Buying a recommended mathematics textbook.							
5. Calculating which is the cheapest method of getting somewhere by public transport.							
6. Dividing a five digit number by a two digit number in private with pencil and paper.							
7. Finding a street in an A to Z atlas.							
8. Walking into a room before a maths class starts.							
9. Listening to another student explain a maths formula.							
10. Having someone watch you as you total up a column of figures.							
11. Adding up 976 + 777 on paper.							
12. Asking someone to do you a favour.							
13. Listening to a lecture in a maths class.							
14. Totalling up a restaurant bill where you think you are being overcharged.							
15. Walking into school or college and thinking about a maths course.							
16. Choosing an item of clothing.							
17. Reading your P60 (or other statement) showing your annual earnings and taxes.							

	Very relaxed	Relaxed	Fairly relaxed	Neither relaxed nor anxious	A little anxious	Moderately anxious	Very anxious
	1	2	3	4	5	6	7
18. Being asked a question by the teacher in a maths class.							
19. Being responsible for keeping track of the amount of subscriptions collected for an organisation.							
20. Sitting in a mathematics class and waiting for the teacher to arrive.							
21. Deciding which film to go and see by yourself.							
22. Reading a cash register receipt after you have bought something.							
23. Raising your hand in a maths class to ask a question.							
24. Figuring out VAT at 15% on a purchase which costs more than one pound.							
25. Taking an examination for a maths course.							
26. Climbing a ladder.							
27. Working out a concrete, EVERYDAY APPLICATION of mathematics that has meaning to you; e.g. calculating how much you can spend on leisure activities after paying other bills.							
28. Realising that you have to do a certain number of maths classes in order to complete your degree.							
29. Raising your hand to ask a question in an English class.							
30. Being given a set of numerical problems involving addition to solve on paper.							
31. Getting the result of a maths diagnostic test.							
32. Being asked a question by the teacher in an English class.							
33. Working out your monthly budget.							
34. Getting the results of an English diagnostic test.							
35. Completing a surprise maths quiz.							
36. Talking in a group of strangers (people from a similar social background to yourself but unknown to you).							
37. Doing this questionnaire.							

Appendix 2
Interview Problems for Solution

The interview was introduced as described in Chapter 8, 'Doing the Interviews'. Then a number of 'life history' questions were asked, concerning:

- the subject the student was studying, and whether he/she had changed since entering the previous September
- age on entry
- where the student had studied most recently, and whether they had used maths then, if at all
- level of qualifications, if any, in mathematics
- the sort of paid work done most recently, and how much they had used numbers there, if at all.

Then four to nine problems were posed, depending on the time available: most students completed six. Each problem was introduced by contexting question C, and rounded off by contexting question R. The details of the first six questions are given below; for the remainder, see Evans (1993: Apps I1 and I2).

Problem 1

(Contexting question C) Does this remind you of any of your current activities? *[Show Figure A2.1.]*

Looking at this 'pie' chart, which do you think uses more water: households or industry with meters?

(Contexting question R) Now, could you tell me about any sorts of earlier experiences it reminds you of, or feelings it brings up?

Problem 2

(Contexting question C) Does this remind you of any of your current activities? *[Show the following question.]*

What is 10% of 6.65?

(Contexting question R) Does this remind you of any earlier experiences?

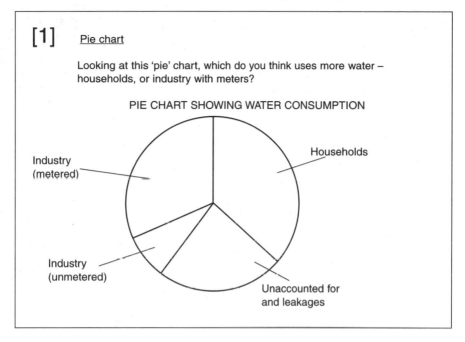

[1] Pie chart

Looking at this 'pie' chart, which do you think uses more water – households, or industry with meters?

PIE CHART SHOWING WATER CONSUMPTION

Figure A2.1 Presentation of Interview Problem 1
Source: Sewell (1982)

Problem 3

(Contexting question C) Does this remind you of any of your current activities? *[Show Figure A2.2.]*

(3A) Which part of the graph shows where the price was rising fastest?

(3B) What was the lowest price that day?

(Contexting question R) Could you tell me about any sorts of earlier experiences it reminds you of, or feelings it brings up?

Problem 4

(Contexting question CA) Do you ever go to a restaurant with a menu anything like this? *[Show the facsimile menu in Figure A2.3.]*

(Contexting question CB) Would you please choose a dish from this menu?

(4A) Suppose the amount of 'service' that you leave is up to the customer: what would you do?

(4B) Could you tell me what a 10 per cent service charge would be?

(Contexting question R) Does this remind you of any earlier experiences?

[3]   Graph

This graph shows how the price of gold varied in one day's trading in London. Which part of the graph shows where the price was rising fastest? What was the lowest price that day?

The London Gold Price – January 23rd 1980

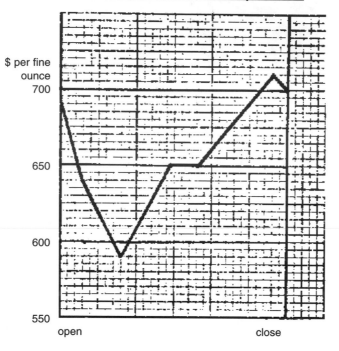

This graph shows how the price of gold (in dollars per fine ounce) varied in one day's trading in London.

Figure A2.2 Presentation of Interview Problem 3
Source: Sewell (1982)

Problem 5

(Contexting question C)   Have you ever received a slip like this? *[Show the copy of a payslip in Figure A2.4.]*
Jennifer is expecting a rise of 9 % on her gross pay.

(5A)   About how much will that be?
(5B)   [If Qu 5A is answered] Can you work it out exactly?

(Contexting question R)   Does this remind you of any earlier experiences?

[4]

CHICKEN MARYLAND	Served with sweet corn, banana fritter, bacon, fresh tomato, whole French beans, jacket baked potatoes with sour cream and chives or French fried potatoes.
	Roll and butter.
	Ice cream, or a selection from our cheese board, biscuits and butter.
	£3.75
SEA FOOD PLATTER	Served with tartare sauce, whole French beans, jacket baked potatoes with sour cream and chives or French fried potatoes.
	Roll and butter.
	Ice cream, or a selection from our cheese board, biscuits and butter.
	£3.53
GRILLED TROUT 10 OZ	Served with tartare sauce, whole French beans, jacket baked potatoes with sour cream and chives or French fried potatoes.
	Roll and butter.
	Ice cream, or a selection from our cheese board, biscuits and butter.
	£3.81
Coffee	Special blend black or with cream **27p**
Connoisseur Coffees	Served in large goblet glass with cream: Irish *(Irish whiskey)*, Caribbean *(Rum)* Russian *(Vodka)*, Parisienne *(Brandy)* Calypso *(Tia Maria)* Highland *(Scotch whisky)* Mine Hosts *(Cointreau)*
	Connoisseur coffees include sugar unless otherwise requested **67p**

Figure A2.3 Presentation of Interview Problem 4

Source: Sewell (1982)

[5]

STAFF NO.	DEPT.		NAME		TAX CODE	DATE	PERIOD	BASIC SALARY/PENSION
	331788		SMITH JENNIFER		116L	07/10/79	28	66.560

GROSS PAY TO DATE	FREE PAY TO DATE	GROSS TAXABLE TO DATE	TAX DUE TO DATE	TAX THIS PERIOD R=REFUND		MISC. DEDUCTIONS		SAYE
1335.450	629.440	681.480	183.800	11.900				

GRAD. PEN.	NAT. INSURANCE	M.S.A./B.U.P.A.	SOC. CLUB	L.A.M.P.S.	R.I.P.	NAT. SAVINGS	TRUSTEE SAV.	DENTAL PAYM'T
	3.150							

	HRS	AMOUNT			
STANDARD PAY	38.75	66.560	NIGHT ALL		GROSS PAY THIS PERIOD
D/TIME MON/SAT			SICK BEN		66.560
O/TIME SUN/BHOL			MISC PAY		TOTAL DEDUCTIONS INCLUDING TAX / ADJ.
LOST TIME PAY			X-BONUS		16.700 / .830
LOST TIME NO PAY			PEN FUND	1.650	NET PAYMENT
HOLIDAYS					49.80
W/E CLEANING				CASH	

Jennifer is expecting a rise of 9% on her gross pay

Figure A2.4 Presentation of Interview Problem 5
Source: Sewell (1982)

[6] The larger bottle in this picture holds 30 oz and costs 69p.
The thinner bottle holds 20 oz and costs 52p.

Figure A2.5 Presentation of Interview Problem 6
Source: Sewell (1982)

Problem 6

(Contexting question CA) Do you ever go shopping for food? Where would you normally go?

(Contexting question CB) Do you ever buy ketchup (or jam, etc.) ?

(Contexting question CC) If you were buying ketchup or [other food mentioned] and several jars were available, how would you decide which one to buy? *[Show the picture in Figure A2.5.]*

The larger bottle in this picture holds 30 oz. and costs 69p. The thinner bottle holds 20 oz. and costs 52p.

(6A)   Which of these two bottles would you buy? Why?
(6B)   Which is better value for money?
(Contexting question R)   Does this remind you of any earlier experiences?

To conclude, I asked:

- if there was anything they would like to ask about any of the questions
- how they felt about the way they were 'able to use numbers these days'
- if there was anything they would like to say about the way that the interview took place.

I then asked each student not to discuss the questions with other people on the course, 'so that people won't come to the interview with any preconceptions'. Finally, I asked for their permission to quote things they had said 'in a suitably anonymised form – in things I may later write', promised to offer them a copy of the transcript when it was ready, and thanked them.

Notes

1 Introduction: Mathematics, the Difficult Subject

1 Another problem that surfaces periodically is that of an allegedly poor performance of UK students in international mathematics studies, compared with students from other countries; however, this does not apply for example to practical problem solving items (Brown et al. 1998).

2 Though this requirement appears reasonable, it can lead to 'double binds' for certain groups – for example, girls – and to devaluation of their correct performance (Walkerdine et al. 1989).

3 Such as the Maths in Work Project, directed by Mary Harris (1991).

4 See also Ernest (1994a), Restivo (1992), and Noss and Hoyles (1996b).

5 Cline-Cohen, in her history of the spread of numeracy in the USA and Britain, cautions against using a strict threshold in defining numeracy either for an individual or for society. She suggests that 'the most interesting question is not how crude numeracy rates have changed over time, but rather how the domain of number has changed and expanded' (Cline-Cohen 1982: 11).

2 Mathematical Thinking in Context among Adults

1 The terms 'practical mathematics' and 'numeracy' are used as broadly equivalent in the first half of this book. The use of the second term in, for example, the 'National Numeracy Strategy' in the 1990s will be discussed in Chapter 11.

2 For further on this pedagogic strategy compared with several others, see Straesser et al. (1989: section 2.2).

3 On parallels and links between literacy and numeracy, see for example Baker and Street (1993) and Hamilton and Stasinopoulos (1987).

4 The terms 'proficiency' and 'functional' are used by Brown et al. (1998), who also discuss a third approach – numeracy as *social practice* (see Chapter 11). The first two types correspond broadly to the 'industrial trainer' and 'technological pragmatist' ideologies in Ernest (1991).

5 Some characteristics of sample members are given in Table 2.1 (p. 18).

6 For the performance questions, see ACACE (1982) or Evans (1989a). For the re-numbering of the ACACE questions used in the Polytechnic study, see Table 2.2 (p. 20).

7 These results are subject to sampling variation (ACACE 1982); thus, any difference between the male and female subgroups of 4 per cent or less would not be very impressive, or 'statistically significant'. This is because it would not be greater than the 'margin of error', and therefore could be expected to occur 19 times out of 20, due only to chance, rather than because of any effect due to (or correlated with) gender.

8 For further discussion of 'critical citizenship', see Chapter 11 and Evans and Thorstad (1995).

9 The NCDS is based on all individuals born in one week of 1958; for a fuller description,

see Fogelman (1983). The 1970 British Cohort Study is based on a longitudinal study of all individuals born in one week in 1970. Along with the 1946 ('Douglas') study, these three longitudinal studies form a valuable resource for studying many aspects of life in Britain.

10   The three cohorts of the survey were entrants to the Polytechnic in 1983, 1984 and 1985. During this period, the DipHE, broadly equivalent to two years of a three-year degree, was part of an informal modular degree scheme, which was formalised in 1985.

11   As had happened often in Sewell's (1981) recruitment of interviewees.

12   More recently, Basic Skills Agency (1997), reporting on the 1996 International Numeracy Survey, notes that the percentage refusing to participate in a short face to face problem-solving interview was 13 per cent in the UK, more than twice that in any of the other six participating countries. Most of the twelve problems posed suggest a 'proficiency' notion of numeracy (see Conceptions of Practical Mathematics and Context).

13   This left aside Questions 22 and 23, as they required the exercise of judgement, and the explication of it, that might have been difficult for respondents to produce, especially when these items were positioned towards the end of a timed set of self-response items.

14   Any attempt to make more precise comparisons between the samples is constrained by the fact that the national survey (ACACE/Gallup) used the 'market research' (IPA) indicator (Reid 1981: 56–7), whereas I used the Registrar General's; these lead to slightly different results (Evans 1993, Chapter 5).

15   The Assessment of Performance Unit (APU) had as its remit the national monitoring of school attainment from the mid-1970s. Its mathematics programme based at the National Foundation for Educational Research produced wide-ranging studies of the 'effects of context', via differently formulated test items, on performance, for different groups of students (e.g. Foxman et al. 1985). This research also developed the idea of context (through work on wording and format of items, and through attempting to characterise and distinguish contexts – for example the 'money context') beyond just naming them.

Importantly, they also studied performance in different *modes*, besides that of large-scale written testing and questionnaires (for attitudes), by using one-to-one practical interviews and small group practical interviews. This variation in the social interaction of the research setting suggests different ways of thinking about the context of mathematical performance.

16   For a breakdown of respondents and a rough estimate of non-response, see Appendix 1.

17   So as to minimise 'mathematical fatigue' (ACACE 1982: 7).

18   These interviewers would be more used to market research and opinion polling, aimed at eliciting product or policy preferences. For them, the ACACE performance questions would be a new sort of task: a precise numerical answer was sought, and was meant to be produced without the usual sorts of prompts and probes.

19   The level of non-response, which had never been more that 3 per cent for any performance item up to Question 15, reached 15 per cent for Question 16 (a complex 'abstract' sum, involving decimals) and was never below 6 per cent or 7 per cent from that point on.

20   However, if the PM and SM scores were likely to change with a re-ordering of the questions, or with a removal of the time limit, such changes would attenuate the correlation. Since estimates of reliability were not produced for the scales used in this study, all correlations reported should be considered as upper bounds.

21   These restrictions were necessary, since only the cohort 3 version of Question 8 was considered comparable mathematically to Question 5, and since the cohort 2 version of Question 9 contained a misprint, which rendered it slightly easier.

3   Mathematics Performance and Social Difference

1   Ethnic group differences are not considered here, for several reasons. The literature on this issue was scant (even in the USA) when I began work on this study. Further, in the UK, the difficulties of measuring ethnic background were well known: the validity of parental birthplace as an indicator was declining as time passed, and there was resistance – which

threatens response rates – to alternative questions among respondents in the run-up to the 1981 Census (though not for 1991). This judgement seemed to be vindicated in my survey's fieldwork, when there were refusals to answer questions on parental occupation, and even gender.

2 Recent reviews of the findings of such studies include Kimball (1989), Fennema and Leder (1990), Hyde et al. (1990), Hanna (1994), Grevholm and Hanna (1995), Fennema (1995) and Keitel et al. (1996).

3 The CSE in each school subject was introduced as an alternative qualification at age 16 to the more 'academic' GCE O-level in the 1960s in the UK. The two were replaced by the single GCSE 16-plus qualification in the late 1980s. A-level remains a very specialised exam at age 18-plus.

4 Here I include statistics with 'mathematics'.

5 Market researchers use a slightly different scale, the IPA / JICNARS, which nevertheless produces 6 categories A/B/C1/C2/D/E, which can be considered as broadly parallel to those (I/II/IIIN/IIIM/IV/V) of the Registrar General.

6 The Registrar General's social class scale was changed in 1998, to include seven major groupings with 'small employers and own account workers' inserted between IIIN and IIIM.

7 In addition, the effect of family aspirations, beliefs and so on are investigated using less structured interviews with a subsample of the students in the second part of the study (see Chapters 8–10). Thus the review of the social variables (Evans 1993: App. V1) was useful in sensitising me to issues that might arise in the interviews.

8 This is because the variable for student's own social class showed small (and anomalous) uncontrolled differences, and did not have sufficient 'predictive power' to be included in the regression models; see also Chapter 4.

9 This comparison of the size of performance differences could be made, using the numbers of *standard deviations*, rather than absolute units, to allow for differences in difficulty between the two scales, if necessary.

10 Since the ratio of the estimate of the gender difference (0.57) to its standard error (0.24) is close to the 'critical value' of 2 (or 1.96) for the statistical significance test.

11 The estimates for the standard errors of each effect are either given by the computer package (SPSS), or calculated from results given by it.

12 The uncontrolled results given in Tables 3.4 and 3.5 differ slightly from those for the whole sample given in Tables 3.2 and 3.3 respectively, because of the latter pair's larger effective sample sizes.

13 Since there are interaction effects of gender with age in this model, the estimate of the gender effect is specific to the age subgroup.

14 This means that, if we were to take samples repeatedly in the same way as this sample was produced, we would find that the estimate for this gender difference would range between 0.50 and –0.18 in 95 per cent of the sampling occasions. (This assumes that my sample could be considered acceptably representative of the working population (see pp. 16–17).) Then, if we wished to test the statistical significance of this difference, we could note the fact that this 95 per cent confidence interval includes zero, which means that we cannot reject the 'null-hypothesis' that the gender difference for younger students is zero, at the 5 per cent (i.e. 100–95 per cent) level of significance ($p < .05$).

Given that a 95 per cent confidence interval estimate for such an effect can always be found (Moore 1976) by calculating an interval of width 2 (more precisely, 1.96) standard errors on either side of the 'point estimate' – here, 0.16, I shall present confidence intervals in this analysis only for the most discussed estimates.

15 I would use an R-squared value of 30 per cent or more as a rough criterion of an 'acceptably powerful' regression model, based on cross-sectional data of this kind. The value of R-squared for the model for PERFS for Cohort 3 only was higher, at 39.7% (see pp. 38–9). The reason for this is related less to the inclusion of social class variables in the Cohort 3 model – for social class adds little to the value of R-squared in this study – than to the fact that one year cohort of students is more homogeneous, and therefore records less variation, on its own – than when merged with two others, as in the 'whole sample' models.

16 This table also excludes results for thirty-eight students in the uncontrolled comparison, and twenty-three for the modelling, who were categorised as 'mixed' social class (one parent classified as middle class, and the other as working class; see Appendix 1).

4 Affect and Mathematics Anxiety

1 McLeod (1989a) also emphasises two further dimensions: (4) the person's level of awareness (which may depend on available attention); and (5) the level of control. These are relevant to learning to manage emotions, e.g. in problem solving.

2 Fennema and Sherman (1976), on the other hand, used a measure of both confidence and anxiety in their battery of scales for attitudes towards mathematics. Because they reported a high negative correlation ($r = -0.89$) between the two, and because confidence was more highly correlated with performance than any other affective variable ($r = 0.40$) (Fennema 1979: 395), the work of Fennema and her colleagues has subsequently used confidence, in preference to mathematics anxiety.

3 What came to be known as the 'Yerkes-Dodson Law' was first proposed in 1908 as a relationship between the strength of electric shocks (presumably producing fear) and simple learning in mice. It has since been applied to simple human learning, as related both to what would now be called 'transitory' anxiety, and also to 'anxiety-proneness'; see Levitt (1968: 144ff.). Theoretical support for the Yerkes-Dodson Law was claimed, first from the 'drive theory' (behaviourist) programme (see Evans 1993, Chapter 3) and later from D.O. Hebb's more cognitive work (1955), by which anxiety might be seen as related to a more encompassing notion of 'arousal'. However, in the 1950s and 1960s, the empirical results, for the most part, were contradictory, and over the longer period, the conceptual basis of the 'Law' seems to be somewhat fluid, depending on differences in focus (for example on emotion, rather than motivation) among research programmes, and on changes in the theoretical base of psychology generally (Teigen 1994).

4 The Mathematics Attitude Scales were conceived in order to 'measure some important, domain specific, attitudes which have been hypothesised to be related to the learning of mathematics by all students and/or cognitive performance of females' (Fennema-Sherman 1976: 1). The other scales include:

- Attitude to Success in Math, based on ideas of women's fear of success in 'male' intellectual areas; see Chapter 7 and Horner (1972).
- Effectance Motivation: persistent exploratory behaviour, active seeking of challenge, similar to a 'problem-solving attitude'.
- Confidence in Learning Mathematics.
- Usefulness of Mathematics.
- Math as a Male Domain.
- Mother's, Father's, and Teacher's Attitudes, as *perceived* towards oneself as a learner of mathematics.

5 Richardson and Suinn (1972) checked on the test's validity (whether it measures what it purports to measure) in a way that was in keeping with their interest in intervention programmes: they noted whether there was a decrease in maths anxiety scores when it was administered before and after behaviour therapy given for mathematics anxiety. (This validation of the MARS as a measure of mathematics anxiety was highly circular, since the change in MARS scores after attending the intervention programme was presumably *also* being used to evaluate the therapy itself!) For the questions in the 1972 version of the MARS, see Evans (1993: App. P4).

6 See p. 58 for an explanation of this criterion of the 'importance' of a factor.

7 A more developed approach to including affective factors comes from the 'expectancy-value' model, so called because the 'expectancies of success' and the 'perception of task value' are seen as operating together to produce the probability of choosing to do another mathematics course (or that of other 'achievement behaviours'). Chipman and Wilson (1985) argue that this approach allows us to subsume most of the 'affective' factors previ-

ously discussed under 'expectancy' (confidence, perceived difficulty, causal attributions) or under 'value' (enjoyment, perceived usefulness of mathematics).

8   Some results for the other affective variables are reported in Evans (1993).

9   No systematic comparisons of the *levels* of mathematics anxiety expressed were attempted between this research and the US research discussed in 'Mathematics Anxiety, Measures and Relationships' earlier. Some of the difficulties were cultural: this was the first time to my knowledge that the MARS had been used outside the USA, and it was uncertain whether the meanings of the descriptions of the situations for each item, or of the words attached to each scale value, could be considered as comparable across the two cultures. Also the original MARS included ninety-eight items, whereas we had selected a sub-sample of twenty-six items for this study (see Appendix 1).

10   On an alternative 'confirmatory' approach, see McDonald (1985).

11   TCA, NA and general anxiety are averages of thirteen, thirteen, and ten items, respectively, from the Situational Attitude Scale (the third part of the questionnaire), and confidence is an average of six items from the Experience Scale (Question 15). The items for the first two were based on a selection of MARS items (see Section 3), and those for the second two were constructed; see Appendix 1.

12   The larger correlation of scores on the general anxiety scale with numerical anxiety than with maths test/course anxiety suggests that NA might indeed be tapping anxiety related to practical contexts, since the contexts of the general anxiety items were also practical, except for the three items constructed to measure what might be called 'English test/course anxiety' (questions 29, 32, and 34 on the Situational Attitude Scale; see Appendix 1).

13   These were Principal Axes Factor Analysis, Maximum Likelihood / Canonical Factor Analysis, and Alpha Factor Analysis. The choice among them is to some extent a technical issue (for example McDonald 1970: 11–15).

14   These were varimax, and direct oblimin (with parameter = 0), respectively. These are the most commonly used rotations of each type, and they are also the rotations used by Rounds and Hendel (1980).

15   Determined by the number of 'characteristic roots' of the correlation matrix that were greater than or equal to one.

16   For each analysis, an 'acceptable' loading (correlation) of an item with a factor was considered to be 0.4 or greater; this is slightly more demanding than the usual criterion (factor loadings 0.3 or greater), and leads to a somewhat more 'simple structure' for the results.

17   From this point onwards, a mathematics anxiety item from the Situational Attitude Scale of the questionnaire will normally be designated 'TC' or 'N', according to its classification by Rounds and Hendel as a 'maths test/course anxiety', or a 'numerical anxiety' item, respectively (see 'Mathematics Anxiety, Measures and Relationships'). Thus, for example, Question 24, 'Figuring out VAT at 15% on a purchase which costs more than one pound' becomes 'N24'.

18   It is conceivable that more than one factor will load on a particular item, as with N24 and N30 here, but the aim in factor analysis generally is to minimise this, so as to promote a 'simple structure' solution.

19   However, both the models for TCA and NA, and the models for performance run with the two anxiety factors as predictor variables, were checked for stability of results when N10, N24 and N30, the three 'maverick' numerical anxiety items singled out in the previous section's analysis, were included in TCA, rather than NA. This change did not lead to any appreciable differences in the values of the regression coefficients, nor in R-squared, so I am not presenting these alternative results for comparison.

20   Multiple regression (MR) analysis aims to explore the relationships of an outcome variable with a set of 'predictor' variables which are considered (normally for theoretical reasons) to 'explain', or at least to predict, the outcome variable. Multiple regression is based on the pattern of correlations among the set of variables, like factor analysis. However, MR also *requires distinguishing between outcome variables and predictor variables.* It then allows the user to consider the 'effect' on the outcome variable of changes in each predictor while all the others are simultaneously held constant. This is

done in a 'virtual' way, by 'statistical control', rather than by the 'physical control' used in experiments.

The main measure of the 'explanatory power' of a regression model is R^2 (or 'R-squared'), the proportion of variation in the outcome variable 'explained' or accounted for by (its correlation with) the set of predictor variables. Other things being equal, the more predictor variables included in the model, the higher the value of R-squared. However, there is a limit on the number of predictor variables that can be included in any particular model for technical reasons (Evans 1993: 145). For further on regression modelling, see Draper and Smith (1980).

21 The models discussed in this section also tested for inclusion particular affective variables as predictors: for the TCA model, 'difficulty with previous maths courses'; and for NA, 'difficulty with using numbers (or maths) in everyday life', 'difficulty with numbers in work', and 'use of numbers in work'. For the wording of the relevant questions (one for each variable), see the Questionnaire in Appendix 1.

22 I used a procedure based on the classical analysis of variance (ANOVA) model with unequal numbers in the cells, since, in my sample, there were different numbers of students in each 'cell' of the research design, for example different numbers of older 'high quali-fied' males, older 'high qualified' females, younger 'high qualified' males, and so on. I decided which predictor variables to include by using an hierarchical strategy (Aitkin 1977, 1979: 200–10) to build up an ANOVA model in the following stages:

1   Any predictor variable considered (theoretically) for inclusion in the model (including variables representing extraneous differences based on year of entry and course) was tested for 'explanatory power' in a one-way ANOVA, at the 5% level of statistical sig-nificance, a stringent requirement.

2   In the process of deciding on the basic set of predictor variables to be included, a number of orderings of the variables selected in (1) were tried, because of the possi-bility of drawing erroneous conclusions in the event that the main effects were highly correlated (Aitkin 1977).

3   A series of models was then produced, each one based on the variables in the model produced by the preceding stage, plus one of the two-way interactions suggested as rel-evant, either by theoretical considerations or by the tables of observed means for the relevant subgroups (see Chapter 3, 'Survey and Modellling Results for Gender Differences in Performance') – so as to select interactions which had 'explanatory power', again using the criterion of statistical significance.

4   Those interactions which passed this 'explanatory power' test were entered into an overall model, beginning with those involving the qualification in maths variable (because of the emphasis in the literature on having taken courses in mathematics as an influence, especially on performance in mathematics), and those interactions which remained significant in this new model were included in an overall model.

5   This overall model was checked 'top-down', for example for three-way interactions that were both statistically significant and theoretically interpretable, by running a full ANOVA model with all effects included, so as to decide on the final set of predictor variables to be included.

6   Those predictors based on more than two categories (e.g. age classified as 18–20, 21–24, or 25+) were re-represented for the MR model as two or more 'dummy vari-ables' (each with one degree of freedom), and the contribution of each dummy variable was tested for inclusion separately, using 'stepwise' procedures.

7   Regression coefficients were estimated, and estimates produced of the size of the 'effect' for each independent variable, so as to allow the calculation of gender differ-ences, say, *with other factors controlled for*.

23 The estimates for the standard errors of each effect are either given by the computer package (SPSS), or calculated from results given by it.

24 For the regression models produced for cohort 3 only, R-squared for the TCA model was 38 per cent, and for NA, it was 30 per cent. This appears better than the values for the

models based on the whole sample, but the R-squared values for the Cohort 3 models were only 19 per cent and 13 per cent respectively, when only social variables (including qualification in maths) were included. At subsequent steps, the modelling procedure selected for inclusion certain affective variables: 'difficulty with previous maths courses' for the TCA model; and 'difficulty with using numbers (or maths) in everyday life', plus 'difficulty with numbers in work', for the NA model.

25 If we isolate out the relevant parts of the equations, we have:

SM performance = 5.64 + . . . [other variables] . . . + 0.43 TCA – 0.06 TCA-squared
PM performance = 7.98 + . . . [other variables] . . . + 0.47 NA– 0.12 NA-squared

Very approximately, by using the usual procedures for finding extreme values for a function (namely, setting the first derivative equal to 0), we could interpret these equations to represent quadratic relationships which reach a maximum (or 'peak') at a value of TCA = 3.6 in the case of SM performance, and at a value of just about NA = 2, for PM performance. Given that the anxiety items required responses of between 1 and 7 on a 'symmetrical' scale (see the questionnaire in Appendix 1), TA = 3.6 would represent an average response (to the thirteen maths test anxiety items) slightly closer to 'neither relaxed nor anxious' than to 'fairly relaxed'; NA = 2 would represent an average response to the thirteen numerical anxiety items of 'relaxed'.

Thus, the 'peak' or maximum value estimated for the relationship between SM performance and TCA was located at a point close to the theoretically 'reasonable' neutral point (4 = 'neither relaxed nor anxious'). This provided additional support for the idea of a quadratic relationship between school maths performance, and TCA. The evidence for a quadratic relationship was less convincing in the case of PM performance, since the peak was estimated to be some distance from the neutral point. This latter anomaly concerning the peak of the inverted U relating PM and NA, and the low R-squared value for the combined contribution of NA and its square recall the decision not present further results from modelling PM performance in Chapter 3.

26 This interpretation is given further support by interchanges in interviews numbers 11 and 16 ('Harriet'); see the latter's case study in Chapter 10.

27 Of course, any decision to highlight this finding would assume that it is robust, that is, can be replicated. One methodological difference that may be relevant is my decision to produce anxiety items each of which offered a *symmetric* scale of seven responses, unlike the previous studies, done in the USA (see 'Mathematics Anxiety, Measures and Relationships').

5 Reflections on the Study So Far

1 In the sense that including each of these affective variables made a statistically significant contribution to the 'explanation' of the relevant performance variable in the model; see pp. 64–5.

2 For example, only 'at the borderline' of statistical significance. Here I use 'borderline' as a rough criterion of confirmation, meaning statistically significant at the 5 per cent level, but not at 1 per cent. The supportability of each finding also depends on whether the difference or estimate is *substantial* in size, whether the measures involved can be considered as valid (see below), and whether appropriate controls were used (see Chapters 3 and 4).

3 Also, reporting anxiety is not necessarily a valid indicator for 'having' or experiencing it (see Chapters 7–10).

4 Possible exceptions are Question 14 on inflation and Questions 21–24 on opinion polling (see the Questionnaire in Appendix 1).

5 The wording of the rubric for the mathematics anxiety items was:

please indicate to what extent you would generally feel either relaxed or anxious in the situations [. . .] describe[d].

6 These models assume that the outcome variable(s) can be clearly differentiated from the

predictor variables so that the functional relationship represents productivity in only one direction (see Chapter 4, note 20).

6 Rethinking the Context of Mathematical Thinking

1 What is called the 'preparation' problem in Chapter 1 might itself be reformulated as a problem of 'transfer' of learning between practices of secondary school and university mathematics (Iben Christiansen, personal communication 1999).

2 Several of my interviews illustrate how a researcher's desire to privilege 'mathematical' approaches to a problem can hinder understanding of other people's cognition; see Chapter 10, especially 'Alan's story'.

3 Various meanings of 'ethnomathematics' have been propounded over the last twenty years, but they have been converging (Barton 1996). Ethnomathematics began as a commitment to describing, and to supporting the development of

> the mathematics which is practised among *identifiable cultural groups*, such as national-tribal societies, labour groups, children of a certain age bracket, professional classes, and so on. Its identity depends largely on focuses of *interest*, on *motivation*, and on certain *codes and jargons* which do not belong to the realm of academic mathematics. We may ... include much of the mathematics which is currently practised by engineers, mainly calculus, which does not respond to the concept of *rigour* and formalism developed in academic courses of calculus.... And builders and well-diggers and shack-raisers in the slums also use examples of ethnomathematics.
> (D'Ambrosio 1985: 45; my emphasis)

The proponents of ethnomathematics have included those committed to the acknowledgement of the achievements of others – for example,. non-European mathematics (Joseph 1991) or women's design work (Harris 1997) – and particularly to post-colonial development. Gerdes (1985) sees an emancipatory mathematics education as necessary for 'problemising reality' (Freire 1970, Coben 1998), so that social practices can be scrutinised and possibly changed; much of this critical activity requires reasoning that can be shown to be mathematical. Gerdes (1986) provides a range of examples from practical everyday activities carried out by Mozambicans that appear able to be 'harnessed' for the purposes of mathematics education. Gerdes's views thus represent an attempt to emphasise the *continuities* between what is often seen as 'European mathematics' and indigenous ethnomathematics.

4 In a reflection on Vygotskian approaches to teaching and learning in the information age, Kathryn Crawford (1996) distinguishes between actions and operations on the somewhat different basis that the former are conscious, purposeful and available for review, while the latter are usually unconscious and automated. See also Bartolini-Bussi (1996).

5 Thus testing sessions (and lab experiments) exhibit the following features:

(1a) The physical environment is highly constrained.

(1b) The person being evaluated is instructed to respond to certain specific features of the environment.

(1c) The tester constructs the stimuli and thereby is empowered to specify (correctly, or incorrectly) the relevant stimuli.

(2) The person evaluated is told the domain of behaviour that will be observed (making it highly probable – but not certain – that such behaviour will be emitted).

(3) The domain of behaviour is chosen to produce hypothetical relations with the stimuli presented (Cole and Traupmann 1979).

6 One of the early concerns of some sociocultural researchers was the 'ecological validity' of many cognitive psychological research findings: that is, the extent to which the findings could be generalised beyond the laboratory contexts, and the specially designed tasks, on

which they have been based, to less contrived (less controlled) everyday settings; see Cole et al. (1978) for an extended discussion. It will be apparent that there are parallels between the methodological issue of the possibilities of generalising from research to everyday settings, and the substantive issue of transfer of learning from formal educational tasks and settings to everyday ones; see Lave (1988: 100ff.).

7 The concept of activity, central to Lave's earlier work (Lave et al. 1984), gave way, to some extent, to 'practice' in Lave (1988); while still using 'activity' (see the text), the latter minimises the explicit links with Soviet psychologists such as Vygotsky, and draws theoretical support from sociologists and anthropologists (for example Giddens 1984, Bourdieu 1977).

8 This moralism around being able to do mathematics 'in the proper way' helps to explain teachers' impatience with students who can't do it that way, and students' feelings of unpleasantness and guilt; see the discussion of case studies in Chapter 10.

9 The differences in methods used between everyday and school-type contexts are discussed more fully by Nunes et al. (1993) researching in Brazil at about the same time (see pp. 88–93).

10 In poor communities in northeast Brazil, children commonly attend only a few years of school, or not at all; thus, the variation in the number of years of schooling attended is much greater than in most European or North American countries .

11 On the importance of adaptability, see also Carraher and Schliemann (1998) and Boaler (1997).

12 See the discussion of case studies and illustrations in Chapters 9 and 10, for example that of 'Peter'.

13 Similar 'syncretic' procedures have been reported among African children by Brenner (1985), and with New Guinean and Brazilian children by Saxe (1991a, 1994).

14 The mathematical basis for this is that, for any amount y, 92 per cent of y (i.e. y discounted at 8 per cent) is less than the present value of an investment at 8 per cent annually that will yield y at the end of one year, namely: y * 100/(100 + 8).

15 Walkerdine's distinction between calculations based on material necessity and those based on symbolic control parallels the distinction made by Reed and Lave (1979) between 'manipulation of quantities' and 'manipulation of symbols' approaches in mathematical thinking (see previous section)

16 The limits and determinations of subjectivity are still very much a focus of debate in the social sciences and philosophy, but see for example Hollway (1989), Habermas (1992), Giddens (1991), Walkerdine (1997), Evans and Tsatsaroni (1994, 1998); and recent issues of *Radical Philosophy*.

17 Note that a practice may be constituted by different discourses; for example, school mathematics may be constituted by transmission learning, or by problem-solving/investigative approaches. These different discourses will make available different versions of the teacher and the pupil positions. This point cannot be developed here, but, on different discourses of gender relations and corresponding subject-positions, see Hollway (1989).

18 For somewhat different approaches, see Carreira (2001), who uses Peirce's, rather than Saussure's linguistics, and Whitson (1997), who aims to reconcile the two.

19 This discussion may suggest the basic form of a strategy for 'teaching for transfer'; see Chapter 11.

20 Walkerdine does not suggest that the teacher's purpose in playing the shopping game was to produce transferable skills (from school to shopping), nor to 'harness' children's (limited) experience with shopping for pedagogic purposes. But it was to give the children experience of action on money, or tokens, which could later be 'disembedded' in the process of producing abstract mathematical knowledge.

21 The learning of a second (or subsequent) language provides another example. Inevitably, the learner attempts to find equivalences, or similarities, between a word in the new language, and one (or more) in the familiar language. If the similarity holds, as it does basically between 'mathematics' in English and '*mathématiques*' in French, fine; if a difference is discovered, as between 'didactics' and '*didactique*' , then a list of differences can begin to be compiled. In any given case, repeated attempts to transfer, or to translate, may be necessary.

22 Or skills, which shows that one of the 1990s buzzwords, 'transferable skills', is fundamentally misguided.

23 However, it would be wrong to imply total neglect. For example, Brenner (1985) describes local African teachers' sensitivity to what might be threatening to their pupils. And Cole and Traupmann (1979) refer to the distress of 'Archie', a participant in the cooking club, and that in others around him, when he is having difficulties solving a problem. So the affective aspects have not gone unnoticed, but they have usually not been at all central to the analysis.

7 Rethinking Mathematical Affect as Emotion

1   DeBellis and Goldin (1997) propose to add *values*, as a fourth type of affect, but do not discuss its positions on McLeod's dimensions of intensity and stability (see Chapter 4); Cobb et al. (1989) emphasise *norms* as closely related to beliefs and emotional acts; see Model B, next section.

2   Winter (1992) and Maxwell (1989) both prefer the term 'mathophobia' / 'mathephobia' to 'mathematics anxiety', since the former connotes both fear and dislike.

3   The German *wunsch* and the French *désir* have tended to be translated into English as 'wish', especially before Lacan's work became known in English (Lacan 1977). Now, the term 'desire' is used more widely, but also rather broadly, for example in the work of Walkerdine and others (discussed later) as something similar to 'libidinal energy' (Laplanche and Pontalis 1973: 481–3 and 239–40).

4   Freud gives the example of a young child's play with a cotton reel, which he argues provides the child with a fantasy of being able to control the mother, masking the child's powerlessness and dependency (Urwin 1984).

5   A similar dynamic can be seen to underlie the process of 'abstraction' depicted in Figure 6.1; see also Whitson (1997), which generally elucidates many of Walkerdine's ideas.

6   For discussion of the difference between the home and the school as 'arenas for children's daily life', see Mayall (1998).

7   These discourses also hold sway in those contexts – beyond the research contexts, but connected – such as Will's ongoing relationship with Beverley (Hollway 1989), in which the activity relating to the decisions discussed, is played out.

8   See also Said's (1978) study of Western views of 'the Orient', Rustin's (1991) study of racism, and Henriques et al. (1984: Ch. 2).

9   Grieb and Easley (1984) has affinities with Walkerdine's description of boys' fantasies of Reason's dream, and focuses on the double-binds to which 'pale, male math mavericks' and their teachers are subject.

10 The first part of the argument requires showing that particular sets of ideas, on the gendering of reason and mathematical thought, were salient, indeed hegemonic, in particular historical periods. This would require using primary sources to examine 'who supported and opposed what educational and psychological moves and in what terms' and 'a conjunctural analysis of the balance of social forces' (Walkerdine 1984: 200, note 20; see also Walkerdine 1997). The sort of evidence needed is not available in Walkerdine's account – though her conclusions seem plausible and consistent with other writers, including those using primary historical sources, for example Clements (1979). In addition, her own research in schools (Walkerdine et al. 1989), using participant observation and interviews with teachers, provides evidence that these sorts of views were subscribed to by the teachers in those particular schools at that particular time.

The second part of the argument appears to be even more difficult to assess, since it seems to require judging the presence of certain types of collective anxieties or fears in particular cultural practices. However, such discourses generally produce texts (for example Colonial Office manuals, orientation programmes for new officials) which can be read.

The third part of the argument could be corroborated by interview material displaying widespread fantasising of Reason's dream. For example, Nimier quotes an 18 year-old female science student describing a boy she tutors in mathematics: '[he] truly follows his fantasy' (1978: 170).

11   Horner (1968) considered the motive to avoid success along with other motives: the motive (or 'need') to achieve (McClelland et al. 1976), with its positive consequences such as pride, and the motive to avoid failure, with its negative consequences such as shame.

8   Developing a Complementary Qualitative Methodology

1   My interviews aimed firmly at a producing research material, rather than at a interventions, to do, say, with overcoming mathematics anxiety, or with mathematics 'consciousness-raising' (e.g. Tobias 1978). Nevertheless, the interviews were designed and conducted to be generally affirmative of the subjects (and certainly not undermining), and there were positive, though largely unplanned, outcomes in 'consciousness-raising' terms; see Harriet's case study in Chapter 10.

2   Hunt uses the term 'transference' to describe the unconscious reactions, to the other, of either subject or researcher, since the latter is, in psychoanalytic terms, also 'lay' (unlike the analyst in a therapeutic relationship).

3   For the sixteen interviewees in cohort 3, social class information came from the items on parents' occupations in the previous autumn's questionnaire. For the nine interviewees in cohort 2, there was no such information, so social class categorisation was based on the student's mention of his/her parents' occupations, or else inferred on the basis of accent, or expressed views, for example on money.

4   It will be recalled that cohorts 1, 2 and 3 were the 1983-4, the 1984-5 and the 1985-6 entrants, respectively.

5   The nine subjects in cohort 2 comprised: five chosen by simple random sampling at a a lecture, one volunteer, and three men known to me, whom I asked because I considered too few men were selected up to that point. For cohort 3, from both genders, two age groupings, and three parental social classes (middle class, working class and 'mixed', see Chapter 3), twelve categories were formed and four students selected from each to be invited for interview, via the student pigeonholes. There were twenty-eight replies, of whom three refused (on grounds of 'work' or 'too busy'); the rest were offered a specific time, and sixteen were finally interviewed.

6   In an attempt to assess the possibility of transference onto myself as researcher, I rated each of the twenty-five students interviewed according to whether I 'knew' them fairly well, was 'acquainted with' them, or 'did not know' them at the time of the interview. About one third were in each category.

7   I consider Miles and Huberman's (1994) approach to be based on principles similar to those used in the analysis of survey data.

9   Reconsidering Mathematical Thinking and Emotion in Practice

1   For some purposes, we might consider more than one set of practices *within* what I have designated as 'college mathematics' here, for example practices concerned with teaching / tutorials, and with testing / assessment. Similarly, we might consider two (competing) forms of the research interview – which could be called the structured interview and the semi-structured interview – with respect to the difference in positions available to 'interviewer' and 'interviewee'; compare, for example, the discourses of interviewing in Moser and Kalton (1971) and in Hammersley and Atkinson (1983).

2   See Chapter 6, note 6.

3   Interviewees from Cohort 2 were numbered 1 to 9, and those from Cohort 3 were numbers 10 to 25. Since Jean's was one of the seven case study interviews, she is given a (fictitious) name, and the transcript for her interview is in Evans (1993).

4   One man (number 8) was not asked to attempt the question; see 'Theme 5' later.

5   However, since some respondents may have done written calculations for Question 18 but not have handed them in with the questionnaire, the absence of written calculations does not necessarily mean that the respondent was positioned in a discourse other than school

maths. Therefore the numbers classified in 'practical maths' for Question 18 in the questionnaire must be regarded as an upper bound.

6 Two items on reading graphs in the questionnaire (Questions 19–20), similar to Question 3, were categorised as PM, rather than SM.

7 Performance on Question 3 was surprisingly error-prone: only eleven of twenty-three students were classed as correct on both parts of the question.. For part A, an error made by 7 of 23 involved choosing the 'afternoon' price change, rather than the one 'before lunch'. (This is despite the apparent 50% chance of *guessing* correctly.) For part B, the error (made by nine of twenty-three) was in misreading the point two subdivisions (on the graph) below 600 as 580 ($ per ounce), rather than as 590. Such errors might have been precipitated by the slightly imperfect photocopy used. This is given some credence by noting that the level of performance (fourteen of twenty-three, or 61 per cent) on part B, is well below that for a similar problem, Question 20 in the survey (87 per cent).

8 Thus the five students classed as 'mixed' social class (see Chapter 3) were reclassified, with one exception, number 25, whose father was a joiner and mother a nurse. Since this student was so anxious at the interview that she answered only Question 1, her exclusion from social class analyses was inconsequential.

9 Using Fisher's exact test for association in a 2x2 cross-tabulation (Blalock 1972).

10 The Learners' Stories

1 There were no social class items in the questionnaire for Jean's year of entry but I judged her to be working class on the basis of her regional accent, and an often-voiced concern about never having enough money.

2 The CSE was introduced as an alternative qualification at age 16 to the more 'academic' GCE O-level in the 1960s in the UK. The two were replaced by the single GCSE 16-plus qualification in the late 1980s. A-level remains a very specialised exam at age 18-plus.

3 In addition, for school mathematics practices, the way Jean was placed in different 'sets' at school is likely to relate to social class (e.g. Ball 1981). At school, Jean was in the top set for the first three years of comprehensive school, then was moved to the third set. She seems ambivalent about losing the top set mathematics teacher who was 'very strict', for the third-set teacher who was 'too soft' and was 'walked all over' by the students. The change also involved leaving students who were 'more intelligent', though also perhaps 'snobs', to go to the set where her friends were, though they 'talked too much' and 'didn't work in class'. Most of her team-mates from the hockey team were in top set. Overall, it was better to be with her friends in the third set, because the top set teacher was so frightening, but she might have got O-level Mathematics if she had stayed there. This account of part of her 'life history' gives some insight into the differing cultures of different sets at her school. These relate in turn to differing experiences in learning maths.

4 Further research is needed into the issue of whether and how this might occur, as it is for the question of how anxiety 'originating' in connection with other ideas may become 'attached' to mathematical concepts and calculations.

5 Note that the final part of this 'statement' is reminiscent of one reported by Nimier (1978) where the 'rigour' of mathematics was celebrated (see Chapter 7).

6 Indeed, in what Ellen says, her position within the relationship is referred to by a quantitative, quasi-mathematical term. This recalls the literature on the commodification of relationships, particularly in the context of women's economic dependency on men; see for example Barrett (1980).

7 This shift also can lead to an enhanced awareness regarding the assumptions about the nature of mathematics made by mathematics teachers and researchers, for example, that mathematics is unequivocally a *closed system* of representation; see Tsatsaroni and Evans (1994) and Lerman (1994).

8 We use the regression model for school mathematics performance produced in Chapter 4 to calculate a 'predicted' performance score for each student by substituting the values for his/her maths qualifications, age, maths test/course anxiety score and so on into the

equation for the model. The 'residual' is obtained by subtracting the predicted score from the observed performance score. If the residual is positive and 'large' – say, greater than the standard error of estimate for the 'whole sample' model (about 1.4 questions) – I will call the student an 'over-performer'. If the residual is negative and large, I call the student an 'under-performer' – relative to other students of the same gender, age, maths qualification and so on. By this standard, in the sample of seven case studies, Alan was an over-performer and Harriet an under-performer in the questionnaire SM performance scale the previous October. Fiona, with a negative residual of -½ a question was a 'slight under-performer' (Evans 1993: Ch. 11).

9   Compare the much less subtle attempt by two young boy pupils to resist their (female) teacher's instructions by shifting to a sexist discourse through playing with words; see Walkerdine et al. (1989: 65–6).

10   This was not in my mind while doing the interview, since I had not retrieved and re-examined her questionnaire at that time.

11   This recalls the finding that some teachers did not want to enter diffident-appearing female students for O- and A-level Mathematics, for fear it would be 'too much' for the girls (Walkerdine et al. 1989).

12   Rather than the 'problem-solving' or investigational work promoted in mathematics education in recent years, not surprisingly since she was at secondary school in the 1970s.

13   Further, one explanation for her difficulties with percentages (about which she felt 'uncomfortable' in the past, and which she resists in the interview) might be that straightforward formulae for percentage problems are not so readily available in textbooks (as they are for, say, gradients or statistics).

14   As does interviewee number 9, another middle class man who had mathematics problems at (fee-paying) school, who wishes he had his calculator and notes in the interview (see p. 158).

15   This seems to contrast with Fiona's stated views, perhaps quoting her father: 'Time is money, money is time' and 'I've been taught from an early age not to waste money'. While the first quotation is even-handed, the second seems, in focussing on money, to give priority to it. This may relate to her feelings about not having been given free access to her father's time, or money, or affection. Also, these different ideas may reflect discourses related to different social class positions, within the middle classes.

16   The Methods and Models module comprised mathematics and philosophy of science (and sometimes IT) strands.

17   In another episode (quoted above), Donald seems to emphasise 'connections' between financial maths and college maths, and to avoid agreeing with my suggestion of a 'division' between them.

18   Pea's (1987) discussion of 'economy of cognitive effort' may be relevant here.

19   However, the *emotional quality* of the bar fantasy is relatively satisfactory, compared with the 'panic' felt in imagining a similar situation by interviewee number 8 (see p. 172).

20   The only other interviewee to mention it explicitly was Keith (Evans 1993), whose father, a local authority accountant, also represented mastery in mathematics, but only up to a point.

21   This lower middle class position is also consistent with his account of deciding how much to tip: 'You don't think about how much you've had, more about what the waiter thinks is fair [. . .] or stingy'. This diffidence, this concern about doing what the waiter thinks is fair, this anxiety about what the waiter might do, can be contrasted with, say, the confidence towards waiters shown by Alan ('I've never felt obliged to give tips') and number 9, both upper middle class males.

22   Actually, the *number of cases* analysed in Chapters 10 and 11 compares reasonably with the numbers used by Lave's team (n = 25 in the supermarket and n = 10 in dieting), and by Taylor (n = 4 interviewees). However, in both these other projects, the interaction was more intensive: about forty hours with each subject for Lave et al., and three or four interviews with each for Taylor.

23   The further comparison of Alan's, Fiona's and Peter's 'middle class' views on money with

those of Donald, Jean and Harriet may also allow us to develop Walkerdine's ideas about the difference between 'calculation for survival' and 'calculation as a theoretical exercise' (Walkerdine 1990b: 52).

24   The prominence in these interviews of the theme of 'sudden decline' (Tobias 1978) is perhaps not surprising, in that having such a history would perhaps motivate acceptance of my invitation to interview. Sudden decline was described by men (such as Peter and number 9) as well as women (for example Jean and Harriet), and there was much variation in the process as perceived by subjects. On the idea of a general discontinuity of girls' performance in mathematics around the transition to secondary school, see Walden and Walkerdine (1985).

25. Investigating the 'relational dynamics' in the research setting is recommended by Walkerdine's (1988) discussion of Cole and Traupmann's account (1979) of the difficulties in research involving 'Archie' (see p. 108). In two incidents, Archie was given a task that was 'easier' than other children were. For example, in a formal testing session, the tester decided to let Archie answer the question under less difficult conditions than were set down. Cole and Traupmann do not raise the issue whether her decision might have been made for reasons of which she was not fully conscious – to do with defences against her own anxiety or emotional distress.

11   Conclusions and Contributions

1   What some poststructuralists call 'the uncontrollable flow of meaning through the signifier', or along the chain of signifiers, sounds somewhat mystical. However, it is illustrated by my case studies, and also by Nimier (1978); see Chapter 7. Further, I would conjecture that this flow is invariably sparked by the flow of emotional charge.

2   The finding of the inverted U, because of its mathematical elegance, and its putative status as one of a few psychological 'laws', is especially difficult to give up to the rigours of critical questioning as suggested here; at one point in my work, I called my reluctance to do so 'positivist withdrawal symptoms'.

3   See also the discussion in chapter 7 of Walkerdine's (1988) example of children 'fantasising too much' in the shopping game at school.

4   For a brief discussion of the disappearing male advantage in school mathematics performance, see Chapter 3.

5   These low performances were of course on the basis of my measures, from questionnaire and interview.

6   Secada (1992) notes critically the tendency of many US studies to attempt to test for *the one* social influence from among social class, gender, ethnicity and so on.

7   Illustrations of these associations are found in the responses to contexting questions for Question1 (see pp. 152–5), and in the case studies, e.g. Ellen's and Fiona's (Chapter 10).

8   See Chapter 6, note 21.

9   For one account of the reasons for this, see Millett and Askew (1994).

10   Paulos (1990) and Moore (1990) both emphasise the importance of the idea of chance and of statistical thinking.

11   The 'Methods and Models' course, taught to students – including all those interviewed for this study – at Middlesex Polytechnic/University from 1969 to 1999, incorporated a vision broadly consistent with Crowther's. It included strands in philosophy of science, mathematics and statistics, and (sometimes) IT. For a discussion of its conceptual basis, see Evans (1989b).

12   For more general discussion of the variety of ways of combining qualitative and quantitative research, see e.g. Brannen (1993), Bryman (1988) and Evans (1979, 1983).

13   That is, if the 'fit' of that subject's performance and other scores to the general model was less good than most; see Chapter 10, note 8.

Bibliography

Abreu, G. and Carraher, D. (1989) 'The Mathematics of Brazilian Sugar Cane Farmers', in Keitel, C., Damerow, P., Bishop, A. and Gerdes, P. (eds), *Mathematics, Education and Society, Reports and Papers presented in Fifth Day Special Programme at the 6th International Conference on Mathematics Education, Budapest, 1988*, Paris: UNESCO, Science and Technology Education Document Series no. 35: 68–70.

Adda, J. (1986) 'Fight against Academic Failure in Mathematics', in Damerow, P., Dunkley, M. E., Nebres, B. F. and Werry, B. (eds), *Mathematics for All, Reports and Papers presented in Theme Group I at the 5th International Conference on Mathematics Education, Adelaide, 1984*, Paris: UNESCO, Science and Technology Education Document Series no. 20: 58–61.

Advisory Council for Adult and Continuing Education (ACACE) (1982) *Adults' Mathematical Ability and Performance*, Leicester: ACACE.

Agre, P. (1997) 'Living Math: Lave and Walkerdine on the Meaning of Everyday Arithmetic', in Kirshner, D. and Whitson, J. A. (eds), *Situated Cognition: Social, Semiotic, and Psychological Perspectives*, Mahwah, N.J.: Lawrence Erlbaum Associates, ch. 5.

Aitkin, M. (1977) 'Discussion on Dr. Nelder's Paper', *Journal of the Royal Statistical Society* A 140, 1: 66–7.

—— (1979) 'The Analysis of Unbalanced Cross-Classifications', *Journal of the Royal Statistical Society* A 141, 2: 195–223.

Anderson, J. R., Reder, L. M., and Simon, H. A. (1996) 'Situated Learning and Education', *Educational Researcher* 25, 4: 5–11.

Askew, M., Brown, M., Rhodes, V., Wiliam, D. and Johnson, D. (1997) *Effective Teachers of Numeracy*, London: King's College London.

Atkinson, P (1979) 'Research Design in Ethnography', Part 5 of Block 3B, *DE304: Research Methods in Education and the Social Sciences*, Milton Keynes: Open University Press.

Atweh, B., Bleicher, R. E. and Cooper, T. J. (1998) 'The Construction of the Social Context of Mathematics Classrooms: A Sociolinguistic Analysis', *Journal for Research in Mathematics Education*, 29, 1: 63–82.

Auster, P. (1991) *The Music of Chance*, London: Faber and Faber.

Baker, D. A. and Street, B. (1993) 'Literacy and Numeracy', *International Encyclopaedia of Education*, Oxford: Pergamon: 3453–9.

Ball, S. (1981) *Beachside Comprehensive: A Case-Study of Secondary Schooling*, Cambridge: Cambridge University Press.

Barrett, M. (1980) *Women's Oppression Today: Problems in Marxist Feminist Analysis*, London: Verso.

Bartolini Bussi, M. G. (1996) 'Mathematical Discussion and Perspective Drawing', *Educational Studies in Mathematics* 31: 11–41.

Mathematics Achievement: A Synthesis of the Research', in Chipman, S., Brush, L. and Wilson, D. (eds), *Women and Mathematics: Balancing the Equation*, Hillsdale N.J.: Laurence Erlbaum Associates, ch. 11.

Clements, M. (1979) 'Sex Differences in Mathematical Performance: An Historical Perspective', *Educational Studies in Mathematics* 10: 305–22.

Cline-Cohen, P. (1982) *A Calculating People: The Spread of Numeracy in Early America*, Chicago: University of Chicago Press.

Cobb, P. (1986) 'Contexts, Goals, Beliefs and Learning Mathematics', *For the Learning of Mathematics* 6, 2, 2–9.

—— (1994) 'Where is the Mind? Constructivist and Sociocultural Perspectives on Mathematical Development', *Educational Researcher* 23, 7: 13–20.

Cobb, P. and Bowers, J. (1999) 'Cognitive and Situated Learning Perspectives in Theory and Practice', *Educational Researcher* 28, 2: 4–15.

Cobb, P., Yackel, E. and Wood, T. (1989) 'Young Children's Emotional Acts While Engaged in Mathematical Problem Solving', in McLeod, D. and Adams, V. (eds), *Affect and Mathematical Problem Solving: A New Perspective*, New York: Springer, ch. 9.

Coben, D. (1998) *Radical Heroes: Gramsci, Freire and the Politics of Adult Education*, New York and London: Garland/ Taylor and Francis.

Coben, D. and Thumpston, G. (1995) 'Researching Mathematics Life Histories: A Case Study', in Coben, D. (ed.), *Mathematics with a Human Face, Proceedings of Second International Conference of Adults Learning Maths:A Research Forum*, Exeter: 39–44.

Cockcroft Committee (1982) *Mathematics Counts*, London: HMSO.

Cohen, G. and Fraser, E. (1991) 'Female Participation in Mathematical Degrees at English and Scottish Universities', *Journal of the Royal Statistical Society A*.

Cole, M., Hood, L. and McDermott, R. (1978) *Ecological Niche picking: Ecological Invalidity as an Axiom of Experimental Cognitive Research*, unpublished manuscript, Rockefeller University, New York.

Cole, M. and Traupmann, K. (1979) 'Comparative Cognitive Research: Learning from a Learning DisabledXhild', manuscript; published (1981) in *Minnesota Symposia on Child Development*, vol. 14., Hillsdale N.J.: Lawrence Erlbaum Associates.

Connell, R. W. (1987) *Gender and Power: Society, the Person and Sexual Politics,* Cambridge: Polity.

Cooper, B. (1985) *Renegotiating Secondary School Mathematics*, Basingstoke: Falmer.

Cooper, B. and Dunne, M. (1998) 'Anyone for Tennis? Social Class Differences in Children's Responses to National Curriculum Mathematics Testing', *Sociological Review* 46, 1: 115–48.

—— (2000) *Assessing Children's Mathematical Knowledge: Social Class, Sex and Problem-Solving*, Buckingham: Open University Press.

Crawford, K. (1996) 'Vygotskian Approaches in Human Development in the Information Era', *Educational Studies in Mathematics* 31, 43–62.

D'Ambrosio, U. (1985) 'Ethnomathematics and its Place in the History and Pedagogy of Mathematics', *For the Learning of Mathematics* 5, 1: 44–8.

D'Andrade, R. (1981) 'The Cultural Part of Cognition', *Cognitive Science*, 5, 3: 179–95.

DeBellis, V. A. and Goldin, G. A. (1997) 'The Affective Domain in Mathematical Problem-Solving', in Pehkonen, E. (ed.), *Proceedings of the 21st Conference of the International Group for the Psychology of Mathematics Education*, Lahti, Finland, vol 2: 209–16.

Delamont, S. (1997) 'Fuzzy Borders and the Fifth Moment: Methodological Issues Facing the Sociology of Education, *British Journal of Sociology of Education* 18, 4: 601–6.

De La Rocha, O. (1985) 'The Reorganization of Arithmetic Practice in the Kitchen', *Anthropology and Education Quarterly* 16, 3: 193–8.

Donaldson M. (1978) *Children's Minds*, London: Fontana.

Dow, S. (1999) 'The Use of Mathematics in Economics', paper presented to the ESRC Seminar on Producing a Public Understanding of Mathematics, Birmingham, May. Online, available HTTP: http://www.ioe.ac.uk:80/esrcmaths/sheila1.html (13 May 1999).

Dowling, P. (1991) 'The Contextualizing of Mathematics: Towards a Theoretical Map', in Harris, M. (ed.), *Schools, Mathematics and Work*, Basingstoke: Falmer, ch. 11.

—— (1998) *The Sociology of Mathematics Education*, London: Falmer.

Draper, N. and Smith, H. (1980) *Applied Regression Analysis*, 2nd edn, New York: Wiley.

Dreger, R. and Aiken, L. (1957) 'The Identification of Number Anxiety in a College Population', *Journal of Educational Psychology* 48: 344–51.

Ekinsmyth, C. and Bynner, J. (1994) *The Basic Skills of Young Adults: Some Findings from the 1970 British Cohort Study*, London: Adult Literacy and Basic Skills Unit (ALBSU).

Elias, N. (1987) 'On Human Beings and Their Emotions: A Process-Sociological Essay', *Theory, Culture and Society* 4: 339–61.

Erickson, R. and Goldthorpe, J. (1993) *The Constant Flux: a Study of Class Mobility in Industrial Societies*, Oxford: Clarendon.

Ernest, P. (1991) *The Philosophy of Mathematics Education*, Basingstoke: Falmer.

—— (ed.) (1994a) *Mathematics Education and Philosophy: an International Perspective*, Studies in Mathematics Education Series no.3, London: Falmer.

—— (ed.) (1994b) *Constructing Mathematical Knowledge: Epistemology and Mathematical Education*, Studies in Mathematics Education Series no.4, London: Falmer.

Evans, J. (1979) 'Evaluation of Research Designs', Part 6 of Block 3B, *DE304: Research Methods in Education and the Social Sciences*, Milton Keynes: Open University Press.

—— (1983) 'Criteria of Validity in Social Research: Exploring the Relationship between Ethnographic and Quantitative Approaches', in Hammersley, M. (ed.), *The Ethnography of Schooling*, Driffield. Nafferton Books, ch. 9.

—— (1989a) 'The Politics of Numeracy', in Ernest, P. (ed.), *Mathematics Teaching: the State of the Art*, Lewes: Falmer: 203–20.

—— (1989b) 'Statistics and the Problem of Empiricism', in Blum, W., Niss, M., and Huntley, I. (eds), *Modelling, Applications and Applied Problem Solving: Teaching Mathematics in a Real Context*, Chichester: Ellis Horwood, ch.3.

—— (1992) 'Mathematics for Adults – Community Research and "Barefoot Statisticians"', in Nickson, M. and Lerman, S. (eds), *The Social Context of Mathematics Education: Theory and Practice*, Proceedings of the Group for Research into Social Perspectives of Mathematics Education, London: South Bank Press: 202–16.

—— (1993) *Adults and Numeracy*, unpublished doctoral dissertation, University of London.

—— (1994) 'Quantitative and Qualitative Research Methodologies: Rivalry or Cooperation?', in Da Ponte, J. P. and Matos, J. F. (eds), *Proceedings of the Eighteenth Conference of the International Group for the Psychology of Mathematics Education*, Lisbon, vol II: 320–7. Online, available HTTP: http://www.mdx.ac.uk/www/mathstat/staff/j_evans.html (13 May 1999).

—— (1999) 'Building Bridges: Reflections on the Problem of Transfer of Learning in Mathematics', *Educational Studies in Mathematics* 39: 23–44.

—— (2000) 'Adult Mathematics and Everyday Life: Building Bridges and Facilitating Learning "Transfer"', in Coben, D., O'Donoghue, J. and Fitzsimons, G. (eds), *Perspectives on Adults Learning Mathematics: Research and Practice*, Dordrecht: Kluwer, ch. 16.

Evans, J. and Rappaport, I. (1998) 'Using Statistics in Everyday Life: From Barefoot Statisticians to Critical Citizenship', in Dorling, D. and Simpson, S. (eds), *Statistics in Society: The Arithmetic of Politics*, London, Arnold: 71–7.

Evans, J. and Thorstad, I. (1995) 'Mathematics and Numeracy in the Practice of Critical

Citizenship', in Coben, D. (ed.), *Proceedings of the Inaugural Conference of Adults Learning Maths – a Research Forum*, Birmingham, July 1994: 64–70.

Evans, J. and Tsatsaroni, A. (1994) 'Language and Subjectivity in the Mathematics Classroom', in Lerman, S. (ed.), *The Culture of the Mathematics Class-room*, Dordrecht: Kluwer: 169–90.

—— (1996) 'Linking the Cognitive and the Affective in Educational Research: Cognitivist, Psychoanalytic and Poststructuralist Models', *British Educational Research Journal: Special Issue on Poststructuralism and Postmodernism* 21, 3: 347–58.

—— (1998) 'You Are as You Read: The Role of Texts in the Production of Subjectivity', in Gates, P. (ed.), *Mathematics Education and Society, Proceedings of the First International Mathematics Education and Society Conference*, Centre for the Study of Mathematics Education, Nottingham University: 168–79.

Fairclough, N. (1992) *Discourse and Social Change*, Cambridge: Polity.

Fennema, E. (1979) 'Women and Girls in Mathematics – Equity in Mathematics Education', *Educational Studies in Mathematics* 10: 389–401.

—— (ed.) (1985) 'Explaining Sex Related Differences in Mathematics: Theoretical Models', *Educational Studies in Mathematics* 16: 303–20.

—— (1989) 'The Study of Affect and Mathematics: A Proposed Generic Model for Research', in McLeod, D. and Adams, V. (eds), *Affect and Mathematical Problem Solving: A New Perspective*, New York: Springer, ch. 14.

—— (1995) 'Mathematics, Gender and Research', in Grevholm, B. and Hanna, G. (eds), *Gender and Mathematics Education: an ICMI Study in Stiftsgarden, Akersberg, Hoor, Sweden 1993*, Lund: Lund University Press: 21–38.

Fennema, E. and Leder, G. (eds) (1990), *Mathematics and Gender: Influences on Teachers and Students*, New York: Teachers College Press.

Fennema, E. and Peterson, P. (1985) 'Autonomous Learning Behaviour: A Possible Explanation of Sex Related Differences in Mathematics', *Educational Studies in Mathematics* 16: 309–11.

Fennema, E. and Sherman, J. (1976) 'Fennema-Sherman Mathematics Attitude Scales', *Catalogue of Selected Documents in Psychology* 6.

Fitzsimons, G. E., Jungwirth, H., Maass, J. and Schloeglmann, W. (1996) 'Adults and Mathematics (Adult Numeracy)', in Bishop, A. J. (ed.), *International Handbook of Mathematics Education*, Dordrecht: Kluwer, ch. 20.

Fogelman, K. (ed.) (1983) *Growing up in Great Britain*, London: Macmillan.

Foucault, M. (1977) *Discipline and Punish*, trans. A. Sheridan, Harmondsworth: Penguin.

—— (1979) *The History of Sexuality*, vol. I, trans. R. Hurley, Harmondsworth: Penguin.

—— (1982) 'The Subject and Power', *Critical Inquiry* 8, 4: 777–89.

Foxman, D., Ruddock, G., Joffe, L., Mason, K., Mitchell, P. and Sexton, B. (1985) *Mathematics Development: Review of Monitoring in Mathematics 1978 to 1982*, 2 vols, London: DES, Assessment of Performance Unit (APU).

Frankenstein, M. (1989) *Relearning Mathematics: A Different Third R – Radical Maths*, vol. 1, London: Free Association Books.

Freire, P. (1970) *Pedagogy of the Oppressed*, trans. M. Ramos, Harmondsworth, Penguin.

Freud, S. ([1900] 1965) *The Interpretation of Dreams*, trans. J. Strachey. New York: Avon Books.

—— ([1901] 1975) *The Psychopathology of Everyday Life*, vol. 5, Pelican Freud Library, Harmondsworth: Penguin.

—— ([1916–17] 1974) 'Anxiety', Lecture 25, in *Introductory Lectures on Psychoanalysis*, Vol. 1, Pelican Freud Library, Harmondsworth: Penguin.

—— ([1917] 1953–) *A Difficulty in the Path of Psychoanalysis*, Standard Edition vol. 17, London: Hogarth.

—— ([1926] 1979) *Inhibitions, Symptoms and Anxiety*, reprinted in *On Psychopathology*, vol. 10, Pelican Freud Library, Harmondsworth: Penguin: 229–333.

—— ([1933a] 1973) 'Anxiety and Instinctual Life', Lecture 32 in *New Introductory Lectures on Psychoanalysis*, vol. 2, Pelican Freud Library, Harmondsworth: Penguin.

—— ([1933b] 1973a) 'Femininity', Lecture 33 in *New Introductory Lectures on Psychoanalysis*, vol. 2, Pelican Freud Library, Harmondsworth: Penguin.

Gay, J. and Cole, M. (1967) *The New Mathematics and an Old Culture*, New York: Holt.

Gerdes, P. (1985) 'Conditions and Strategies for Emancipatory Mathematics Education in Undeveloped Countries', *For the Learning of Mathematics* 5, 1: 15–20.

—— (1986) 'How to Recognize Hidden Geometrical Thinking: a Contribution to the Development of Anthropological Mathematics' *For the Learning of Mathematics* 6, 2: 10–17.

Giddens, A. (1984) *The Constitution of Society: Outline of the Theory of Structuration*, Cambridge: Polity.

—— (1991) *Modernity and Self-Identity: Self and Society in the Late Modern Age*, Cambridge: Polity.

Ginsburg, H., Kossan, N., Schwartz, R. and Swanson, D. (1983) 'Protocol Methods in Research on Mathematical Thinking', in Ginsburg, H. (ed.), *The Development of Mathematical Thinking*, New York: Basic Books, ch. 1.

Ginsburg, H. and Asmussen, K. (1988) 'Hot Mathematics', in Ginsburg, H. and Gearhart, M. (eds), *Children's Mathematics, New Directions for Child Development*, no. 41, San Francisco: Jossey-Bass: 89–111.

von Glasersfeld, E. (1998) *Radical Constructivism*, London: Falmer.

Glenn, J. (ed.) (1978) *The Third R: Towards a Numerate Society*, London: Harper and Row.

Gottheil, E. (n.d.) 'Gender and Mathematics: A New Look at "Math Anxiety"', unpublished paper.

Grevholm, B. and Hanna, G. (eds) (1995) *Gender and Mathematics Education: An ICMI Study in Stiftsgarden, Akersberg, Hoor, Sweden 1993*, Lund: Lund University Press.

Grieb, A. and Easley, J. (1984) 'A Primary School Impediment to Mathematical Equality: Case Studies in Rule-Dependent Socialization', in Steinkampf, M. and Maehr, M. (eds), *Advances in Motivation and Achievement*; vol. 2: Women in Science, Greenwich, Conn.: JAI Press: 317–62.

Griffin, P., Cole, M. and Newman, D. (1982) 'Locating Tasks in Psychology and Education', *Discourse Processes* 5: 111–25.

Habermas, J. (1992) *Postmetaphysical Thinking: Philosophical Essays*, trans. W. M. Hohengarten, Cambridge: Polity.

Hamilton, M. and Stasinopoulos, M. (1987) *Literacy, Numeracy and Adults: Evidence from the National Child Development Study*, London: Adult Literacy and Basic Skills Unit (ALBSU).

Hammersley, M. (1979) 'Ethnographic Interviewing', Part 3 of Block 4, *DE304: Research Methods in Education and the Social Sciences*, Milton Keynes, Open University Press.

Hammersley, M. and Atkinson, P. (1983) *Ethnography: Principles and Procedures*, London: Tavistock.

Hanna, G. (1994) 'Should Girls and Boys be Taught Differently?', in Biehler, R. et al. (eds), *Didactics of Mathematics as a Scientific Discipline*, Dordrecht: Kluwer.

—— (ed.) (1996) *Towards Gender Equality in Mathematics Education: An ICMI Study*, Dordrecht: Kluwer.

Harman, H. (1976) *Modern Factor Analysis*, 3rd edn, Chicago: University of Chicago Press.

Harris, C. (1967) 'On Factors and Factor Scores', *Psychometrika* 32: 363–79.

Harris, M. (1991) 'Looking for the Maths in Work', in Harris, M. (ed.), *Schools, Mathematics and Work*, Basingstoke: Falmer, ch. 13.

—— (1997) *Common Threads: Women, Mathematics and Work*, Stoke-on-Trent: Trentham.

Hart, L. (formerly Reyes) (1989) 'Describing the Affective Domain: Saying What We Mean', in McLeod, D. and Adams, V. (eds), *Affect and Mathematical Problem Solving: A New Perspective*, New York: Springer, ch.3.

Hawkes, T. (1977) *Structuralism and Semiotics*, London: Methuen.

Hebb, D. O. (1955) 'Drives and the CNS', *Psychological Review* 62: 243–54.

Hembree, R. (1990) 'Nature, Effects and Relief of Mathematics Anxiety', *Journal for Research in Mathematics Education* 21, 1, 33–46

Henriques, J., Hollway, W., Urwin, C., Venn, C. and Walkerdine, V. (1984) *Changing the Subject: Psychology, Social Regulation and Subjectivity*, London: Methuen.

Hinshelwood, R. D. (1991) *A Dictionary of Kleinian Thought*, London: Free Association Books.

HMI (1989) *Girls Learning Mathematics; Education Observed no.14*, London: HMSO.

Holland, J. (1981) 'Social Class and Changes in Orientations to Meanings', *Sociology* 15, 1: 1–18.

Hollway, W. (1984) 'Gender Difference and the Production of Subjectivity', in Henriques, J. et al., *Changing the Subject: Psychology, Social Regulation and Subjectivity*, London: Methuen, ch. 5.

—— (1989) *Subjectivity and Method in Psychology: Gender, Meaning and Science*, London: Sage.

Horner, M. (1968) *Sex Differences in Achievement Motivation and Performance in Competitive and Non-Competitive Situations*, unpublished doctoral dissertation, University of Michigan.

—— (1972) 'Toward an Understanding of Achievement-related Conflicts in Women', *Journal of Social Issues* 28: 157–75.

Howson, G. (1983) *Curriculum Development and Curriculum Research. A Review of Research in Mathematical Education, Part C*, Slough: NFER-Nelson.

Hoyles, C., Morgan C. and Woodhouse G. (eds) (1999) *Rethinking the Mathematics Curriculum*, London: Falmer.

Hunt, J. (1989) *Psychoanalytic Aspects of Fieldwork*, London: Sage.

Hyde, J. S., Fennema E. and Lamon, S. J. (1990) 'Gender Differences in Mathematics Performance: A Meta-analysis', *Psychological Bulletin* 107, 2: 139–55.

International Statistical Institute (1985) *Declaration on Professional Ethics*, 45th Session, Amsterdam.

Irvine, J., Miles, I. and Evans, J. (eds) (1979) *Demystifying Social Statistics*, London: Pluto.

Jaques, E. (1977) 'Social Systems as a Defence Against Persecutory and Depressive Anxieties', in Klein, M. et. al. (eds), *New Directions in Psychoanalysis*, London: Maresfield Reprints.

Jaworski, B. (1994) *Investigating Mathematics Teaching*, London: Falmer.

Joseph, G. G. (1991) *The Crest of the Peacock: Non-European Roots of Mathematics*, London: Tauris.

Keitel, C., Luelmo, M.-J., and Grevholm, B. (eds.) (1996) *20 Years of Cooperative Research on Gender and Mathematics: Where we Are, Where we Go; Programs and Abstracts for IOWME Sessions / Working Group 6: Gender and Mathematics, ICME 8, Seville 1996*, Madrid, International Organisation of Women in Mathematics Education (IOWME).

Kimball, M. (1989) 'A New Perspective on Women's Math Achievement', *Psychological Bulletin* 105, 2: 198–214.

Kirshner, D. and Whitson, J. A. (eds) (1997) *Situated Cognition: Social, Semiotic, and Psychological Perspectives*, Mahwah N.J.: Lawrence Erlbaum Associates.

Laboratory of Comparative Human Cognition (1978) 'Cognition as a Residual Category in Anthropology', *Annual Review of Anthropology* 7: 51–69.

Lacan, J. (1977) *Ecrits*, trans. A. Sheridan, London: Tavistock.

Laplanche, J. and Pontalis, J.-B. (1973) *The Language of Psychoanalysis*, London: Institute of Psychoanalysis and Karnac Books.

Lave, J. (1977) 'Tailor-made Experiments and Evaluating the Intellectual Consequences of Apprenticeship Training', *Quarterly Newsletter of the Institute for Comparative Human Development* 1, 2: 1–3.

—— (ed.) (1985) 'The Social Organisation of Knowledge and Practice: A Symposium', *Anthropology and Education Quarterly* 16, 3: 171–213.

—— (1988) *Cognition in Practice: Mind, Mathematics and Culture in Everyday Life*, Cambridge: Cambridge University Press.

—— (1996a) 'Teaching as Learning, in Practice', *Mind, Culture, Activity* 3, 3: 149–64.

—— (1996b) Seminar, Oxford University Dept. of Educational Studies, Oxford, 3 May.

—— (1997) 'The Culture of Acquisition and the Practice of Understanding', in Kirshner, D. and Whitson, J. A. (eds), *Situated Cognition: Social, Semiotic, and Psychological Perspectives*, Mahwah N..J: Lawrence Erlbaum Associates: 17–35.

Lave, J., Murtaugh, M. and De La Rocha, O. (1984) 'The Dialectic of Arithmetic in Grocery Shopping', in Rogoff, B. and Lave, J. (eds), *Everyday Cognition: Its Development in Social Context*, Cambridge, Mass.: Harvard University Press: 69–97.

Lave, J. and Wenger, E. (1991) *Situated Learning: Legitimate Peripheral Participation*, Cambridge: Cambridge University Press.

Lawler, R. (1981) 'The Progressive Construction of Mind', *Cognitive Science* 5: 1–30.

Lazarus, R. S. (1982) 'Thoughts on the Relations between Emotion and Cognition', *American Psychologist* 37, 2: 1019–24.

Lee, L. (1992) 'Gender Fictions', *For the Learning of Mathematics* 12, 1: 28–37.

Legault, L. (1987) 'Investigation des facteurs cognitifs et affectifs dans les blocages en mathematiques', in Bergeron, J., Herscovic, N. and Kieran, C. (eds), *Proceedings of the Eleventh International Conference, Psychology of Mathematics Education*, Montreal, vol. I: 120–5.

Lerman, S. (1994) 'Changing Focus in the Mathematics Classroom', in Lerman, S. (ed.), *The Culture of the Mathematics Classroom*, Dordrecht: Kluwer: 191–213.

Levitt, E. (1968) *The Psychology of Anxiety*, New York: Bobbs-Merrill.

Liebert, R. and Morris, L. (1967) 'Cognitive and Emotional Components of Test Anxiety: A Distinction and some Initial Data', *Psychological Reports* 20: 975–8.

Llabre, M. and Suarez, E. (1985) 'Predicting Math Anxiety and Course Performance in College Women and Men', *Journal of Counselling Psychology* 32, 2: 283 7.

London Mathematical Society (LMS), with Institute for Mathematics and its Applications and Royal Statistical Society (1995) *Tackling the Mathematics Problem*, London: LMS.

McClelland, D. C., Atkinson, J. W., Clark, R. and Lowell, E. (1976) *The Achievement Motive*, New York: Appleton-Century-Crofts.

McDonald, R. (1970) 'The Theoretical Foundations of Principal Factor Analysis, Canonical Factor Analysis and Alpha Factor Analysis', *British Journal of Mathematical and Statistical Psychology* 23, 1: 1–21.

—— (1985) *Factor Analysis and Related Methods*, Hillsdale, N.J.: Lawrence Erlbaum Associates.

Mcintosh, A. (1981) 'When Will They Ever Learn?', in Floyd, A. (ed.), *Developing Mathematical Thinking*, London: Addison-Wesley/Open University.

McLeod, D. (1989a) 'The Role of Affect in Mathematical Problem Solving', in McLeod, D. and Adams, V. (eds), *Affect and Mathematical Problem Solving: A New Perspective*, New York: Springer, ch. 2.

—— (1989b) 'Beliefs, Attitudes, and Emotions: New Views of Affect in Mathematics Education', in McLeod, D. and Adams, V. (eds), *Affect and Mathematical Problem Solving: A New Perspective*, New York: Springer, ch. 17.

—— (1992) 'Research on Affect in Mathematics Education: A Reconceptualisation', in Grouws, D. A. (ed.), *Handbook of Research in Mathematics Education Teaching and Learning*, New York: Macmillan: 575–96.

—— (1994) 'Research on Affect and Mathematics Learning in the JRME: 1970 to the Present', *Journal for Research in Mathematics Education* 25, 6: 637–47.

Maier, E.(1980) 'Folk Mathematics', *Mathematics Teaching* 93: 21–3.

Mandler, G. (1989a) 'Affect and Learning: Causes and Consequences of Emotional Interactions', in McLeod, D. and Adams, V. (eds), *Affect and Mathematical Problem Solving: A New Perspective*, New York: Springer, ch. 1.

—— (1989b) 'Affect and Learning: Reflections and Prospects', in McLeod, D. and Adams, V. (eds), *Affect and Mathematical Problem Solving: A New Perspective*, New York: Springer, ch. 16.

Marsh, C. (1982) *The Survey Method: The Contribution of Surveys to Sociological Explanation*, London: Allen and Unwin.

Marx, K. ([1852] 1968) 'The Eighteenth Brumaire of Louis Napoleon', in Marx, K. and Engels, F., *Selected Works*, London: Lawrence and Wishart: 96–179.

Masingila, J. O., Davidenko, S. and Prus-Wisniowska, E. (1996) 'Mathematics Learning and Practice in and out of School: A Framework for Connecting these Experiences', *Educational Studies in Mathematics* 31: 175–200.

Maxwell, J. (1989) 'Mathephobia', in Ernest, P. (ed.), *Mathematics Teaching: The State of the Art*, Lewes: Falmer, ch.19.

Mayall, B. (1998) 'Children, Emotions and Daily Life at Home and School', in Bendelow, G. and Williams, S. J. (eds), *Emotions in Social Life: Critical Themes and Contemporary Issues*, London: Routledge, ch. 8

Mead, M. (1949) *Male and Female*, New York: Morrow.

Meehl, P. (1954) *Clinical vs. Statistical Prediction: A Theoretical Analysis and a Review of the Evidence*, Minneapolis: University of Minnesota Press.

Mellin-Olsen, S. (1987) *The Politics of Mathematics Education*, Dordrecht: Reidel.

Menzies, I. (1960) 'A Case-Study in the Functioning of Social Systems as a Defence Against Anxiety: A Report on a Study of the Nursing Service of a General Hospital', *Human Relations* 13: 95–121.

Metcalfe, A. and Humphries, M. (eds) (1985) *The Sexuality of Men*, London: Pluto.

Miles, M. and Huberman, M. (1994) *Qualitative Data Analysis*, 2nd edn, London: Sage.

Millett, A., and Askew, M. (1994) 'Teachers' Perceptions of Using and Applying Mathematics', *Mathematics Teaching* 148: 3–7.

Ministry of Education (1959) *15 to 18: A Report of the Central Advisory Council for Education (England), vol. I: Report* (the Crowther Report), London: HMSO.

Molyneux, S. and Sutherland, R. (1996) *Mathematical Competencies of GNVQ Science Students: The Role of Computers*, Report, Graduate School of Education, University of Bristol.

Moore, D. W. (1976) *Statistics: Concepts and Controversies*, San Francisco: W. H. Freeman.

Moore, P. (1990) 'The Skills Challenge of the Nineties', *Journal of Royal Statistical Society A 153*, Part 3.

Morris, L., Kelleway, D., and Smith, D. (1978) 'Mathematics Anxiety Rating Scale: Predicting Anxiety Experiences and Academic Performance in Two Groups of Students', *Journal of Educational Psychology* 70: 589–94.

Morgan, C. (1998) *Writing Mathematically: The Discourse of Investigation*, London: Falmer.

Moser, C. and Kalton, G. (1971) *Survey Methods in Social Investigation*, 2nd edn, London: Heinemann.

Muller, J. and Taylor, N. (1995) 'Schooling and Everyday Life: Knowledges Sacred and Profane', *Social Epistemology* 9, 3: 257–75.

Murphy, P. (1989) 'Assessment and Gender', *NUT Education Review* 3, 2: 37–41.

Murtaugh, M. (1985) 'The Practice of Arithmetic by American Grocery Shoppers', *Anthropology and Education Quarterly* 16, 3: 186–92.

Newman, D., Griffin, P., Cole, M., (1984) 'Social Constraints in Laboratory and Classroom Tasks', in Rogoff, B. and Lave, J. (eds), *Everyday Cognition: Its Development in Social Context*, Cambridge, Mass.: Harvard University Press, ch. 8.

—— (1989) *The Construction Zone: Working for Cognitive Change in School*, Cambridge: Cambridge University Press.

Nimier, J. (1977) 'Mathematique et Affectivité', *Educational Studies in Mathematics* 8: 241–50.

—— (1978) 'Mathematique et Affectivité', *Revue Francaise de Pédagogie* 45: 166–72.

—— (1993) 'Defence Mechanisms against Mathematics', trans. Hoare, C. and Tahta, D., *For the Learning of Mathematics* 13, 1: 30–4.

Noss, R. (1997) *New Cultures, New Numeracies*, Inaugural Professorial Lecture, Institute of Education, University of London.

Noss, R. and Hoyles, C. (1996a) 'The visibility of Meanings: Modelling the Mathematics of Banking', *International Journal for Computers in Mathematics Learning* 1, 1, July: 3–30.

—— (1996b) *Windows on Mathematical Meanings: Learning Cultures and Computers*, Dordrecht: Kluwer.

Noss, R., Hoyles, C. and Pozzi, S. (1998) *Towards a Mathematical Orientation through Computational Modelling Project*, ESRC End of Award Report, Mathematical Sciences Group, Institute of Education, University of London; excerpted and revised as 'Working Knowledge: Mathematics in Use', in Bessot, A. and Ridgway, J. (eds) (2000) *Education for Mathematics in the Workplace,* Dordrecht: Kluwer.

Numeracy Task Force (1998) *The Implementation of the National Numeracy Strategy; The Final Report of the Numeracy Task Force*, London: DfEE.

Nunes, T., Schliemann, A. D., and Carraher, D. W. (1993) *Street Mathematics and School Mathematics*, Cambridge: Cambridge University Press.

Office of Population Censuses and Surveys (OPCS) (1980) *Classification of Occupations*, London: HMSO.

Ortony, A., Clore, G. L. and Collins, A. (1988) *The Cognitive Structure of Emotion*, Cambridge, Cambridge University Press.

Paulos, J. (1990) *Innumeracy: Mathematical Illiteracy and its Consequences*, New York: Vintage.

Pea, R. (1987) 'Socializing the Knowledge Transfer Problem', *International Journal of Educational Research* 11, 6: 639–63.

—— (1990) 'Inspecting Everyday Mathematics: Reexamining Culture-Cognition Relations', *Educational Researcher*, May: 28–31.

Pozzi, S., Noss, R. and Hoyles, C. (1998) 'Tools in Practice, Mathematics in Use' *Educational Studies in Mathematics* 36, 2: 105–22.

Rappaport, I. and Evans, J. (1998) 'Resources for Lay Statisticians and Critical Citizens', in Dorling, D. and Simpson, S. (eds), *Statistics in Society: The Arithmetic of Politics*, London, Arnold: 78–82.

Reed, H. and Lave, J. (1979) 'Arithmetic as a Tool for Investigating Relations Between Culture and Cognition, *American Ethnologist* 6, 3: 568–82.

Rees, R. and Barr, G. (1984) *Diagnosis and Prescription: Some Common Maths Problems*, London: Harper and Row.

—— (1985) *Developing Numeracy Skills*, London: Longman.

Reid, I. (1981) *Social Class Differences in Britain*, 2nd ed., London: Grant McIntyre.

—— (1998) *Class in Britain*, Cambridge: Polity.

Restivo, S. (1992) *Mathematics in Society and History*, Dordrecht: Kluwer.

Reyes, L. H. (1984) 'Affective Variables and Mathematics Education', *Elementary School Journal* 84: 558–1.

Reyes, L. H. and Stanic, G. (1988) 'Race, Sex, Socioeconomic Status, and Mathematics', *Journal For Research in Mathematics Education* 19, 1: 26–43.

Richardson, F. and Suinn, R. (1972) 'The Mathematics Anxiety Rating Scale: Psychometric Data', *Journal of Counselling Psychology* 19: 551–4.

Riley, T. (1984) 'Functional Numeracy', in *Viewpoints 1: Numeracy*, London: ALBSU: 2–4.

Rogers, C. (1971), *Talk to La Jolla Program*, La Jolla, Calif., August.

Rotman, B. (1980) *Mathematics: An Essay in Semiotics*, unpublished monograph, University of Bristol.

Rounds, J. and Hendel, D. (1980) 'Measurement and Dimensionality of Mathematics Anxiety', *Journal of Counselling Psychology* 27: 138–49.

Rustin, M. (1991) *The Good Society and The Inner World: Psychoanalysis, Politics and Culture*, London: Verso.

Said, E. (1978) *Orientalism*, London: Routledge and Kegan Paul.

Salomon, G. (ed.) (1993) *Distributed Cognitions: Psychological and Educational Considerations*, Cambridge: Cambridge University Press.

Santos, M. And Matos, J. F. (1998) 'Learning about Mathematics Learning with Ardinas at Cabo Verde', in Gates, P. (ed.), *Mathematics Education and Society, Proceedings of the First International Mathematics Education and Society Conference*, Centre for the Study of Mathematics Education, Nottingham University: 306–9.

Saussure, F. de (1916/1974) *Course in General Linguistics*, London: Duckworth.

Saxe, G. (1991a) *Culture and Cognitive Development: Studies in Mathematical Understanding*, Hillsdale N.J.: Lawrence Erlbaum Associates.

—— (1991b) 'Emergent Goals in Everyday Practices: Studies in Children's Mathematics', in Furinghetti, F. (ed.), *Proceedings of the Fifteenth Conference of the International Group for the Psychology of Mathematics Education*, Assisi, vol. 3: 230–7.

—— (1994) 'Studying Cognitive Development in Sociocultural Context: The Development of a Practice-Based Approach', *Mind, Culture, and Activity* 1, 3: 135–57.

Schliemann, A. (1995) 'Some Concerns about Bringing Everyday Mathematics to Mathematics Education', in Meira, L. and Carraher, D. (eds.), *Proceedings of the 19th International Conference for the Psychology of Mathematics Education*, Recife, Brazil, vol. I: 45–60.

—— (1999) 'Everyday Mathematics and Adult Mathematics Education', Keynote Address, in van Groenestijn, M. and Coben, D. (eds), *Mathematics as Part of Lifelong Learning*, Proceedings of the fifth international conference of Adults Learning Maths – a Research Forum, Utrecht, July 1998: 20–31.

Schliemann, A. and Acioly, N. (1989), 'Mathematical Knowledge Developed at Work: The Contribution of Practice versus the Contribution of Schooling', *Cognition and Instruction* 6, 3: 185–221.

Schliemann, A. and Carraher, D. (1992) 'Proportional Reasoning in and out of School', in Light, P. and Butterworth, G. (eds), *Context and Cognition: Ways of Learning and Knowing*, Hemel Hempstead: Harvester Wheatsheaf, ch. 4.

Schoenfeld, A. (1985) *Mathematical Problem Solving*, Orlando: Academic Press.

Schon, D. (1983) *The Reflective Practitioner*, New York: Basic Books.

Scribner, S. (1984) 'Studying Working Intelligence', in Rogoff, B. and Lave, J. (eds), *Everyday Cognition: Its Development in Social Context*, Cambridge, Mass.: Harvard University Press, ch. 1.

Scribner, S. (1985) 'Knowledge at Work', *Anthropology and Education Quarterly* 16, 3: 199–206.

Scribner, S. and Cole, M. (1973) 'Cognitive Consequences of Formal and Informal Education', *Science* 182: 553–9.
—— (1978) 'Literacy without Schooling: Testing for Intellectual Effects', *Harvard Educational Review* 48: 448–61.
Scribner, S. and Fahrmeier, E. (1982) 'Practical and Theoretical Arithmetic: Some Preliminary Findings', Working Paper no. 3, Industrial Literacy Project, Graduate Center, City University of New York.
Secada, W. (1992) 'Race, Ethnicity, Social Class, Language, and Achievement in Mathematics', in Grouws, D. A. (ed.), *Handbook of Research in Mathematics Education Teaching and Learning*, New York: Macmillan: 623–60.
Segal, L. (1990) *Slow Motion: Changing masculinities, Changing Men*, London: Virago.
Sewell, B. (1981) *Use of Mathematics by Adults in Everyday Life*, Leicester: ACACE.
Simon, H. (1982) 'Comments', in Clark, M. S. and Fiske, S. (eds), *Affect and Cognition*, Hillsdale, N.J., Lawrence Erlbaum Associates: 333–42.
Singley, M. K. and Anderson, J. R. (1989) *Transfer of Cognitive Skill*, Cambridge, Mass.: Harvard University Press.
Skemp, R. (1971) *The Psychology of Learning Mathematics*, Harmondsworth: Penguin.
—— (1976) 'Relational Understanding and Instrumental Understanding', *Mathematics Teaching 77.*
—— (1979) *Intelligence, Learning and Action*, Chichester: Wiley.
Spielberger, C. (ed.) (1972) *Anxiety: Current Trends in Theory and Research*, vol. I, New York: Academic Press.
—— (1972a) 'Current Trends in Theory and Research on Anxiety', in Spielberger, C. (ed.), *Anxiety: Current Trends in Theory and Research*, vol. I, New York: Academic Press, ch. 1.
—— (1972b) 'Anxiety as an Emotional State', in Spielberger, C. (ed.), *Anxiety: Current Trends in Theory and Research*, vol. I, New York: Academic Press, ch. 2.
Spielberger, C., Gorsuch, R., and Lushene, R. (1970) *Manual for the State-Trait Anxiety Inventory (Self-Evaluation Questionnaire)*, Palo Alto: Consulting Psychologists Press.
Straesser, R., Barr, G., Evans, J. and Wolf, A. (1989) 'Skills versus Understanding', *Zentralblatt für Didaktik der Mathematik, Analyses: Mathematics in Adult Education, including Distance Education*, 21, 6: 197–202; reprinted in Harris, M. (ed.), *Schools, Mathematics and Work*, Basingstoke: Falmer: 158–68.
Suinn, R. (1972) *Mathematics Anxiety Rating Scale* (MARS), Ft. Collins, Colorado: RMBSI.
Suinn, R., Edie, C., Nicoletti, J., and Spinelli, P. (1972) 'The MARS, A Measure of Mathematics Anxiety: Psychometric Data', *Journal of Clinical Psychology* 28: 373–75.
Tahta, D. (1991) 'Understanding and Desire', in Pimm, D. and Love, E. (eds), *Teaching and Learning in School Mathematics*, London: Hodder and Stoughton, ch. 19.
Tahta, D. (ed.) (1993) *For the Learning of Mathematics, Special Issue on Psychodynamics in Mathematics Education 13, 1.*
Taylor, J. (1953) 'A Personality Scale of Manifest Anxiety', *Journal of Abnormal and Social Psychology* 48: 285–90.
Taylor, N. (1989) 'Let Them Eat Cake: Desire, Cognition and Culture in Mathematics Learning', in Keitel, C., Damerow, P., Bishop, A. and Gerdes, P. (eds), *Mathematics, Education and Society, Reports and Papers presented in Fifth Day Special Programme at the 6th International Conference on Mathematics Education, Budapest 1988*, Paris: UNESCO, Science and Technology Education Document Series no. 35: 161–3.
—— (1990a) *Making Sense of Children Making Sense: Imagery, Educational Television and Mathematical Knowledge*, unpublished doctoral thesis, Faculty of Education, University of the Witwatersrand, Johannesburg.
—— (1990b) 'Picking Up the Pieces: Mathematics Education in a Fragmenting World', in

Noss, R., Brown, A., Dowling, P., Drake, P., Harris, M., Hoyles, C. and Mellin-Olsen, S. (eds) *Political Dimensions of Mathematics Education: Action and Critique; Proceedings of the First International Conference*, Dept. of Mathematics, Statistics and Computing, Institute of Education, University of London: 235–242; revised as (1991) 'Independence and Interdependence: Analytical Vectors for Defining the Mathematics Curriculum in Schools in a Democratic Society', *Educational Studies in Mathematics 22*.

Teigen, K. H. (1994) 'Yerkes-Dodson: a Law for all Seasons', *Theory and Psychology* 4, 4: 525–47.

Thom, M. (1981) 'The Unconscious Structured as a Language', in MacCabe, C. (ed.), *The Talking Cure: Essays in Psychoanalysis and Language*, London: Macmillan, ch. 1.

Thorstad, I. (1992) 'Adult Numeracy and Responsible Citizenship', *Adults Learning* 4, 4: 104–5.

Tobias, S. (1978) *Overcoming Math Anxiety*, San Francisco: Houghton Mifflin.

Tobias, S. and Weissbrod, C. (1980) 'Anxiety and Mathematics: An Update', Harvard *Educational Review* 50, 1: 63–9.

Tolson, A. (1977) *The Limits of Masculinity*, London: Tavistock.

Triandis, H. (1971) *Attitudes and Attitude Change*, New York: Wiley.

Tsatsaroni, A. and Evans, J. (1994) *Mathematics; the Problematical Notion of Closure*, in Ernest, P. (ed.), *Mathematics, Education and Philosophy: An International Perspective*, London: Falmer, ch. 7..

Urwin, C. (1984) 'Power Relations and the Emergence of Language', in Henriques, J. et al., *Changing the Subject: Psychology, SocialRegulation and Subjectivity*, London: Methuen, ch. 6.

Vergnaud, G. (1988) 'Multiplicative Structures', in Hiebert, M. and Behr, J. (eds), *Number Concepts and Operations in the Middle Grades*, National Council of Teachers of Mathematics.

Volmink, J. (1995) 'When we say Curriculum Change, How Far are we Prepared to Go as a Mathematics Community?', in Julie, C., Angelis, D. and Davis, Z. (eds), *Curriculum Reconstruction for Society in Transition, Political Dimensions of Mathematics Education, Second International Conference*, Cape Town: National Education Coordinating Committee Mathematics Commission and Maskew Miller Longman: 122–9.

Vygotsky, L. S. (1962) *Thought and Language*, trans. E. Hanfmann and G. Vakar, Cambridge, Mass.: MIT Press.

Walden, R. and Walkerdine, V. (1985) *Girls and Mathematics: From Primary to Secondary Schooling*, Bedford Way papers 24, Institute of Education, University of London.

Walkerdine, V. (1982) 'From Context to Text: a Psychosemiotic Approach to Abstract Thought', in M. Beveridge (ed.), *Children Thinking through Language*, London: Edward Arnold, ch. 6.

—— (1984) 'Developmental Psychology and the Child-Centred Pedagogy: The Insertion of Piaget into Early Education', in Henriques, J. et al., *Changing the Subject: Psychology, Social Regulation andSubjectivity*, London: Methuen, ch. 4.

—— (1985) 'Science and the Female Mind: The Burden of Proof', *PsychCritique* 1, 1: 1–20; reprinted in Walkerdine, V. (1990a), *Schoolgirl Fictions*, London: Verso: 61–81.

—— (1988) *The Mastery of Reason: Cognitive Development and the Production of Rationality*, London: Routledge.

—— (1990a) *Schoolgirl Fictions*, London: Verso.

—— (1990b) 'Difference, Cognition and Mathematics Education', *For the Learning of Mathematics* 10, 3: 51–5.

—— (1997) 'Redefining the Subject in Situated Cognition Theory', in Kirshner, D. and Whitson, J. A. (eds), *Situated Cognition: Social, Semiotic, and Psychological Perspectives*, Mahwah, N.J.: Lawrence Erlbaum Associates, ch. 4.

Walkerdine, V. and Girls and Mathematics Unit (1989) *Counting Girls Out, London: Virago.*

Walkerdine, V. and Lucey, H. (1989) *Democracy in the Kitchen*, London: Virago.

Wedege, T. (1999) 'To Know or not to Know – Mathematics, that is a Question of Context', *Educational Studies in Mathematics* 39: 205–27.

Weiner, B. (1986) *An Attributional Theory of Motivation and Emotion*, New York: Springer-Verlag.

Wenger, E. (1998) *Communities of Practice: Learning, Meaning and Identity,* Cambridge: Cambridge University Press.

White, K. R. (1982) 'The Relation between Socioeconomic Status and Academic Achievement', *Psychological Bulletin* 91: 461–81.

Whitson, J. A. (1997) 'Cognition as a Semiosic Process: From Situated Mediation to Critical Reflective Transcendence', in Kirshner, D. and Whitson, J. A. (eds), *Situated Cognition: Social, Semiotic, and Psychological Perspectives*, Mahwah, N.J.: Lawrence Erlbaum Associates, ch. 7.

Willis, J. (1984) 'Who are These People with Numeracy Problems?' in *Viewpoints 1: Numeracy*, London: ALBSU: 19–21.

Willis, S. (ed.) (1990) *Being Numerate: What Counts?* Hawthorn, Victoria: Australian Council for Educational Research.

Winter, R. (1992) 'Mathophobia, Pythagoras and Roller-Skating', in Nickson, M. and Lerman, S. (eds), *The Social Context of Mathematics Education: Theory and Practice, Proceedings of the Group for Research into Social Perspectives of Mathematics Education,* London: South Bank Press: 81–93.

Withnall, A. (1981) *Numeracy and Mathematics for Adults*, vol. 7 in *Review of Existing Research in Adult and Continuing Education.*

Wolf, A. (1991) 'Assessing Core Skills: Wisdom or Wild Goose Chase?', *Cambridge Journal of Education* 21, 2: 189–201.

Yerkes, R. and Dodson, J. D. (1908) 'The Relationship of Strength of Stimulus to Rapidity of Habit-Dormation', *Journal of Comparative Neurology and Psychology* 18: 459–82.

Zajonc, R. B. (1984) 'On the Primacy of Affect', *American Psychologist* 39, 2: 117–23.

Zappert and Stansbury (1984) *In the Pipeline: A Comparative Analysis of Men and Women in Graduate programs in Science, Engineering and Medicine at Stanford University*, Palo Alto: Stanford University.

Zaslavsky, C. (1975) 'What is Math for?', *Urban Review* 8, 3: 232–40.

Zevenbergen, R. (1996) 'Constructivism as a Liberal Bourgeois Discourse', *Educational Studies in Mathematics* 31: 95–115.

Index